HANDBOOK OF STRATA-BOUND AND STRATIFORM ORE DEPOSITS

Volume 3
SUPERGENE AND SURFICIAL ORE DEPOSITS; TECTURES AND FABRICS

HANDBOOK OF STRATA-BOUND AND STRATIFORM ORE DEPOSITS

Edited by
K.H. WOLF

I
PRINCIPLES AND GENERAL STUDIES

1. Classifications and Historical Studies

2. Geochemical Studies

3. Supergene and Surficial Ore Deposits ; Textures and Fabrics

4. Tectonics and Metamorphism

II
REGIONAL STUDIES AND SPECIFIC DEPOSITS

5. Regional Studies

6. Cu, Zn, Pb, and Ag Deposits

7. Au, U, Fe, Mn, Hg, Sb, W, and P Deposits

ELSEVIER SCIENTIFIC PUBLISHING COMPANY
Amsterdam — Oxford — New York   1976

HANDBOOK OF STRATA-BOUND AND STRATIFORM ORE DEPOSITS

I. PRINCIPLES AND GENERAL STUDIES

Edited by

K.H. WOLF

Volume 3

SUPERGENE AND SURFICIAL ORE DEPOSITS ; TEXTURES AND FABRICS

ELSEVIER SCIENTIFIC PUBLISHING COMPANY
Amsterdam — Oxford — New York   1976

ELSEVIER SCIENTIFIC PUBLISHING COMPANY
335 Jan van Galenstraat
P.O. Box 211, Amsterdam, The Netherlands

Distributors for the United States and Canada:

ELSEVIER/NORTH-HOLLAND INC.
52, Vanderbilt Avenue
New York, N.Y. 10017

ISBN: 0-444-41403-7

Printed in The Netherlands

LIST OF CONTRIBUTORS TO THIS VOLUME

B. BOULANGE
Office de la Recherche Scientifique et Technique, Outre Mer (ORSTOM), Boudy, France

G. GRANDIN
Office de la Recherche Scientifique et Technique, Outre Mer (ORSTOM), Boudy, France

J.R. HAILS
Department of Environmental Studies, The University of Adelaide, Adelaide, S.A., Australia

F. LELONG
Laboratoire de Géologie Appliquée et Institut de Recherches sur les Ressources et Matériaux Minéraux, Université d'Orleans, Orleans (CEDEX), France

W. SCHOTT
Bundesanstalt für Geowissenschaften und Rohstoffe, Hannover, German Federal Republic

O. SCHULZ
Abteilung Geochemie und Lagerstättenlehre, Institut für Mineralogie und Petrographie der Universität Innsbruck, Innsbruck, Austria

M. SCHWARZBACH
Geologisches Institut der Universität Köln, Köln, German Federal Republic

Y. TARDY
Centre de Sédimentologie et de Géochimie de Surface (C.N.R.S.), Strasbourg (CEDEX), France

J.J. TRESCASES
Office de la Recherche Scientifique et Technique, Outre Mer (ORSTOM), Boudy, France

J. VEIZER
Department of Geology, University of Ottawa, Ottawa, Ont., Canada

H. WOPFNER
Geologisches Institut der Universität Köln, Köln, German Federal Republic

R.A. ZIMMERMAN
Mineralogisch-Petrographisches Institut der Universität Heidelberg, Heidelberg, German Federal Republic

P. ZUFFARDI
Cattedra di Giacimenti Minerari, Università de Milano, Milano, Italy

CONTENTS

Chapter 3. PEDOGENESIS, CHEMICAL WEATHERING AND PROCESSES OF FORMATION
OF SOME SUPERGENE ORE DEPOSITS
by F. Lelong, Y. Tardy, G. Grandin, J.J. Trescases and B. Boulange

Chapter 4. KARSTS AND ECONOMIC MINERAL DEPOSITS
by P. Zuffardi

Chapter 5. PLACER DEPOSITS
by J.R. Hails

Chapter 6. MINERAL (INORGANIC) RESOURCES OF THE OCEANS AND OCEAN
              FLOORS: A GENERAL REVIEW
by W. Schott

Chapter 7. TYPICAL AND NONTYPICAL SEDIMENTARY ORE FABRICS
by O. Schulz (Translation: I. Vergeiner)

Chapter 8. RHYTHMICITY OF BARITE-SHALE AND OF Sr IN STRATA-BOUND DEPOSITS
OF ARKANSAS
by R.A. Zimmermann

EVOLUTION OF ORES OF SEDIMENTARY AFFILIATION THROUGH GEOLOGIC HISTORY; RELATIONS TO THE GENERAL TENDENCIES IN EVOLUTION OF THE CRUST, HYDROSPHERE, ATMOSPHERE AND BIOSPHERE

JÁN VEIZER

INTRODUCTION

Economic geology, like any other field of the geosciences, is at present in a turbulent state of development. One obvious trend is evident in attempts to relate economic deposits, particularly of the endogenic type, to the "new global tectonics" with all its consequences. The second feature, currently very strongly emphasized, is the treatment of economic deposits not as a special entity, but as an integral part of their geological environment. Economic geology is becoming more and more a geology of rocks, with ores being only one example of them (Stanton, 1972, p. 34). This is a very important step, in particular for "ores of sedimentary affiliation", which may have a direct relationship to the surrounding "country rocks". Their actual genesis is treated in detail in various parts of this multi-volume publication and is not a subject of the present review. On the other hand, if such ores are intimately related to their sedimentary environment, and this environment evolved during geologic history, the apparent evolution should be reflected also by the "ores of sedimentary affiliation".

An attempt to formulate pertinent questions in this direction was hampered by the overwhelming influence of the magmatic theory of the origin of ores, and even more by the lack of knowledge of the Precambrian and its stratigraphy. These two defects have been rectified to a considerable degree during the last 10–15 years and as a result, first attempts at a comprehensive approach to the problem were formulated. Of particular importance were the publications of Strakhov (1964, 1969), Ronov (1964), Ronov and Migdisov (1971), Garrels and Mackenzie (1971) and Stanton (1972). These authors set the stage for the treatment of ore genesis and sedimentary rocks as a complementary system.

A review of such a broad subject is extremely difficult and this chapter is no more than an attempt to indicate the general problem and to expose possible interrelations and mutual consistency of the evolution of the outer spheres of the earth. The author is deeply convinced that concentrated research into the questions tackled in the present review would be extremely rewarding and beneficial and may change to a considerable degree the picture presented below.

ORES OF SEDIMENTARY AFFILIATION AND THEIR TRENDS IN GEOLOGIC HISTORY

Limiting the ores of sedimentary affiliation to those described below, it is possible to discuss their major types and distribution during geologic history. The available data vary in detail and reliability and therefore discussion of the problem is rather uneven. Nevertheless, this description may serve as a basis for outlining general features and the direction of future research and interpretation.

Residual deposits and weathering crusts[1]

Depending on the climate and bedrock composition, weathering may lead to formation of *laterites* of either iron- or alumina-rich type. These, in some cases, may serve as

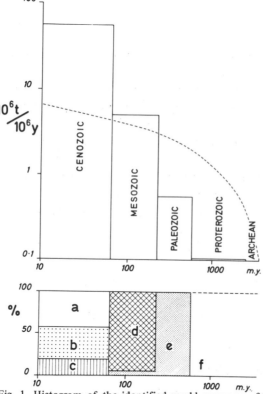

Fig. 1. Histogram of the identified world resources of bauxites and the proportion of various facies during geologic history. (Based on the data of Valeton, 1972, and compilation of resources by Paterson, 1967.)

a = gibbsite bauxites on igneous and metamorphic rocks; *b* = gibbsite bauxites on clastic rocks; *c* = gibbsite bauxites on karst; *d* = boehmite bauxites on karst; *e* = diaspore bauxites on karst; *f* = corundum schists.

The dotted line represents the slope for decrease in total world sedimentary mass with increasing age (cf. Fig. 7).

---

[1] Editor's note: see also in this Volume, Chapter 3 by Lelong et al.

economic accumulations of iron, manganese and alumina (bauxites). Ferruginous laterites also contain a high proportion of known nickel resources (Cornwall in Brobst and Pratt, 1973). Their potential for extraction of other metals as byproducts is not yet fully realized (cf., Krauskopf, 1955), but the ferruginous laterites could be a potential source of Co, Ni, As, Be, Cr, Mo, Cu and possibly also Sc, Se, V, and the alumina-rich laterites of Be, Ga, Ti and possibly also Mo, Zn, Nb, Zr and others.

The origin and classification of *bauxites* was discussed recently by Valeton (1972). From Fig. 1 it is evident that the total reserves as well as the facies of deposits were highly variable during geologic history. The mineralogical composition changes in the order gibbsite → boehmite → diaspore → corundum with increasing age. This trend very likely represents diagenetic and/or metamorphic dehydratation (cf., Valeton, 1972 p. 179). The rapid decrease in total reserves of bauxites with increasing age, as well as the higher proportion of karst-type deposits in Mesozoic and Paleozoic sequences, will be discussed later in the text.

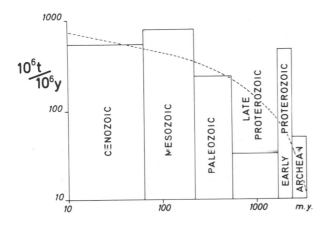

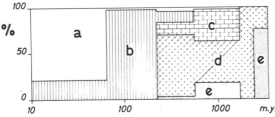

Fig. 2. Histogram of the identified world resources of sedimentary iron ores and the proportion of various facies during geologic history. (Based on the *United Nations Survey of World Iron Ore Resources*, 1970.)
a = laterites, gossans, placers, etc.; *b* = Minette type; *c* = Clinton type; *d* = Superior type; *e* = Algoma type.
The dotted line as in Fig. 1.

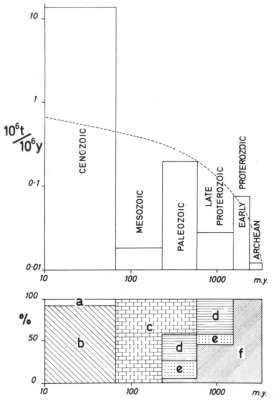

Fig. 3. Histogram of the identified world resources of sedimentary manganese ores and the proportion of various facies during geologic history. (Based on the summary of Varentsov, 1964.)
a = residual deposits; *b* = orthoquartzite-glauconite-clay association; *c* = carbonate association; *d* = orthoquartzite-siliceous shale-manganese carbonate association; *e* = volcano-sedimentary type; *f* = "jaspilitic" type.
The dotted line as in Fig. 1.

The rapid decrease in iron- (and manganese)-rich laterites in sequences older than Cenozoic is evident from Figs. 2 and 3. This decrease is more evident for iron than manganese. However, this is due mainly to the enormous tonnage of the Mn deposits of the orthoquartzite–glauconite–clay association (cf., Fig. 3). In fact the absolute economic manganese reserves of lateritic type in Cenozoic formations are almost equal to the total Paleozoic reserves.

Deposits associated with fluviatile to deltaic clastic sediments

(*1*) According to *United States Mineral Resources* (Brobst and Pratt, 1973), *placer and palaeoplacer deposits,* whether of elluvial, alluvial or beach type, are at present the world's major source of Nb, Ta, Sn, W, Ti, Th, Zr and Hf. They are, or were, the source of

TABLE I

Summary of characteristic features of the ores of fluviatile-deltaic association (in part a modification of Roscoe's, 1969, table XI)

	Placers and palaeoplacers[1]	Conglomerate—gold—uranium—pyrite type[2]	Sandstone—uranium—vanadium—copper type[3]
Host rocks	mostly oligomictic sandstones and sands, less conglomerates and pebbles	oligomictic conglome-rates; greenish argilla-ceous sandstones, arkoses, shales	sandstones, arkoses, siltstones, mudstones of variegated, pre-dominantly red, colours (red beds), conglomerates, lignites and organic debris, carbonates
Sedimentary environment	continental-marine: elluvial, alluvial, beach placers	continental: fluviatile-deltaic	continental: mostly fluviatile
Tectonic setting		platform cover	platform cover and rigid sialic blocs
Age	mostly Quaternary and Tertiary	mostly Early Proterozoic	Late Paleozoic to Quaternary
Shape of deposits	lenses and blankets (concordant with sedimentary features)	stratiform, concordant in detail with sedimen-tary features	peneconcordant tabular bodies, elongate irregular bodies; discordant C-shaped bodies (rolls)
Controls of localization	primary features favourable for concentra-tion of detrital heavy minerals; palaeotopography	(often dissemination in arkosic matrix)	facies variation, large-scale sedi-mentary features and tectonic structures affecting permeability and distribution of organic material
Minerals	"heavy minerals"; monazite, chromite, zircon, columbite, tantalite, native Au, Pt, cassiterite, wolframite, scheelite, magnetite, hematite	uraninite, uranotho-rite, brannerite, thucholite, pyrite, gold, "heavy minerals"	uraninite, coffinite, secondary U minerals; roscoelite, montroseite, V-clay; chalcocite, covellite, chalcopyrite, bornite, malachite, azurite; pyrite, marcasite, galena, sphalerite
Ore mineral textures	discrete grains	small discrete sub-rounded to rounded grains	very fine-grained impregnations and replacements
Metals recovered or concen-trated	Nb, Ta, Sn, W, Ti, Th, Zr, Hf, Au, Pt-group, Sc, Y, Cr, REE	Au, U, Th, Y, REE, Pt-group (Pb, Ti, Zr, Co, Ni, Cu, Zn, Ag)	U, V, Cu, Mo, Ag (Se, As, Ni, Co)
Sulphur		little isotopic fractionation	extreme isotopic fractionation
Examples	tin belt of SE Asia; Florida beach sand for Ti (U.S.A.)	Witwatersrand (S.A.); Blind River-Elliot Lake area (Canada); Serra de Jacobina (Brazil); Tarkwa (?) (Ghana); Aldan Shield (U.S.S.R.)	Colorado Plateau (U.S.A.); Ferghana (?) (U.S.S.R.); Hamr (Czechosl.)

[1,2,3] Editor's notes: [1] see Chapter 5 by Hails; [2] Chapters 1 and 2, in Vol. 7, by Pretorius, describe the Witwatersrand deposits; [3] see, in Vol. 7, Chapter 3 by Rackley.

about 25% of the world's Au, and also yield Sc, Y, REE, Cr and the Pt-group. The quantitative statistical distribution of deposits of this type through geologic history is not known to the present author, but deposits older than Cenozoic are rare (cf. also Table I).

(2) Considering the absence of typical palaeoplacers in sequences older than about 65 m.y., it is rather surprising that a modified type of palaeoplacers, the *conglomerate–gold–uranium–pyrite* association (Witwatersrand–Blind River type of Stanton, 1972, p. 564), is statistically important in the range 1800–≥2500 m.y. (cf., e.g., Salop, 1969). Deposits of this type are characterized by association of fluviatile to deltaic quartz conglomerates with gold, thucholite, detrital uraninite and pyrite, Pt-group and other minor metals and "heavy"-mineral assemblages (cf., Ramdohr, 1958; Pretorius, 1966; Roscoe, 1969).

According to *United States Mineral Resources*, this type of deposit is the major source of the world's gold and contains about 40% of the western world's identified resources of uranium, 40% of Th and 20% of REE.

(3) The third type of deposits associated with fluviatile-deltaic clastic sediments is the so-called *sandstone–uranium–vanadium–copper (silver)* type (cf., Stanton, 1972, p. 553) known also as Colorado Plateau-type deposits. The ores are generally associated with "bleached" parts of red-bed sequences and particularly with organic matter present in host rocks. This type differs from the previously discussed deposits not only by its association with finer clastic rocks of red colours and the presence of organic matter, but also by distinct features of (dia)epigenetic origin.

According to *United States Mineral Resources,* such deposits contain about 30% of the western world's known resources of uranium, and are an important source of Cu and V, and also of Ag and Se as byproducts. Deposits of this type, at least the U–V-rich varieties, are mostly in sequences of Late Paleozoic to Recent age (Fischer in Brobst and Pratt, 1973). The Cu-rich end members grade into the Cu red-bed types discussed below.

Base-metal strata-bound deposits of marine and marine-volcanogenic associations[1]

Cu–Pb–Zn stratiform and strata-bound deposits form a highly variable but continuous series of genetic types. Consequently, their classification is complex. Subdivision based on the discussions by Gilmour (1971), Stanton (1972, pp. 495–563), Hutchinson (1973) and Sangster (1972, and personal communication, 1973) will be used here. According to these authors the economic concentrations of base metals can be subdivided into deposits of:

(A) (1) volcanogenic and volcano–sedimentary associations; (2) sedimentary associations of: (a) basinal type; (b) red-bed type;

(B) carbonate–Pb–Zn deposits (known also as Mississippi Valley, Silesian, Alpine and Karatau types).

The general features of various types are summarized in Table II. Group-A deposits are

---

[1] Editor's note: refer to Vol. 6, the chapters by Sangster and Scott, and by Vokes, and to Vol. 1, the chapter by Gilmour.

probably of syngenetic to diagenetic origin. The carbonate—Pb—Zn association is often discordant to sedimentary features and is most likely of epigenetic (and diagenetic ?) origin.

Strata bound base-metal deposits contain about 30% of identified resources of world's Cu, ~70% Zn, ~50% Pb and an appreciable amount of Au and Ag (Laznicka and Wilson, 1972; Brobst and Pratt, 1973). The distribution of genetic types during geologic history can be treated only qualitatively. Quantitative data, although commercially available (Laznicka, 1970) are not within financial reach of the author. The volcanogenic and volcano—sedimentary type, although known in eugeosynclinal sequences of all ages, seems to be dominant in the Archean (cf. Sangster, 1972), often as a distal basinal facies of the Algoma-type iron deposits (cf., Goodwin and Ridler, 1970; Gross, 1973). Hutchinson (1973) subdivided this type of deposits into Zn—Cu—pyrite, Pb—Zn—Cu—pyrite and Cu—pyrite subtype; with the first one being very abundant in the Archean, the second common in the Early Proterozoic and Phanerozoic and the less frequent Cu—pyrite subtype present in the Phanerozoic. The basinal-type deposits do not indicate any definite age pattern. The red-bed sedimentary deposits of Cu (Co) are known from about 2000 m.y. onwards (Udokan Series; Salop, 1968) However, the latter type is most abundant from about 1000 m.y. ago. The carbonate—Pb—Zn deposits are typical of the Phanerozoic although some Proterozoic deposits in the Grenville Province and elsewhere could be included in this category (cf., Stanton, 1972, p. 543; Wedow et al., 1973).

The production values for Canada (cf., Lang et al. in Douglas, 1970) indicate that the Precambrian Shield supplied about 37% of Cu, about 20% of Zn and negligible Pb, whereas the younger (Late Proterozoic and Phanerozoic) eastern and western parts of the country produced about 16% of Cu, 60% of Zn and almost all Pb. Such secular variation in composition of Canadian base-metal deposits was also observed by Sangster (1972) and world-wide by Laznicka and Wilson (1972). Although these data are only a rough indication of a general progression, the pattern agrees well with the distribution of genetic types during geologic history and may be therefore significant.

Sedimentary iron and manganese deposits

In contrast to the previously described base-metal and some of the fluviatile-deltaic economic concentrations, the syngenetic nature of stratified iron and manganese deposits was not seriously questioned.

Iron ores. In his introduction to the compilation of the world iron ore deposits Gross (1970) recognized the following categories of iron ores associated with sedimentary rocks:

(*A*) Bedded: (*1*) Algoma-type; (*2*) Superior-type; (*3*) Clinton-type; (*4*) Minette-type; (*5*) blackband ores and non-oolitic beds; (*6*) ferruginous sandstones and shales (palaeoplacers); (*7*) placers; (*8*) ferruginous carbonate beds;

(*B*) Massive: (*9*) Bilbao-type;

TABLE II

Summary of characteristic features of strata-bound base-matel deposits (modified after Gilmour, 1971; Stanton, 1972; Sangster 1972, and personal communication, 1973; and Hutchinson, 1973)

	Carbonate–lead–zinc type	Ores of sedimentary affiliation		Ores of volcanogenic and volcano-sedimentary affiliation		
		red-bed type	basinal type	Pb–Zn–Cu–pyrite	Zn–Cu–pyrite	Cu–pyrite
Host rocks	carbonates, mostly dolomitized reefal sequences	arkoses, sandstones, quartzites, mudstones, carbonates less frequent	black shales and dark coloured carbonates rich in C_{org}	intermediate-felsic calc-alkaline volcanics; greywackes, Mn-shales, graphitic shales and argillites, silt-stones, cherts, iron formations	intermediate tholei-itic and calc-alkaline volcanics; greywackes, cherts, iron formations, pyroclastics	ophiolitic suites and tholeiitic basalts; cherts, iron formations
Sedimentary environment and tectonic setting	marine; miogeosynclinal and stable platform setting	marine, but in parts also fluviatile-deltaic; stable platforms, aulacogenes	marine-epicontinental to miogeosynclinal; lagoons and intermontaine basins of post-orogenic geosynclinal stages	late eugeosynclinal stage	early eugeosynclinal stage	early stage of continental plate rifting and separation
Age	mostly Phanerozoic	the oldest 2000 m.y. ago, but abundant since 1000 m.y. ago	possibly throughout the whole geologic column (?)	dominant in Early Proterozoic and Phanerozoic	all ages, but dominant in Archean	Phanerozoic
Shape of deposits	irregular bodies, peneconcordant and discordant tabular bodies	irregular peneconcordant bodies	well-layered stratiform bodies	──────── stratiform bodies and lenses ──────── →		
Control of localization	facies variations, large-scale sedimentary features and tectonic structures affecting permeability	primary facies variations, permeability and distribution of organic matter	primary sedimentary facies variations and distribution of organic matter	──────── primary stratification ──────── →		

Minerals	galena, sphalerite > barite, pyrite, marcasite, tetrahedrite	chalcopyrite, bornite, chalcotite > galena, sphalerite, covellite	sphalerite, galena > pyrite	sphalerite, galena, pyrite, tetrahedrite, argentite, stephanite, pyrrhotite	sphalerite, galena, pyrite, chalcopyrite > tetrahedrite, argentite, pyrargirite, stephanite, pyrrhotite	pyrite, chalcopyrite > sphalerite
Ore mineral textures	massive and disseminated	fine grained disseminations in bleached parts of red beds	massive		massive and disseminated	
Metals recovered or concentrated	Zn, Pb > Ni, Co, Cu, Ag, Cd, Ge	Cu > Pb, Co, Zn, V, Se	Zn, Pb > Cu, Ag	Pb, Zn, Cu > Ag	Zn, Cu > Au, Ag	Cu > Au
Total sulphide	~15%	~4–8%	≲50%	≲75%	≲75%	≲75%
Examples	Pine Point (Can.); SE Missouri, Tri State (U.S.A.); Silesia (Poland); Karatau (U.S.S.R.); Bleiberg (Austria)	Katangan-Zambian Copper Belt (?); Udokan, Dzezkazgan (U.S.S.R.)	Menesuch shale (U.S.A.); Kupferschiefer (?) (FRG); HYC, Mt. Isa (?) (Austr.); Sullivan (?) (Can.)	Bathurst, Errington, Vermilion (Can.); E. Shasta (U.S.A.); Kuroko (Japan)	Timmins, Noranda, Rambler (Can.); United Verde, W. Shasta (U.S.A.)	Cyprus; W. Newfoundland; Philippines

(*C*) Residual: (*10*) laterites and gossans; (*11*) river-bed deposits and bog-iron types.

The basic features of these types are summarized also in James (1966). Their major properties[1] are illustrated in Tables III and IV.

Among these types the placers, palaeoplacers, ferruginous carbonates, blackbands and the river-bed—bog-iron ores are not very important economically on a worldwide basis. The massive Bilbao-type deposits are epigenetic replacements of reefal and reef-detrital carbonates and are mostly of Phanerozoic age. They often show a close spatial relation with the carbonate—Pb—Zn deposits (Gross, 1970), but their reserves are <1% of the total world iron. The residual accumulations are quite important economically, but they are mostly due to surface enrichment on particularly the Proterozoic Superior type. Surface leaching of silica (chert layers) leads to an upgrading of the iron ore.

As seen from Fig. 2, the distribution of sedimentary iron ores during geologic history is of a bimodal type with peaks in the Early Proterozoic and Mesozoic and a minimum of reserves per time unit during the Late Proterozoic. There are also definite variations in the proportion of various types of iron ores during geologic history. The Algoma-type deposits are dominant during Archean, Superior during Early Proterozoic, Clinton in Late Proterozoic and Early Paleozoic, Minette during Late Paleozoic and Mesozoic, and the residual deposits during Cenozoic (cf., also James, 1966; Gross, 1970; Garrels et al., 1973). The high proportion of the Superior-type deposits in the Paleozoic is due to one huge region (Mutun-Urucum; Brazilian—Bolivian M. Grosso) (cf., Alvarado, 1970; Dorr, 1973). Its Cambro-Ordovician (?) age and classification is uncertain. If this deposit were excluded, the Minette and Clinton types would each be about 1/2 of the total tonnage, with Minette being confined to the Late and Clinton to the Early Paleozoic. This discussion, of course, does not exclude the presence of other types of iron ores during different periods, but their share of the total reserves is negligible.

In general, one may observe a shifting of the site of iron deposition from an Archean "eugeosynclinal" setting into a miogeosynclinal-unstable shelf environment in the Proterozoic, continental to marine (often paralic) in the Carboniferous, and terrestrial in the Cenozoic. This indicates a general landward movement of iron-rich facies with decreasing age (cf., also Strakhov, 1964, 1969) accompanied by chemical and mineralogical changes evident from Tables III and IV.

Manganese ores.[2] Stratified manganese deposits can be classified into several categories on the basis of their lithological association and genetic significance. The division of manganese deposits follows a modified scheme based on the reviews of Varentsov (1964), Roy (1968) and Stanton (1972, pp. 454—494), and recognizes the following types:

Editor's notes: [1] see also, in Vol. 7, Chapter 4 by Eichler, and Chapter 5 by Dimroth, for treatment of Precambrian banded iron-formations, and oolitic ores, respectively; [2] Cf., in Vol. 7, Chapter 9 by Roy, and Chapter 8 by Callender and Bowser, for details on ancient and Recent manganese deposits, respectively.

TABLE III

Chemical compositions of major types of sedimentary iron ores (after Gross in Douglas, 1970, p. 173; courtesy Geol. Surv. Can.)

	Algoma mainly oxide facies	Superior oxide facies	silicate - carbonate f.	Clinton composite (oxide mostly)	Minette composite
Total Fe	33.52	33.97	30.23	*51.79*	30.97
SiO_2	*47.90*	*48.35*	*49.41*	11.42	28.06
Al_2O_3	0.90	0.48	0.68	*5.07*	*5.79*
Fe_2O_3	31.70	45.98	16.34	61.83	29.81
FeO	14.60	2.33	24.19	11.00	13.08
CaO	1.45	0.10	0.10	*3.32*	*1.92*
MgO	*1.80*	*0.32*	*2.95*	0.63	1.54
Na_2O	0.20	0.33	0.03	ND	0.33
K_2O	0.32	0.01	0.07	ND	0.53
H_2O^+	0.47	2.00	5.20 }	1.94	*13.10*
H_2O^-	0.10	0.04	0.38		
TiO_2	0.05	0.01	0.01	0.02	0.18
P_2O_5	0.10	0.04	0.08	*1.96*	*1.59*
MnO	0.05	0.03	0.65	0.17	0.16
CO_2	ND	0.03	0.22	2.15	2.89
S	ND	0.01	0.05	0.02	ND
C	ND	0.08	0.15	ND	ND
Location	Timagami	Knob Lake	Knob Lake	Wabana	Peace River

Italicized values indicate typical chemical features of a particular facies type.

(A) Bedded deposits of:
   (1) sedimentary affiliation: (a) marsh and lake deposits; (b) orthoquartzite—glauconite—clay association; (c) orthoquartzite—siliceous shale—Mn carbonate association; (d) carbonate association;
   (2) volcano—sedimentary affiliation;
   (3) association with Precambrian iron formations and their metamorphosed equivalents ("jaspilitic" type);
(B) residual deposits.

Although the relation of various types of manganese deposits (cf. Table V and Fig. 3) with their tectono-sedimentary regime is not as clear as in the case of iron, several features are still recognizable. Fig. 3 indicates a decrease in economic resources per time unit with increasing age. Whether the decrease with time is more regular than this diagram indicates is difficult to judge. The histogram may simply reflect insufficient sampling. For example, it is probable that the evaluated resources for "jaspilitic" type are underestimated (cf., Kalahari; Dorr et al., 1973). On the other hand inclusion of manganese nodules into this picture would enhance very strongly the Cenozoic peak. Therefore, the only definite

TABLE IV

Summary of characteristic features of major types of sedimentary iron ores (based on descriptions of Gross, 1970, 1973, and Stanton, 1972)

	Bilbao type	Laterites, gossans and residual accumulations	Bog and river deposits	Blackbands and nonoolitic beds	Minette type	Clinton type	Superior type	Algoma type
Host rocks	carbonates, mostly dolomitized reefal sequences (often with carbonate–Pb–Zn type)	concentrations on preexisting ores, sulphides or ultrabasics	detrital sediments with carbonaceous material and organic debris	dark fine-grained shales, greywackes, volcanics, coal formations, limestones	carbonates, coal measures, dark shales, clastic rocks	carbonates, clastic rocks, carbonaceous shales	siliceous rocks, shales, carbonates, greywackes, volcanics and pyroclastics	tholeiitic and calc-alkaline volcanics; pyroclastics, greywackes, black graphitic schists (often with base metals)
Sedimentary environment and tectonic setting	marine, unstable terrestrial; shelf to miogeosyncline	terrestrial; mostly stable platform	terrestrial in humid climate; tectonic setting variable	paralic-marine; unstable shelf to geosyncline	paralic-marine; unstable shelf to miogeosyncline	shallow marine; unstable shelf to miogeosyncline	marine (?); stable shelf to miogeosyncline	marine-eugeosyncline
Age	mostly Phanerozoic	mostly Cenozoic	mostly Cenozoic	Late Paleozoic Quaternary	mostly Late Paleozoic–Tertiary	mostly Cambrian-Devonian	dominant during Proterozoic (particularly 1.8–2.5 b.y. interval)	all ages but dominant in Archean
Shape of deposits	irregular bodies	blankets	small lenses	massive lenses	lenses and lenticular beds of broad regional extent (X0 km; ≲20 m thick)		layers of broad regional extent (X0-X00 km; to X000 m thick)	en echelon lenses (X km; X00 m thick)
Control of localization	facies variations, large-scale sedimentary features and tectonic structures effecting permeability	(paleo) topography and sedimentary features			primary sedimentary stratification and sedimentary conditions often cyclic sedimentation		lower part of the transgressive sequence	

Minerals	hematite and goethite above and siderite below the water level	hematite, goethite, limonite	goethite, limonite, siderite	siderite; hematite and goethite; siderite, chamosite and goethite	siderite, chamosite ≫ limonite, hematite	hematite > chamosite, glauconite > siderite	magnetite; hematite > siderite, ankerite, Fe dolomite > greenalite (minnesotaite, stilpnomelane, cummingtonite, grunerite) > pyrite	magnetite, hematite ≫ siderite > pyrite, pyrhotite
Ore mineral textures	massive	earthy masses, bothryoidal textures	pisolites, oolites, botryoidal textures earthy masses	massive	oolites, pisolites common; concretions		oolites and granules in hematite; banded	oolites and granules absent; banded
Chemical characteristics	high Ca, Mg (Pb, Zn, Cu, As), variable Mn, low Si, P	high Al, Ti, Ni, Cr, (H_2O)	high Mn	high Si, Al, Ca, Mn, P, S, (As, Cu, Pb, Ni, Zn)	low Si; high Al, P; Mg/Ca < 1; high H_2O	high Al, P; Mg/Ca < 1; high total Fe Na, Ca, trace metals	high Si, Fe^{2+}/Fe^{3+}, Mn; low Al, P, Na, Ca, trace metals; Mg/Ca > 1	
Total iron (as Fe)	20–60%	variable but generally low		< 25%	~31%	~52%	~30–33%	~33%
Examples	Bilbao (Spain); Cumberland (U.K.); Erzberg (Austria)	Conakry (Guinea)	Robe River (Austr.); Kustansi (U.S.S.R.)	Gross Ilsede, Peine (F.R.G.); Vareš (Yugosl.)	Lorraine basin (France); Salzgitter (F.R.G.)	Birmingham (U.S.A.); Wabana (Can.)	Messabi Range (U.S.A.); Minas Gerais (Braz.); Hamersley R. (Austr.)	Steep Rock Lake (Can.); Soudan Mine (U.S.A.)

TABLE V

Summary of characteristic features of major types of sedimentary manganese ores (based on descriptions of Varentsov, 1964, and Stanton, 1972)

	Marsh and lake deposits	Orthoquartzite—glauconite—clay association	Siliceous shale—orthoquartzite—Mn carbonate association	Carbonate association	Volcano-sedimentary type	"Jaspilitic" type
Host rocks	marsh and lake sediments	sandstones, siltstones, shales, glauconitic beds, marls, conglomerates and coal beds	orthoquartzite, siliceous shales, dark carbonates	marine limestones and dolomites (red beds, volcanics, evaporites, black shales)	volcanics, pyroclastic rocks, siliceous rocks, greywackes	banded iron formations, argillaceous shales, siliceous rocks, orthoquartzites, carbonates (in places their metamorphosed equivalents)
Sedimentary environment and tectonic setting	continental; mostly on stable platform in subarctic climate	shallow-marine-estuarine; stable basement (platform, posttectonic intermontane troughs)	marine; stable platform - (miogeosyncline)	marine; eugeosyncline—miogeosyncline—stable platform	marine; mostly eugeosyncline	eugeosyncline < miogeosyncline > stable platform
Age	mainly Cenozoic	mostly Tertiary and Mesozoic	Late Proterozoic–Paleozoic	Paleozoic-Mesozoic	Precambrian–Phanerozoic	Precambrian
Shape of deposits	lenses (8–10 acres, ≤5 m thick)	thin lenses and continuous beds (≤8 m thick)		lenses and continuous beds	small lenses with frequent abundance continuous beds	extensive lenses and continuous beds
Control of localization	(palaeo) topography and sedimentary features	in transgressive sequences	primary sedimentary stratification and sedimentary conditions in transgressive sequences →			
Minerals	amorphous hydrous oxides, after burial higher oxides	higher oxides ≫ carbonates and mixed zones	carbonates > oxides	higher and lower oxides and/or carbonates	lower and higher oxides > silicates; carbonates	lower oxides ≫ silicates
Ore mineral textures	oolites, pisolites, earthy masses	concretionary nodules, rounded earthy masses	oolites, laminae			banded laminae

Chemical characteristics	Mn/Fe = 0.1–2.5	Mn/Fe = 8–10 (often higher)		Mr/Fe < 1–10 (variable), low Al_2O_3, P_2O_5	Mn/Fe ~ 0.3, high P_2O_5 and trace metals	Mn/Fe ~ 2–5, low P_2O_5, sometimes high Al_2O_3 (as in gondites)
Total manganese (as Mn)	5–20%	~20–50%	low-grade	<1–10%	low-grade (~5–10%)	~35%
Examples	Kaslo, Dawson settlement (Can.); Scandinavia	Nikopol, Labinsk, Chiatura (U.S.S.R.); Groote Eylandt (?) (Aust.)	Hsiangtàn (China); Uzbekistan (U.S.S.R.)	Irmini (Marocco); Usinsk (U.S.S.R.)	Kokuriki, Nikura (Japan); Karadzhal (U.S.S.R.)	Postmasburg, Kalahari (S.A.); Minas Gerais, Matto Grosso (Brazil); Darwar (India)

conclusion is the exceptionally high rate of manganese fractionation into economic deposits during the Cenozoic. Comparing the distribution of various types of manganese ores with age, one may observe (as in the case of iron) the dominant role of "jaspilitic" types during the Archean and Early Proterozoic. This group is probably equivalent to the combined Algoma and Superior types of iron deposits, and its tectonic setting might therefore cover the whole range from "eugeosyncline" to stable shelf. The volcano—sedimentary ores are prominent during Late Proterozoic and Paleozoic and mainly represent eugeosynclinal sequences. The orthoquartzite—siliceous shale and carbonate associations are typical for uppermost Proterozoic and Paleozoic—Mesozoic sequences and are associated with (eu)-miogeosynclinal-stable shelf regions. The orthoquartzite—glauconite—clay facies, predominantly Tertiary and economically the most important, was deposited in a shallow-marine to estuarine environment on a stable basement. The marsh, lake and residual accumulations are terrestrial and mostly Quaternary. Therefore, as in the case of iron, zone of the maximum manganese accumulation is shifting landwards with decreasing age (cf. also Strakhov, 1964, and Varentsov, 1964).

This change in tectonic setting is accompanied by mineralogical changes with higher oxides of Mn being dominant in younger and lower oxides in older sequences. The older sequences also contain a higher proportion of Mn silicates and a lower proportion of Mn carbonates. These mineralogical changes may be related to the diagenetic and metamorphic trends discussed in detail by Roy (1968; and in this publication Chapter 9, Vol. 7).

Potential resources

The previously discussed ores of sedimentary affiliation are mined as direct sources of their respective metals. Their utilization reflects the present state of technology, which dictates the "cut off" boundary for ores on the basis of economic considerations. In future, with improvement of technology and increasing demand and/or depletion of natural resources, many sources considered at present as "country rocks" or non-metallic resources may be economically exploited. With the exception of the already discussed laterites, such sources for potential exploitation (mostly as byproducts) could be (cf. Krauskopf, 1955; Brobst and Pratt, 1973):

(*1*) black shales for *Cu, Ag, U,* As, Mo, Au, V, Ni, Pb, Zn, Cd, Se;
coal for *U, Ge, Mo,* Ag, Ni, Au, Bi (As, Sb, Cd, Tl, Se, Te, B, Zn, Be, Ga, Sc);
asphalt and petroleum for *V, Mo, Ni, Pb, Cu,* Zn, Cr, Co, Sn, (Ag, As, REE);

(*2*) phosphorites for *U, V, Mo, Pb, REE,* Ag, Nb, As, Ni, Zn, Cr, Co, Cd, Sc, (Be, Se, Zr, Hf);

(*3*) manganese nodules for *Cu, Co, Mn, Pb, Ni, Te,* Tc, Tl, Ti, Zn;

(*4*) carbonates for Zn and disseminated Au;

(*5*) evaporites for Li, Sr, Zn, (Ba, Cs, Rb).

The discussion of the age distribution of those potential host rocks is included in the paragraph related to sedimentary rocks.

Interim summary

The discussion of ores of sedimentary affiliation revealed apparent evolutionary trends during geologic history. These apparent trends are well documented for iron and manganese and are also evident for deposits associated with fluviatile-deltaic clastic sediments. The trends for base metal deposits are suggestive, but as yet not well documented quantitatively.

If these trends are genuine, the question arises, as to whether they are secondary features (related to the level of erosion, diagenesis, metamorphism, etc.), "primary" phenomena (related to the evolution of the earth's crust, hydrosphere and atmosphere) or a combination of both. It is probable that we are dealing with a combination of primary and secondary factors. However, recognition of possible primary features in the apparent trends necessitates a general discussion of the evolution of the earth's crust and its sedimentary cover during geologic history.

EVOLUTION AND COMPOSITION OF THE CONTINENTAL CRUST

The formation of continental crust by fractionation from the upper mantle is a generally accepted feature of the terrestrial evolution (cf., e.g., Engel, 1963; Taylor, 1964; Ringwood, 1969; Wyllie, 1971, chapter 8 and others). The disputed questions are related mainly to the ways and rates of this differentiation.

The most comprehensive recent summary of "age" determinations of continental basement was published by Hurley and Rand (1969) (cf. Fig. 4). The diagram indicates that the crustal area per time unit is decreasing with increasing age. This decrease on a semi-

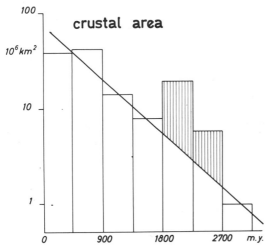

Fig. 4. Histogram of the present-day areal distribution of the world continental basement per 450 m.y. time units. (Based on the summary of Hurley and Rand, 1969.)

logarithmic scale is roughly linear for all time units, except for the 2700–1800 m.y. interval. This shows that the decrease in crustal area with time may be defined by the exponential equation:

$$M_t = M_T \cdot e^{-\lambda t}$$

where: M_t = crustal area (or mass, volume, etc.) at time t; M_T = total area of the crust; λ = "decay constant" defined by the slope of the line; t = time before present.

Possible models generating the pattern in Fig. 4 include:

(1) An exponentially increasing rate of formation of the upper continental crust towards the present (Hurley and Rand, 1969).

(2) Total fractionation of the whole upper continental crust at or before 3000 m.y. ago; and afterwards its continuous recycling either through the mantle (Armstrong, 1968; Armstrong and Hein, 1973) or within the crust.

(3) Linear growth of the upper continental crust combined with recycling, if the rate of recycling is faster than the rate of growth (cf. Garrels and Mackenzie, 1971, chapter 10 for mathematical treatment of the model). If the rate of recycling is ⩾5 times faster than the linear growth rate, the exponential curves for models (2) and (3) will be identical (cf. Garrels and Mackenzie, 1971, chapter 10).

It is obvious that models (1) and (2) are limiting cases and model (3) lies between the two extremes.

Since the continental crust and upper mantle differ by about one order of magnitude in their Rb/Sr ratio, the generation of radiogenic ^{87}Sr will be faster in the continental crust than in the mantle. Strontium from weathering of the upper continental (and oceanic) crust is carried into marine basins and accumulates in sea water and its precipitates. Therefore, the ^{87}Sr/^{86}Sr evolutionary curve for sea water may indicate one of the alternatives. Fig. 5 presents the generalized data for sea water as presently known (Veizer, 1971; Veizer and Compston, 1973) and the calculated evolutionary curves for upper mantle and continental crust. The sea-water data indicate a sudden influx of radiogenic ^{87}Sr between 2500 and 2000 m.y. ago. This correlates well with the peak in Fig. 4.

The calculated ^{87}Sr/^{86}Sr evolutionary curves for model (1) are not consistent with available estimates for the present day upper mantle and also do not explain the sharp rise in ^{87}Sr/^{86}Sr of sea water during the Early Proterozoic. This model is also difficult to reconcile with the probable higher heat production during the early history of the earth (cf. Birch, 1965; Lubimova, 1969), which would rather favour a decelerating rate of the crustal differentiation towards the present.

Model (2) with recycling through the mantle might satisfy all constraints with respect to Rb/Sr and Sr isotopes. However, Ringwood (1969) maintains that the fractionation of the continental crust is an irreversible process and the elements forming this crust would be the first to enter the melt upon subduction and so would fractionate again to the surface. The small difference in ^{87}Sr/^{86}Sr between oceanic tholeiites and island arc

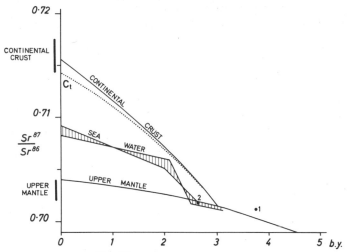

Fig. 5. Calculated $^{87}Sr/^{86}Sr$ models for the continental crust and upper mantle and a *generalized trend* of Sr-isotopic evolution of sea water (after Veizer, 1971). The Rb and Sr budget for the mantle and continental crust is from Armstrong (1968) and the model is compatible with higher rates of (upper ?) continental crust fractionation during the Late Archean and Early Proterozoic. Explanation:
C_t = crustal evolutionary curve for the model of·Hart and Brooks (1970) calculated by W. Compston; *1* = initial $^{87}Sr/^{86}Sr$ for the Amîtsoq gneisses of Greenland (Moorbath et al., 1972); *2* = initial $^{87}Sr/^{86}Sr$ of the Archean mantle-derived volcanics (Hart and Brooks, 1970).

andesites (cf., e.g., Gill, 1970) is indeed a strong point against a large crustal, sedimentary and sea water contribution to the mantle. Moreover, this model implies a steady-state situation within and between the mantle-crust systems. If so, the composition and volumetric relations of rocks within the continental crust should be the same over the entire period since its formation. This seems to be contradicted by the chemistry of clastic rocks. The most revealing example of such secular variations is the K_2O/Na_2O ratio (Fig. 6). It is likely that this and other trends are not of epigenetic origin (cf. Engel, 1963; Ronov and Migdisov, 1971; Veizer, 1973) and if so, reflect secular changes related to the evolution of the continental crust.

Similarly the model of recycling within the continental crust would imply steady-state conditions. Moreover, entire fractionation of the continental crust at or before 3000 m.y. ago would produce crustal $^{87}Sr/^{86}Sr$ ratio ⩾0.7095 (cf. Armstrong, 1968) which is too high to be compatible with the measured data (cf. Veizer and Compston, 1974 and Fig. 5).

This discussion favours model (3), with recycling mainly within the crust, as the most likely approximation to the real situation. However, the rates of crustal formation might have been somewhat higher during the early history of the Earth which might indicate a deviation in the direction of model (2). In particular, it seems that the Late Archean and Early Proterozoic were periods of exceptionally high rates of formation of (upper ?)

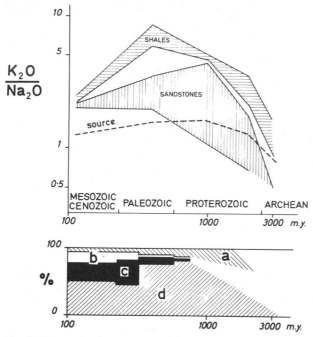

Fig. 6. Changes in the average K_2O/Na_2O ratio and mineralogical composition of clastic sedimentary rocks during geologic history. The chemical composition is based on the summary of Ronov and Migdisov (1971) and the data include the Russian Platform and North America. The curve marked as "source" indicates changes in average chemical composition of source regions and was computed by Ronov (1971b). The mineralogical variation of North American shales is from the data of Weaver (1967). The initial increase in K_2O/Na_2O ratio is related, at least in part, to a partly open-system diagenetic alteration of the kaolinite – "montmorillonite" – "degraded" illite assemblage into a stable assemblage of illite and chlorite.
a = chlorite; b = kaolinite; c = "montmorillonite"; d = illite.

continental crust as indicated by Figs. 4 and 5 and many other independent observations (cf. Ronov, 1964; Goodwin, 1970; Condie, 1973; Goldich and Mudrey, 1973; Veizer, 1973; and others).

The $^{87}Sr/^{86}Sr$ ratios for Archean sea water are not only compatible with the upper mantle data (Fig. 5) but they also define a Rb/Sr slope of 0.02–0.04, which is similar to typical ratios for present-day and ancient island arcs (0.03–0.05) (Gill, 1970; Hart et al., 1970; Jakeš and White, 1971). This is in good agreement with geological observations (cf. Goodwin, 1968; Baragar and Goodwin, 1969; Glickson, 1971; etc.), which show rather conclusively the compositional similarity of Archean "greenstone belts" to present-day island arcs. If so, the Archean "protocrust" was formed predominantly by volcanics of tholeiitic and calc-alkaline associations and their derivative sediments (greywackes). The felsic differentiates, if present, would mostly be sodium-rich tonalites (cf. Glickson, 1971; Glickson and Sheraton, 1972; Fyfe and Brown, 1972; Glickson and Lambert, 1973) similar to those described by Gill (1970) from Fiji, where they form the last stage of island arc development.

During the Late Archean and Early Proterozoic K-rich granites became a very prominent phase in the upper continental crust. The sudden widespread evolution of K-granites is a remarkable feature of Precambrian geology. This process started about 3000 m.y. ago in South Africa (Kröner et al., 1973) and at about 2700 m.y. ago on other continents (cf. Ronov, 1964; Glickson, 1971; Goldich and Mudrey, 1973; Condie, 1973; and others). This "event(s)" may mark the time when protocontinental crust became thick enough to undergo large-scale reprocessing. This would be at ~25 km at present geothermal gradients (cf. Winkler, 1967, p. 4 and chapter 16), but due to the already mentioned probability of a higher temperature gradient in the early history of the earth, the depth might have been proportionally less. Afterwards, the fractionation of new continental crust and particularly its stratification into "granitic" and "basaltic" layers by reprocessing ($\approx$recycling) probably continued at a slower rate.

This sudden change at the Archean–Proterozoic boundary thus denotes the time of formation of the felsic nuclei of stable platforms and emergence of an early "geosynclinal" tectonics with a facies association similar to Phanerozoic belts.

For latest, added information, see "Note added in proof" (p. 36).

EVOLUTION AND COMPOSITION OF SEDIMENTARY ROCKS

The most recent estimates of the total mass of sediments vary between 1.7 and $3.2 \cdot 10^{18}$ t (Kuenen, 1950, p. 386; Poldervaart, 1955; Ronov and Yaroshevski, 1969; Garrels and Mackenzie, 1971, p. 257). The age distribution of this sedimentary mass is presented in Fig. 7. The slope of the decrease of sedimentary mass per time unit (dotted) is almost parallel to the slope for crustal areas in Fig. 4 (note the logarithmic scale in Fig. 7). The major differences in the estimates of the total sedimentary mass are due to uncertainties in the value for the Precambrian. Since the Precambrian estimate of Garrels and Mackenzie is the highest, the general slope for sediments is either as presented here or steeper.

As in the case of the continental crust, *the most probable model for the generation of this age pattern may be a modified linear accumulation model* of Garrels and Mackenzie (1971). These authors concluded that the time distribution of sediments (Fig. 7) is consistent with a total deposited sedimentary load of about 5 times the present mass and its permanent cycling through geologic history. Therefore, *the fraction of sediment remaining is inversely proportional to its age.* Due to the rate of cycling, which is much faster than addition of the new sedimentary mass, the favoured model will be essentially cannibalistic. As a consequence, the "mixing" of sediments will be enhanced and the average composition of clastic rocks will therefore reflect the general evolutionary trend rather than the local igneous source area composition.

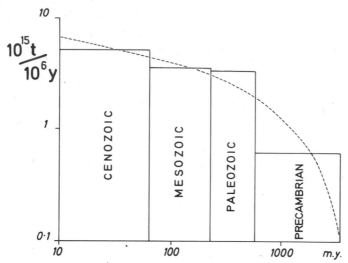

Fig. 7. Present-day distribution of sediments as a function of their geologic age, based on the data of Garrels and Mackenzie (1971). (Due to its construction as mass/time unit, the histogram does not represent an absolute mass of a given era.)

Unanswered by this discussion is the question of whether the composition of this sedimentary mass was constant through geologic time, even if its remaining mass was not. The accumulated data (cf. Rutten, 1962, 1966; Engel, 1963; Strakhov, 1964, 1969; Ronov, 1964, 1968, 1971a; Cloud, 1968, 1971, 1972; Garrels and Mackenzie, 1971; Veizer, 1973; etc.) show quite conclusively that this was not the case. In the present state of art the treatment of this subject can be at best only semiquantitative. However, the general trends are evident even from the limited data, and a short summary is presented below.

(1) Clastic sediments (cf. Ronov, 1964). Graywackes were dominant sediments during the Archean. Since then, their proportion in sediments decreased by a factor of 4. Arkoses on the other hand were dominant during the Early and Middle Proterozoic. Conglomerates and orthoquartzites were scarce in the Archean, but fairly abundant since the Early Proterozoic. Detrital siltstones and mudstones of red-bed type are known since about 2000—1800 m.y. ago (cf. also Cloud, 1968). The proportion of shales is more or less stable, although their composition is not.

(2) Carbonates (cf. also Kuenen, 1950, p. 392; Strakhov, 1964; Veizer, 1973). Calcium and magnesium carbonates are rare in the Archean and their relative proportion in sedimentary sequences is increasing with decreasing age. Iron and manganese carbonates were discussed in the part dealing with their ores. The oldest sequences, particularly in the Early and Middle Proterozoic, were dominated by lagoonal and littoral—sublittoral early diagenetic (and primary ?) dolomites. Since about 1000 m.y. ago the dolomites were

superseded by inorganic (and biochemical) limestones of similar depositional milieu. During Paleozoic time the most important carbonate rocks were littoral—neritic organogenic and organodetrital limestones and late diagenetic dolomites formed at their expense. Since the Jurassic the dominant carbonate sediments are bathyal and abyssal (?) (deep sea) limestones (Twenhofel in Kuenen, 1950, p. 392). Thus the trend for carbonates is just the opposite of the one observed for iron and manganese ores. Whereas the axis of maximum sedimentation for Fe—Mn ores shifted landwards with decreasing age, the trend for carbonates was seawards.

(3) Sulphates. With the exception of a few localities (cf. Engel, 1963; Veizer, 1973), which may be as old as 1300 m.y., the majority of gypsum and anhydrite deposits are of Phanerozoic age (Strakhov, 1964, 1969; Ronov, 1964; Cloud, 1968).

(4) Phosphorites. Phosphorites in economic concentrations are known only since Middle Riphean and Middle Sinian (Strakhov, 1964; Bushinskii, 1966) and are therefore very likely younger than about 1000 m.y.

(5) Salts. Although pseudomorphs after NaCl are known from the whole Proterozoic column, the economic deposits of sodium chloride as well as K—Mg salts are confined to the Phanerozoic (cf. Strakhov, 1964; Ronov, 1964). However, collapse and solution breccias, which might possibly be related to the original evaporites are known from sequences as old as 2700 m.y. (Veizer, 1973; Mackenzie and Garrels, 1974).

(6) Siliceous rocks. No definite trend for siliceous rocks with age is known. The chert content of iron ores and carbonates is increasing with age (cf. Fig. 16). However, since Mesozoic times pelagic siliceous rocks have been widespread (cf., e.g., Kuenen, 1950, p. 361) and it may be that the actual trend is only a shift of the major formation of siliceous rocks into the deep sea with decreasing age. However, this inference must be tested by more definite data.

(7) Manganese nodules. Although by contrast this is not a distinct group of sedimentary rocks its great economic potential merits a comment. The concentration of manganese nodules so typical in Recent pelagic areas (cf., e.g., Mero, 1965) is found only rarely in older sediments. Similar "condensation horizons" are rare in the Mesozoic and even more so in the Paleozoic (cf., e.g., Aubouin, 1965, pp. 126—129).

(8) Organics. Formation of organics is directly related to the evolution of organic life and its decay. The amount of organic matter in sediments is increasing towards the present (Fig. 8) paralleling the growth of organic life. Bituminous shales are known since the Archean and no age trend is observed. Oil, gas and asphalt accumulations are known principally in the Phanerozoic, with 14% of identified world oil resources in the Paleo-

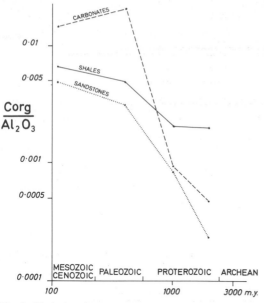

Fig. 8. Variation in average C_{org} of sediments during geologic history, based on the summary of Ronov and Migdisov (1971) for the Russian Platform. Recently, Cameron and Jonasson (1972) reported C_{org} concentrations from Archean and Early Proterozoic shales of the Canadian Shield which are higher, by a factor of 2–3, than those reported previously for the Precambrian sediments of the Russian Platform and North America (cf. table 2 in Ronov and Migdisov, 1971). If so, it might indicate that the trends in this figure are not representative. Also the high C_{org}/Al_2O_3 ratio of carbonates is due to the low Al_2O_3 content and not to the high concentration of C_{org}.

zoic, 57% in the Mesozoic and 29% in Cenozoic rocks (McCulloh in Brobst and Pratt, 1973). This is probably related directly to the considerable increase in biomass observed at the Precambrian–Cambrian boundary (Fig. 8). Coal-like deposits are known from the Early Proterozoic (Strakhov, 1964; Cloud, 1968), but the important economic deposits date from the Late Paleozoic. According to Volkov (1968) there is, at least in the U.S.S.R., a progressive shift in the site of coal formation from geosynclinal (mostly paralic) to terrestrial environments with decreasing age.

The evolution of organics is directly related to the evolution of organic life, with appearance of Procaryota prior to 3200 m.y. ago (Cloud, 1968; Engel et al., 1968), Eucaryota between 2000 and 1000 m.y. ago, and the evolution and "explosion" of Metazoa at about 600 m.y. ago (Cloud, 1968; Glaessner, 1968). This was followed by migration of the life onto the continents in Silurian–Devonian.

CONSEQUENCES

To interpret the apparent age trends discussed in the first part of this paper, it is inevitable to consider them together with the general trends for crust and sediments.

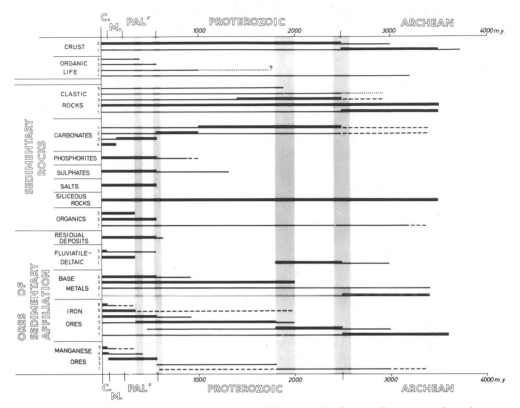

Fig. 9. Summary of variations in the composition of the crust, biosphere, sedimentary rocks and ores of sedimentary affiliation during geologic history.

Crust: *1* = Archean "protocrust"; *2* = thick continental crust with "granitic" upper layer.

Organic life: *1* = Procaryota; *2* = Eucaryota; *3* = Metazoa; *4* = terrestrial life.

Clastic rocks: *1* = greywackes; *2* = shales and slates; *3* = arkoses; *4* = oligomictic conglomerates and orthoquartzites; *5* = red beds.

Carbonates: *1* = early diagenetic dolomites; *2* = inorganic and/or biochemical (?) limestones; *3* = littoral-neritic organogenic and organodetrital limestones and late diagenetic dolomites; *4* = deep-sea limestones.

Organics: *1* = bituminous shales; *2* = oil and gas; *3* = coal.

Fluviatile-deltaic deposits: *1* = conglomerate-U-Au-pyrite type; *2* = sandstone-U-V-Cu type; *3* = placers and palaeoplacers.

Base metal deposits: *1* = volcanogenic and volcano-sedimentary type; *2* = sedimentary basinal type; *3* = sedimentary red-bed type; *4* = carbonate-Pb-Zn type.

Iron ores: *1* = Algoma type; *2* = Superior type; *3* = Clinton type; *4* = Bilbao type; *5* = Minette type; *6* = river bed, bog iron, laterite and similar types.

Manganese ores: *1* = "jaspilitic" type; *2* = volcano-sedimentary and orthoquartzite-siliceous shale-Mn carbonate types; *3* = carbonate association; *4* = orthoquartzite-glauconite-clay association; *5* = marsh and lake deposits.

From comparison of the slope for total identified reserves per time unit with the slope for total sediments and crust (dotted line in Figs. 1—3 and 7) it can be inferred that the residual deposits (cf. bauxites) and probably also manganese ores are decreasing at a faster rate than the total of sediments with time. The iron ores on the other hand are decreasing at a slower rate. The data are not sufficient as yet for other types of ores or sediments, but from the general lack of several types of sedimentary rocks in older sequences (cf. Fig. 9) it can be inferred that the slope of their decrease with time will steepen in the order: carbonates < phosphorites < sulphates < salts < organics. On the other hand, the higher proportion of clastic rocks in older sequences would indicate a shallower slope of their decrease with time than the slope observed for total sediments (cf. also Ronov, 1964; Garrels and Mackenzie, 1971, p. 260). In other words, *the relative proportion of sedimentary types is a function of age* and therefore variable.

There are two basic schools of interpretation of the observed secular trends. The first one (Rutten, 1962, 1966; Ronov, 1964; Strakhov, 1964, 1969; Cloud, 1972) considers the majority of the variations as a result of an irreversible evolution of the earth's crust, hydrosphere, atmosphere and biosphere. The second group (Garrels and Mackenzie, 1971; Garrels et al., 1972) considers *this pattern as basically due to recycling in which the cycling of various types of rocks proceeds at different rates, depending on their resistance to chemical and mechanical destruction.* A model based on the latter reasoning may explain very elegantly the general pattern of temporal variations for major groups of rocks (cf. Garrels and Mackenzie, 1971, chapter 10). Thus the "appearance" of carbonates, mostly since 2500 m.y., Ca-sulphates since about 1300 m.y. and salts since ~600 m.y. ago may be due to their increasing rate of cycling related to the increase in solubility from carbonates to salts. However, there are difficulties with phosphorites, clastics and the lack of carbonates in the Archean (Cameron and Baumann, 1972; Veizer, 1973), and also the confinement of coal and oil to the Phanerozoic is probably a primary feature.

On the other hand the chemical destruction of residual ores and sediments is negligible, yet they are confined to the younger periods of the earth's history. However, since the statistical probability of mechanical erosion of terrestrial deposits is higher than that of the marine ones, the confinement of residual ores to Phanerozoic may be expected. This in fact may also explain the longer preservation of karst-type bauxites than of the residual bauxite accumulations on igneous and clastic rocks (Fig. 1). An alternative explanation for the lack of bauxites in Precambrian would have to be based on a postulate that the type of weathering for those two periods was different or on some other evolutionary features, although some of the bauxite deposits might have been metamorphosed to sillimanite—kyanite—andalusite schists with corundum (Bushinskii, 1958, in Allen, 1972).

The statistical probability of mechanical destruction should decrease from terrestrial to geosynclinal environments. Therefore within-group variations should indicate a progression of facies from continent to geosyncline with increasing age. Such a feature can indeed be observed in the case of detrital rocks, iron and manganese ores, organics and possibly base metals whereas carbonates (and siliceous rocks?) indicate just the reverse. Moreover, as already mentioned, the temporal distribution of organics is strictly related to the evolution of organic life and the chemical, mechanical and tectonic factors are of secondary importance only. In addition, in the case of iron and manganese ores it may be also stated that the Precambrian iron and manganese formations do not have Phanerozoic equivalents of similar magnitude.

Summarizing the above discussion it seems feasible to conclude, that *the differential cycling of Garrels and Mackenzie is probably the decisive factor for secular variations among major groups of sediments and ores, whereas the secular facies differences within groups were influenced by this factor only to a smaller degree. Such facies differences,* as well as the occurrence of organics and possibly phosphorites, *would demand additional factors for their explanation* very likely *related to the general evolution of the earth's crust, hydrosphere, atmosphere and biosphere.* Considering the great dispersal rate of fluviatile-deltaic sediments, the high frequency of Witwatersrand—Blind River-type ores in the Early Proterozoic is a distinct anomaly and this supports again the conclusion that the facies variations within groups are of a "primary" nature.

An attempt to integrate all these variables into a unified system will be presented in the following sections. From the review of the ores of sedimentary affiliation it is evident that, excluding strictly mechanical accumulations (placers) with their suite of economic elements, the metals of interest are mainly U, V, Au, Cu, Pb, Zn, Fe and Mn. Considering sedimentary rocks as a possible source of these metals, it is of great interest to know the temporal variations in chemistry of sediments with respect to those elements. Such data are however limited.

DISCONTINUITIES IN THE GEOLOGIC RECORD

The apparent evolutionary features discussed previously are strongly interrelated. The summary of those variations is presented in a qualitative to semiquantitative manner in Fig. 9. This figure shows at least four marked discontinuities in the overall trends. These are the Archean—Proterozoic boundary (2500 m.y. ago), the period between 2000 and 1800 m.y., the Precambrian—Cambrian boundary and the Late Paleozoic. There may be other discontinuities (1000 m.y.; Mesozoic—Cenozoic boundary), but they are not as yet clearly defined.

The 2500-m.y. discontinuity is marked by a decrease in importance of the "continental protocrust" (greenstone belts), greywackes, Algoma-type iron (and manganese ?) ore formations and also of volcanogenic and volcano—sedimentary base metal concentrations. The same period marks a growth of a thick continental crust with its "granitic" upper layer, increase in importance of arkoses, carbonates, orthoquartzites, conglomerates, Superior-type iron (and manganese ?) ore deposits and of fluviatile-deltaic deposits of the Witwatersrand—Blind River type. All *these features would be consistent with the forma-*

tion of continental nuclei of felsic composition from the Archean "protocrust" at this time interval.

The period between 2000 and 1800 m.y. was characterized by disappearance of common detrital pyrite and uraninite from the fluviatile-deltaic deposits (cf., e.g., Rutten, 1962; Holland, 1973) by the sharp decrease in proportion of iron (and manganese ?) ores of Superior type as well as of detrital siderite; and by the appearance of red beds, red-bed copper deposits and Clinton-type iron ores. Since all these changes are related to poly-valent metals (Fe, Mn, U, Cu) it is reasonable to suppose, that *this period marked a change in the oxidation state of the contemporaneous sedimentary environments.*

The beginning of *the Phanerozoic is associated with the appearance of Metazoa* and therefore an increase in the $C_{org.}$ of sediments and appearance of fossil fuels and organo-detrital carbonates with their associated Pb–Zn, Fe (Bilbao) and Mn ores. It is conceiv-able that the "appearance" of phosphorites mainly in this period may be related to the same factor. On the other hand the "appearance" of bauxites, sulphates and salts is probably related to recycling as discussed previously.

The *Late Paleozoic* discontinuity *is marked by the migration of life onto the conti-nents* and this caused the accumulation of organic material in terrestrial environments (coal). The high proportion of the Minette-type iron ores, which are often associated with coal seams, and U–V–Cu fluviatile-deltaic deposits (Colorado Plateau) are also consistent with this phenomenon.

It appears therefore that the last two discontinuities were basically caused by organic evolution. Since the increase in biomass and the oxygen and carbon dioxide budgets are directly related through photosynthesis (cf., e.g., Berkner and Marshall, 1964; Tappan and Loeblich, 1970; Tappan, 1971; and others) it is reasonable to suppose, that the last three discontinuities were related essentially to the evolution of the earth's atmosphere and hydrosphere, whereas the main cause of the Archean–Proterozoic discontinuity was related to the evolution of the crust.

GENERAL TENDENCIES IN THE EVOLUTION OF THE EARTH'S CRUST, ATMOSPHERE, HYDROSPHERE AND BIOSPHERE – A DISCUSSION

Evolution of the crust

As already discussed, the unstable Archean "protocrust" was composed of tholeiitic and calc-alkaline volcanic rocks with their orogenic flysch-type sediments (greywackes) and basins and trenches with random sublinear distribution (greenstone belts), as well as by some granulitic and "granitic" regions of tonalitic composition (cf. Goodwin, 1968, 1970, 1973; Salop, 1969; Anhauser et al., 1969; Viljoen and Viljoen, 1969; Glickson, 1971; etc.). Its average chemical composition was probably similar to, or somewhat more mafic than, andesite (cf. Baragar and Goodwin, 1969). As shown later, the segregation of metals into economic accumulations of sedimentary affiliation could be related to general evolution of the outer spheres of the earth. However, the Archean may be somewhat exceptional in this respect, since the extensive volcano-exhalative activity may provide an additonal source of metals. This may help to explain the high frequency of volcanogenic and volcano-sedimentary base metal deposits and of the Algoma-type iron formations.

The Proterozoic and younger stable continental crust is characterized by compositionally varied non-sequential volcanism, mature shelf-type sediments and tectonically by stable platforms, broad basins and long linear "geosynclinal" features (e.g. Goodwin, 1970; Rogers and McKay, 1972). Its average chemical composition is similar to, or somewhat more mafic than, andesite, but its upper layer has an average composition similar to granodiorite (cf. Wedepohl, 1969).

The crustal evolution, if correct, should cause secular variations in average composition of clastic sedimentary rocks. This on the other hand may have a direct bearing on the consideration of sedimentary rocks as a source of metals in ores of sedimentary affiliation.

Consequences for the chemistry of clastic rocks. Accepting the modified linear accumulation model as a first-order approximation to the actual situation, the total mass of sediments might have been $\geqslant 1/2$ of its present mass at the end of the Archean. Since the new material added continually during the Archean was of similar average composition, no secular trends in chemistry of clastic sediments would be expected. However, with the formation of the thick continental crust of about 2500 m.y. ago the fractionation of the new "basaltic—andesitic" crust in island arcs would be followed by its further fractionation in the island arcs of the Andean type as well as within the crust. If so, the part of the continental crust exposed to weathering would now be the "granitic" layer of average granodioritic composition (cf. Wedepohl, 1969). Consequently, the new material added to the sedimentary mass will displace the average chemical composition of clastic sediments towards this felsic end member. The slope of the change with time for a particular element will increase with increasing difference in concentration between "granodiorite" and "basalt—andesite" and will tend to flatten when approaching the felsic end member. In general, the slope will be very small for major elements and high for trace elements. Comparing, for example, the "andesite" and "granodiorite" of Taylor (1969) the expected secular trends in composition of clastic sedimentary rocks should be negligible for Al and Si; the slope of increasing Total Fe (and Fe^{3+}), Mn, Mg and Ca with age may be very slight, whereas the slope of decreasing K/Na ratio should be considerable (Fig. 6). Similarly, there should be a considerable decrease with increasing age in U, Th, Rb, Ba, Pb, REE and their fractionation pattern and increase in Cu, Co, Ni, V, Cr, Ag, Zn, Au and others. This prediction agrees quite well with the known data (cf. Engel, 1963; Ronov and Migdisov, 1971; Ronov et al., 1972; and others). For present purposes the elements of interest would be Fe, Mn, Pb, Zn, Cu, U, V, Ag, Au. The summaries available to the author are for total Fe, Fe^{3+} (which forms the major proportion of the detrital iron), Mn, U and

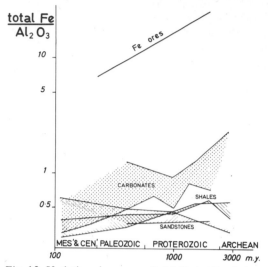

Fig. 10. Variations in average Total Fe (as Fe_2O_3) composition of sedimentary rocks during geologic history. The graphs for clastic rocks and carbonates of North America and Russian Platform are based on the summary of Ronov and Migdisov (1971). The carbonate field includes also the data for Australia (Veizer, 1971) and the trend for iron ores is based on the North American data of Lepp and Goldich (1964).

Zn (cf. Ronov and Migdisov, 1971; Cameron and Jonasson, 1972) and they are indeed following the predicted pattern (cf. also Figs. 10—14). It follows, that for example if the clastic rocks are the source of metals for strata-bound ores, the statistical chance of

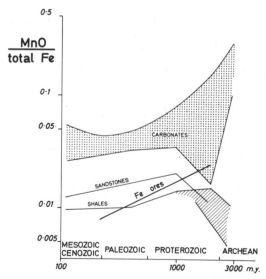

Fig. 11. Variations in average MnO/Total-Fe ratio of sedimentary rocks during geologic history. Sources of the data as in Fig. 10. The trend for iron ores does not seem to be supported by Table III. This casts some doubt on the representative nature of the Lepp and Goldich's (1964) data.

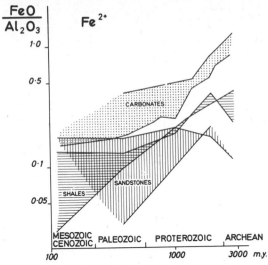

Fig. 12. Variations in average Fe^{2+} concentrations of sedimentary rocks during geologic history. Sources of the data as in Fig. 10.

younger base metal accumulations being rich in Pb may be higher. As already mentioned, such a trend was indeed observed for Canada. Consequently, the general pattern of secular variation in evolution of endogenic ores (cf., e.g., Bilibin, 1968; and summarized recently in the third part of the book edited by Borovikov and Semenov, 1969) may not be entirely fortuitous, because the anatexis of younger sediments with their more "felsic" composi-

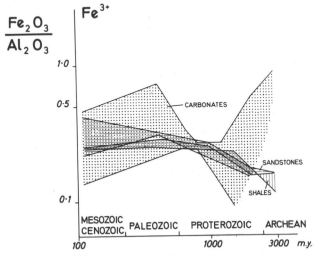

Fig. 13. Variations in average Fe^{3+} concentrations of sedimentary rocks during geologic history. Sources of the data as in Fig. 10.

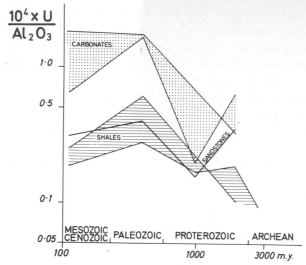

Fig. 14. Variations in average uranium concentrations of sedimentary rocks during geologic history. (Based on the summary of Ronov and Migdisov, 1971, for North America and the Russian Platform.)

tion will favour formation of aureoles enriched in those metals. Therefore at least the high frequency of Ni, Cr, (Au) economic concentrations in the Archean may be to some extent a primary feature, but there is little doubt that the major factor in this apparent evolutionary trend is the level of erosion (cf. Stanton, 1972, chapter 19; Laznicka, 1973).

Evolution of oxygen

Although the evolution of the crust may have some bearing on base metals and Algoma-type Fe ores, other trends for Fe, Mn and Cu—U—V associated with red beds are not entirely compatible with such an explanation. Both iron and manganese ores show definite secular trends (Figs. 2, 3), but the complementary trends in composition of detrital rocks are only slight or, non-existent (cf. also Figs. 10, 11). However, the slopes for "chemical" sediments are increasing markedly with increasing age for Mn, Total Fe and Mn/Fe ratio. This together with the fact of increase in soluble Fe^{2+} for all lithologies hints that the concentration of those metals in their parent solution (sea water ?) was higher in the early history of the earth. Unpublished work of the present author indicates that the trends for carbonates are not epigenetic features and also the lack of the complementary negative slope for Fe^{3+} shows that the Fe^{2+} trend is due only in part to the reduction of ferric iron during diagenesis and epigenesis (Figs. 12, 13).

As already mentioned, Fe, Mn, Cu, V and U can be present in sedimentary environments in several oxidation states, with Fe and Mn being soluble at lower and Cu, V and U at higher states (cf. Garrels and Christ, 1965; Krauskopf, 1967). From theoretical considerations (e.g. Holland, 1964; French, 1966) it is probable, that the P_{O_2} in the atmosphere was increasing towards the present mainly due to photosynthetic activity of the biosphere. Prior to the 1800—2000-m.y. interval, P_{O_2} in the atmosphere was probably controlled by reduced Fe^{2+}, Mn^{2+}, sulphide and organic C buffer, and only after satisfying the first three reservoirs (cf. Sillén, 1965, in Stumm and Morgan, 1970, p. 338; Stumm and Morgan, 1970, p. 335; Garrels et al., 1973; Garrels and Perry, 1975) could the free oxygen start to accumulate in significant quantities. Therefore, the high frequency of iron ores prior to this period (cf. Lepp and Goldich, 1964 and the references in James, 1966) as well as the presence of detrital pyrite, uraninite and siderite could be related to this factor or to a variant of such a model (cf. La Berge, 1973; Cloud, 1973; Holland, 1973; Goodwin, 1973; Drewer, 1973, and others). For example, Holland (1973) and Garrels and Perry (1975) calculated the dissolved iron in such sea water as 3—30 p.p.m. as opposed to about 0.01 p.p.m. at present (cf. Krauskopf, 1967, table 3). Grandstaff (1974) calculated that for uraninite to survive as a detrital mineral, the P_{O_2} must have been less than $\sim 10^{-2}$ to 10^{-6} of its present atmospheric level.

With increasing P_{O_2}, linked with organic C buffer (Garrels and Perry, 1975) and/or with the availability of nitrate and phosphate which may control the growth of the biomass (Holland, 1972), manganese and particularly iron were less mobile and accumulated progressively closer to their continental source, with some iron as hematite remaining in continental-deltaic red beds. The same factor also caused oxidation of the detrital pyrite, uraninite (and siderite) and hence their absence from younger fluviatile-deltaic deposits. The oxidation of U, V and Cu led to their mobilization by oxidized surface and/or diagenetic waters (for example in red beds — cf. Vine and Tourtelot, 1973), although the oldest red bed deposits (Udokan· ~ 2000 m y) still contain some copper of possible detrital origin (cf. Cox et al., 1973). After encountering C_{org}-rich layers these elements could be concentrated into economic accumulations such as the basinal Paleozoic black shales of Sweden for U (Finch et al., 1973), the Kupferschiefer (Wedepohl, 1971) or in the bleached parts of red beds themselves. Only with migration of life into terrestrial environments in Late Paleozoic and subsequent accumulations of C_{org} in such sediments was the stage set for concentration of those metals in terrestrial environments (e.g., Colorado Plateau) (cf. Fisher, in Brobst and Pratt, 1973, p. 683). Moreover, this feature may als explain the concentration of manganese deposits in younger geological sequences. Since the Mn/Fe ratio of Precambrian "chemical" sediments is higher than of their Phanerozoic counterparts (Fig. 11), they contained a high proportion of manganese. Separation of Mn from Fe is achieved by higher oxidation potential of iron (cf. Krauskopf, 1957, 1967, p. 266). Therefore with increasing partial pressure of oxygen and recycling of older sediments manganese was progressively separated into its own deposits.

Evolution of carbon dioxide (?)

A glance back at Fig. 9, shows that the acceptance of crustal and oxygen evolution (complementary with the evolution of organic life) can help to explain the major part of

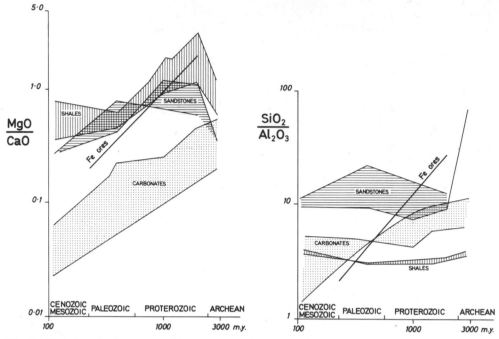

Fig. 15. Variations in average MgO/CaO ratio of sedimentary rocks during geologic history. Sources of the data as in Fig. 10. In addition, the shale field includes also the Australian trend of Van Moort (1973).

Fig. 16. Variations in average SiO_2/Al_2O_3 ratio of sedimentary rocks during geologic history. Sources of the data as in Fig. 10.

the observed trends. Other features, such as the decreasing proportion of residual deposits, salts, sulphates and carbonates may be due to the cyclic nature of the sedimentary process. The apparent trends not affected by the above variables would be the increasingly dolomitic nature of carbonates (Fig. 15) and increasing SiO_2 (chert) content of the "chemical" precipitates with increasing age (Fig. 16) (Daly, 1909; Vinogradov et al., 1952; Chilingar, 1956; Veizer, 1971, 1973). Since the generation of oxygen by photosynthesis is chiefly at the expense of carbon dioxide, it might be expected that this would influence the CO_2 budget. According to Holland (1965) the P_{CO_2} of the atmosphere is controlled by mineral equilibria. In coexistence of chlorite, dolomite, calcite and quartz the variation in P_{CO_2} could have been within $10^{-3.7}$ and $10^{-2.9}$ atm. (Stumm and Morgan, 1970, p. 416). Garrels and Perry (1975) postulate on similar basis P_{CO_2} as high as 10^{-2} to 10^{-1} atm. for the Archean. In any case, the increasing P_{CO_2} with age will cause stabilization of carbonates at the expense of silicates. Due to the absence of silica-secreting organisms, sea water would be saturated with respect to amorphous silica (Siever, 1957) and thus the excess SiO_2 will precipitate as chemical (?) sediment. Also accepting that the pH of sea water is determined by a heterogeneous mineralogical "buffer" system (e.g.,

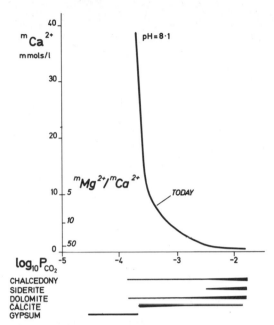

Fig. 17. Stability of mineral phases and variations in $a_{Ca^{2+}}$ with P_{CO_2} in a system with fixed pH. (Modified after Holland, 1965, and the sources listed in Veizer, 1973.)

Sillén, 1961, 1967; Holland, 1965; Mackenzie, 1969; Stumm and Morgan, 1970, p. 418; Wangersky, 1972; and others) the activity of Mg^{2+} in sea water would be controlled by silicate equilibria, whereas the $a_{Ca^{2+}}$ being controlled by carbonate equilibria, will decrease with increasing P_{CO_2} (Holland, 1965). Consequently the higher Mg/Ca ratio of ancient sea water may contribute to more extensive dolomite formation in the early history of the earth. The succession of mineral phases along the decreasing P_{CO_2} path is in agreement with the actual situation observed in the geologic record (cf. Fig. 17).

FINAL COMMENTS

The apparent evolution of ores of sedimentary affiliation as well as of sedimentary rocks and their chemistry during geologic history is probably a product of the irreversible evolution of the outer spheres of the earth combined with cycling phenomena and proceeding in a spiral way. The irreversible evolutionary features may be consistent with several variables such as the evolution of the crust, organic life, P_{O_2} and P_{CO_2} and their mutual interrelations.

The author is aware that the discussion in this chapter is based on limited data only and influenced also by personal, time and space limitations. It would be an irresponsible attitude to maintain that this review is correct in all its points. It is inevitable that a substantial part of the interpretation as well as of the "basic" data will prove erroneous. The chapter should be considered only as an attempt to show the considerable potential of a concentrated research into this problem. Such a concentrated research might prove not only as exciting as the current glamorous projects, but also considerably cheaper and with immediate benefit to mankind. In the meantime it is hoped that the present review will at least help to formulate several pertinent questions for further research.

NOTE ADDED IN PROOF

This review was written in 1973. Recent publications require some additional comments, particularly related to the part dealing with crustal evolution (see pp. 17–21).

There is a broad consensus among students of the Precambrian, that the Early Proterozoic continental crust has features similar to the present-day continental crust. As to the nature of the Archean crust, scientists are about evenly divided in favour of various alternatives of the two major hypotheses (cf., A discussion on the evolution of the Precambrian crust; *Phil. Trans. R. Soc. Lond., Math. Phys. Sci.,* 273). According to the first hypothesis, the continental crust was only of subordinate importance during the Archean, whereas, according to the second one, the continental crust similar to its present-day counterpart in volume and composition, although thinner and distributed more evenly on the global surface, was already in existence in the Archean. Consequently, the Archean–Proterozoic discontinuity may represent a structural as well as compositional change, if the first alternative is accepted, or only a structural change in the case of the second alternative.

It is likely that the actual situation was somewhere between the two extremes. However, the recently published data of Engel et al. (*Geol. Soc. Am. Bull.,* 85: 843) show convincingly that he Archean crust was more mafic than the Proterozoic one. Their secular trend for alkalies is mirrored in detail by the strontium isotopic trend for sea water (generalized in Fig. 5). The Archean sea water/strontium isotopic ratios indicate that crustal layers in contact with meteoric waters, were close to "upper mantle" in their composition. Therefore the near-surface crustal layers, which were the source for sediments, were more mafic than the Proterozoic near-surface continental crust. However, the composition of deeper crustal layers remains open for discussion.

A posssible cause of this compositional and structural discontinuity, based on decreasing temperature gradients during geologic history, was suggested by Green (*Geology,* 3: 15).

ACKNOWLEDGEMENTS

The author would like to acknowledge his gratitude to Mr. E.W. Hearn for preparation of all illustrations; Miss L.G. Dubeau for the typing of the manuscript; Dr. G.A. Armbrust and Mr. M.R. Gibling for the review of the text; Dr. D.F. Sangster for his agreement to incorporate some of his ideas into Table II; Dr. R.J.W. Douglas and the Geological Survey of Canada for permission to republish Table III and Profs. R.M. Garrels, F.T. Mackenzie and K.H. Wedepohl for the permission to publish information from the preprints of their papers and/or comments to several points dealt with in this review. However, this does not necessarily indicate their agreement with the above presentation.

In addition to acknowledgements, the author owes apology for omission of many important older publications related to the subject. Their quotation is mostly included in the reference lists of the quoted papers.

REFERENCES

Allen, J.B., 1972. Aluminum ore deposits. In: R.W. Fairbridge (Editor), *The Encyclopedia of Geochemistry and Environmental Sciences*. Van Nostrand Reinhold, New York, N.Y., pp. 23–27.

Alvarado, B., 1970. Iron ore deposits of South America. In: United Nations (Editor), *Survey of World Iron Ore Resources*. United Nations, New York, N.Y., pp. 302–380.

Anhäusser, R.C., Mason, R., Viljoen, J.G., and Viljoen, R.P. 1969. A reappraisal of some aspects of Precambrian shield geology. *Geol. Soc. Am. Bull.*, 80: 2175–2200.

Armstrong, R.L., 1968. A model for evolution of strontium and lead isotopes in a dynamic earth. *Rev. Geophys.*, 6: 175–199.

Armstrong, R.L. and Hein, S.M., 1973. Computer simulation of Pb and Sr isotope evolution of the earth's crust and upper mantle. Geochim. Cosmochim. Acta, 37: 1–18.

Aubouin, J., 1965. *Geosynclines*. Elsevier, Amsterdam, 335 pp.

Baragar, W.R.A. and Goodwin, A.M., 1969. Andesites and Archean volcanism of the Canadian Shield. *Ore. Dept. Geol. Mining Ind. Bull.*, 65: 121–142.

Berkner, I.V. and Marshall, I.G., 1964. The history of growth of oxygen in the earth's atmosphere. In: J.P. Brancazio and A.S.W. Cameron (Editors), *The Origin and Evolution of Atmospheres and Oceans*. Wiley, New York, N.Y., pp. 102–126.

Bilibin, Yu.A., 1968. *Metallogenetic Provinces and Metallogenetic Epochs*. Queens College Press, Flushing, N.Y., 35 pp.

Birch, F., 1965. Speculations on the earth's thermal history. *Geol. Soc. Am. Bull.*, 76: 133–154.

Borovikov, L.I. and Semenov, A.I., 1969. *Geological Framework of the U.S.S.R.*, 5. Nedra, Moscow, 494 pp. (in Russian).

Brobst, D.A. and Pratt, W.P., 1973. *United States Mineral Resources – Geol. Surv. Prof. Pap.*, 820: 722 pp.

Bushinskii, G.I., 1966. Old phosphorites of Asia and their genesis. *Acad. Sci. U.S.S.R., Geol. Inst. Trans.*, 149. (Translated from Russian: Israel Program for Scientific Translations, Jerusalem, 1969, 266 pp.)

Cameron, E.M. and Baumann, A., 1972. Carbonate sedimentation during the Archean. Chem. Geology, 10: 17–30.

Cameron, E.M. and Jonasson, I.R., 1972. Mercury in Precambrian shales of the Canadian Shield. *Geochim. Cosmochim. Acta*, 36: 985–1005.

Chilingar, G.V., 1956. Relationship between Ca/Mg ratio and geologic age. *Am. Assoc. Petrol. Geologists Bull.*, 40: 2256–2266.

Cloud, P.E., 1968. Pre-metazoan evolution and the origin of the Metazoa. In: E.T. Drake (Editor), *Evolution and Environment*. Yale University Press, New Haven, pp. 1–72.

Cloud, P.E., 1971. The Precambrian (Meeting). *Science*, 173: 851–854.

Cloud, P.E., 1972. A working model of the primitive earth. *Am. J. Sci.*, 272: 537–548.

Cloud, P.E., 1973. Paleoecological significance of the banded iron-formation. *Econ. Geol.*, 68: 1135–1143.

Condie, K.C., 1973. Archean magmatism and crustal thickening. *Geol. Soc. Am. Bull.*, 84: 2981–2992.

Cox, D.P., Schmidt, R.G., Vine, J.D., Kirkemo, H., Tourtelot, E.B. and Fleischer, M., 1973. Copper. In: D.A. Brobst and W.P. Pratt (Editors), *United States Mineral Resources – U.S. Geol. Surv. Prof. Pap.*, 820: 163–190.

Daly, R.A., 1909. First calcareous fossils and the evolution of limestones. *Geol. Soc. Am. Bull.*, 20: 153–170.

Dorr, J.Van N. II., 1973. Iron-formation in South America. *Econ. Geol.*, 68: 1005–1022.

Dorr, J. Van N. II., Crittenden, M.D. Jr. and Worl, R.G., 1973. Manganese. In: D.A. Brobst and W.P. Pratt (Editors), *United States Mineral Resources – U.S. Geol. Surv. Prof. Pap.*, 820: 385–399.

Douglas, R.J.W. (Editor), 1970. *Geology and Economic Minerals of Canada – Geol. Surv. Can., Econ. Geol., Rep.*, 1: 838 pp.

Drewer, J.I., 1973. A geochemical model for the origin of Precambrian banded iron-formations. *Geol. Soc. Am. (Abstr.)*, 5(7): 605.

Engel, A.E.J., 1963. Geological evolution of North America. *Science* 140: 143–152.

Engel, A.E.J., Nagy, B., Nagy, L.A., Engel, C.G., Kemp, G.O.W. and Drew, Ch.M., 1968. Algae-like forms in Onverwacht Series, South Africa – oldest recognized life-like forms on the Earth. *Science*, 161: 1005–1008.

Finch, W.I., Butler, A.P. Jr., Armstrong, F.C. and Weissenborn, A.E., 1973. Uranium. In: D.A. Brobst and W.P. Pratt (Editors), *United States Mineral Resources – U.S. Geol. Surv. Prof. Pap.*, 820: 456–468.

French, B.V., 1966. Some geologic implications of equilibrium between graphics and C-H-O phase at high temperatures and pressures. *Rev. Geophys.*, 4: 223–253.

Fyfe, W.S. and Brown, G.C., 1972. Granites past and present. *J. Earth Sci., Leeds*, 8: 249–260.

Garrels, R.M. and Christ, C.L., 1965. *Solutions, Minerals and Equilibria.* Harper and Row, New York, N.Y., 450 pp.

Garrels, R.M. and Mackenzie, F.T., 1971. *Evolution of Sedimentary Rocks: a Geochemical Approach.* Norton, New York, N.Y., 397 pp.

Garrels, R.M. and Perry, E.A. Jr., 1975. *Cycling of Carbon, Sulphur, and Oxygen through Geologic Time. The Sea, 5.* Interscience, New York (in press).

Garrels, R.M., Mackenzie, F.T. and Siever, R., 1972. Sedimentary cycling in relation to the history of the continents and oceans. In: E.C. Robertson (Editor), *The Nature of the Solid Earth.* McGraw-Hill, New York, N.Y., pp. 93–121.

Garrels, R.M., Perry, E.A. Jr. and Mackenzie, F.T., 1973. Genesis of Precambrian iron formations and the development of atmospheric oxygen. *Econ. Geol.*, 68: 1173–1179.

Gill, J.B., 1970. Geochemistry of Viti Levu, Fiji, and its evolution as an island arc. *Contrib. Mineral. Petrol.*, 27: 179–203.

Gilmour, P., 1971. Stratabound massive pyritic sulfide deposits – a review. *Econ. Geol.*, 66: 1239–1249.

Glaessner, M.F., 1968. Biological events and the Precambrian time scale. *Can. J. Earth Sci.*, 5: 585–590.

Glickson, A.Y., 1971. Primitive Archean element distribution patterns: chemical evidence and geotectonic significance. *Earth Planet. Sci. Lett.*, 12: 309–320.

Glickson, A.Y. and Lambert, I.B., 1973. Relations in space and time between major Precambrian shield units. An interpretation of Western Australian data. *Earth Planet. Sci. Lett.*, 20: 395–403.

Glickson, A.Y. and Sheraton, J.W., 1972. Early Precambrian trondhjemitic suites in Western Australia and northwestern Scotland, and the geochemical evolution of shields. *Earth Planet. Sci. Lett.*, 17: 227–242.

Goldich, S.S. and Mudrey, M.G. Jr., 1973. Long-term tectonic evolutionary model for the Canadian Shield. *Geol. Soc. Am. (Abstr.)*, 5(7): 638.

Goodwin, A.M., 1968. Archean protocontinental growth and early crustal history of the Canadian Shield, *23rd Int. Geol. Congr., Prague*, 1: 69–89.

Goodwin, A.M., 1970. Metallogenetic evolution of the Canadian Shield. In: R.J.W. Douglas (Editor), *Geology and Economic Minerals of Canada – Geol. Surv. Can., Econ. Geol. Rep.*, 1: 156–162.

Goodwin, A.M., 1973. Archean iron-formations and tectonic basins of the Canadian Shield. *Econ. Geol.*, 68: 915–933.

Goodwin, A.M. and Ridler, R.H., 1970. The Abitibi orogenic belt. In: A.J. Baer (Editor), *Symposium on Basins and Geosynclines of the Canadian Shield – Geol. Surv. Can. Pap.*, 70(40): 1–30.

Grandstaff, D.E., 1974. Uraninite oxidation and the mid-Precambrian atmosphere. *Trans. Am. Geophys. Union*, 55(4): 457.

Gross, G.A., 1970. Nature and occurrence of iron-ores. In: United Nations (Editor), *Survey of World Iron Ore Resources*. United Nations, New York, N.Y., pp. 13–31.

Gross, G.A., 1973. The depositional environment of principal types of Precambrian iron-formations. In: UNESCO (Editor), *Genesis of Precambrian Iron and Manganese Deposits. Proceedings of the Kiev Symposium, 1970*. UNESCO, Paris, pp. 15–21.

Hart, S.R. and Brooks, C., 1970. Rb–Sr mantle evolution model. Carnegie Inst., Ann. Rep. D.T.M., pp. 426–429.

Hart, S.R., Brooks, C., Krogh, T.E., Davis, G.L. and Nava D., 1970. Ancient and modern volcanic rocks: a trace-element model. *Earth Planet. Sci. Lett.*, 10: 17–28.

Holland, H.D., 1964. On the chemical composition of the terrestrial and cytherean atmospheres. In: A.E.J. Engel, H.L. James and B.F. Leonard (Editors), *The Origin and Evolution of Atmospheres and Oceans*. Wiley, New York, N.Y., pp. 86–101.

Holland, H.D., 1965. The history of ocean water and its effect on the chemistry of the atmosphere. *Proc. Natl. Acad. Sci. U.S.*, 53: 1173–1182.

Holland, H.D., 1972. The geologic history of sea water – an attempt to solve the problem. *Geochim. Cosmochim. Acta*, 36: 637–651.

Holland, H.D., 1973. The oceans: a possible source of iron in iron-formations. *Econ. Geol.*, 68: 1169–1172.

Hurley, P.M. and Rand, J.R., 1969. Pre-drift continental nuclei. *Science*, 164: 1229–1242.

Hutchinson, R.W., 1973. Volcanogenic sulfide deposits and their metallogenic significance. *Econ. Geol.*, 68: 1223–1246.

Jakeš, P. and White, A.J.R., 1971. Composition of island arcs and continental growth. *Earth Planet. Sci. Lett.*, 12: 224–230.

James, H.L., 1966. Chemistry of iron-rich sedimentary rocks. *U.S. Geol. Surv. Prof. Pap.*, 440 W: 60 pp.

Krauskopf, K.B., 1955. Sedimentary deposits of rare metals. *Econ. Geol.*, Annivers. Vol., I: 411–463.

Krauskopf, K.B., 1957. Separation of manganese from iron in sedimentary processes. *Geochim. Cosmochim. Acta*, 12: 61–84.

Krauskopf, K.B., 1967. *Introduction to Geochemistry*. McGraw-Hill, New York, N.Y., 721 pp.

Kröner, A., Anhausser, C.R. and Vajner, V., 1973. Neue Ergebnisse zur Evolution der präkambrischen Kruste im südlichen Africa. *Geol. Rundsch.*, 62: 281–309.

Kuenen, Ph.H., 1950. *Marine Geology*. Wiley, New York, N.Y., 568 pp.

LaBerge, G.L., 1973. Possible biological origin of Precambrian iron-formations. *Econ. Geol.*, 68: 1098–1109.

Laznicka, P., 1970. The University of Manitoba file of nonferrous metal deposits of the world (MANIFILE). University of Manitoba, Winnipeg (unpublished).

Laznicka, P., 1973. Development of nonferrous metal deposits in geological time. *Can. J. Earth Sci.*, 10: 18–25.

Laznicka, P. and Wilson, H.D.B., 1972. The significance of a copper-lead line in metallogeny. *24th Int. Geol. Congr., Montreal, Sect. 4*, pp. 25–36.

Lepp, H. and Goldich, S.S., 1964. Origin of Precambrian iron-ore formations. *Econ. Geol.*, 59: 1025–1060.

Lubimova, E.A., 1969. Thermal history of the earth. In: P.J. Hart (Editor), *The Earth's Crust and Upper Mantle – Am. Geophys. Union, Geophys. Monogr.*, 13: 63–77.

Mackenzie, F.T., 1969. Chemistry of sea water. In: F.E. Firth (Editor), *The Encyclopedia of Marine Resources.* Van Nostrand Reinhold, New York, N.Y., pp. 106–112.

Mackenzie, F.T. and Garrels, R.M., 1974. Chemical history of the oceans. In: W. Hay (Editor), *Studies in Paleo-Oceanography – Soc. Econ. Paleontologists Mineralogists, Spec. Publ.,* 20.

Mero, J.L., 1965. *The Mineral Resources of the Sea.* Elsevier, Amsterdam, 310 pp.

Moorbath, S., O'Nions, R.K., Pankhurst, R.J., Gale, N.H. and McGregor, V.R., 1972. Further rubidium-strontium age determinations of the very early Precambrian rocks of the Godthaab district, West Greenland. *Nature-Phys. Sci.,* 240: 78–82.

Paterson, S.H., 1967. Bauxite reserves and potential aluminum resources of the world. *U.S. Geol. Surv. Bull.,* 1228: 176 pp.

Poldervaart, A., 1955. Chemistry of the earth's crust. In: A. Poldervaart (Editor), *Crust of the Earth – Geol. Soc. Am; Spec. Pap.,* 62: 119–144.

Pretorius, D.A., 1966. Conceptual geological models in the exploration of gold mineralization in the Witwatersrand Basin. *Univ. Witwatersrand, Johannesburg, Econ. Geol. Resour. Unit, Inf. Circ.,* 33: 39 pp.

Ramdohr, P., 1958. Die Uran- und Goldlagerstätten Witwatersrand-Blind River District – Dominion Reef – Serra de Jacobina: Erzmikroskopische Untersuchungen und ein geologischer Vergleich. *Abh. Dtsch. Akad. Wiss. Berlin, Kl. Chem. Geol. Biol.,* 3: 35 pp.

Ringwood, A.E., 1969. Composition and evolution of the upper mantle. In: P.J. Hart (Editor), *The Earth's Crust and Upper Mantle – Am. Geophys. Union, Geophys. Monogr.,* 13: 1–17.

Rogers, J.J.W. and McKay, S.M., 1972. Chemical evolution of geosynclinal material. In: B.R. Doe and D.K. Smith (Editors), *Studies in Mineralogy and Precambrian Geology – Geol. Soc. Am., Mem.,* 135: 3–28.

Ronov, A.B., 1964. Common tendencies in the chemical evolution of the earth's crust, ocean and atmosphere. *Geochem. Int.,* 1: 713–737.

Ronov, A.B., 1968. Probable changes in the composition of sea water during the course of geological time. *Sedimentology,* 10: 25–43.

Ronov, A.B., 1971a. Evolution of sedimentary rock composition and of geochemical processes in the history of the Earth. *Int. Geochem. Congr. (Abstr.),* Moscow, 1971, pp. 943–946.

Ronov, A.B., 1971b. Allgemeine Entwicklungstendenzen in der Zusammensetzung der äusseren Erdhülle. *Ber. Dtsch. Ges. Geol. Wiss., A, Geol. Paläontol.,* 16(3): 331–350.

Ronov, A.B. and Migdisov, A.A., 1971. Evolution of the chemical composition of the rocks in the shields and sediment cover of the Russian and North American platforms. *Sedimentology,* 16: 137–185.

Ronov, A.B. and Yaroshevski, A.A., 1969. Chemical composition of the earth's crust. In: P.J. Hart (Editor), *The Earth's Crust and Upper Mantle – Am. Geophys. Union, Geophys. Monogr.,* 13: 37–57.

Ronov, A.B., Balashov, Yu.A., Girin, Yu.P., Bratishko, R.Kh. and Kazakov, G.A., 1972. Trends in rare-earth distribution in the sedimentary shell and in the erth's crust. *Geochimija,* 12: 1483–1514 (in Russian).

Roscoe, S.M., 1969. Huronian rocks and uraniferous conglomerates in the Canadian Shield. *Geol. Surv. Can. Pap.,* 68(40): 205 pp.

Roy, S., 1968. Mineralogy of the different genetic types of manganese deposits. *Econ. Geol.,* 63: 760–786.

Rutten, M.G., 1962. *The Geological Aspects of the Origin of the Life on the Earth.* Elsevier, Amsterdam, 146 pp.

Rutten, M.G., 1966. Geologic data on atmospheric history. *Palaeogeogr. Palaeoclimatol. Palaeoecol.,* 2: 47–57.

Salop, L.I., 1968. Proterozoic. In: A.I. Zhamojda (Editor), *Geological Framework of the U.S.S.R., I.* Nedra, Moscow, pp. 111–229 (in Russian).

Salop, L.I., 1969. The ways of establishing a unified stratigraphic division of the Precambrian. In: L.I. Borovikov and A.I. Semenov (Editors), *Geological Framework of the U.S.S.R., 5.* Nedra, Moscow, pp. 63–82 (in Russian).

Sangster, D.F., 1972. Precambrian volcanogenic massive sulphide deposits in Canada: a review. *Geol. Surv. Can. Pap.,* 72(22): 44 pp.

Siever, R., 1957. Silica budget in sedimentary cycle. *Am. Mineralogist,* 42: 821–841.

Sillén, L.G., 1961. The chemistry of sea water. In: M. Sears (Editor), *Oceanography – Am. Assoc. Adv. Sci. Publ.,* 67: 549–581.

Sillén, L.G., 1967. The ocean as a chemical system. *Science,* 156: 1189–1197.

Stanton, R.L., 1972. *Ore Petrology.* McGraw-Hill, New York, N.Y., 713 pp.

Strakhov, N.M., 1964. Stages of development of the external geospheres and formation of sedimentary rocks in the history of the earth. *Int. Geol. Rev.,* 6: 1466–1482.

Strakhov, N.M., 1969. *Principles of Lithogenesis, 2.* Oliver-Boyd, Edinburgh, 609 pp.

Stumm, W. and Morgan, J.J., 1970. *Aquatic Chemistry.* Wiley-Interscience, New York, N.Y., 583 pp.

Tappan, H., 1971. Microplankton, ecological succession and evolution. *Proc. N. Am. Paleontol. Conv., September 1969,* H: 1058–1103.

Tappan, H. and Loeblich, A.R. Jr., 1970. Geobiologic implications of fossil phytoplankton evolution and time-space distribution. *Geol. Soc. Am. Spec. Pap.,* 127: 247–340.

Taylor, S.R., 1964. Trace-element abundances and the chondritic earth model. *Geochim. Cosmochim. Acta,* 28: 1989–1998.

Taylor, S.R., 1969. Trace-element chemistry of andesites and associated calc-alkaline rocks. *Ore. Dept. Geol. Mines. Ind. Bull.,* 65: 43–63.

United Nations, 1970. *Survey of World Iron Ore Resources.* United Nations, New York, N.Y., 479 pp.

Valeton, I., 1972. *Bauxites.* Elsevier, Amsterdam, 226 pp.

Van Moort, J.C., 1973. The magnesium and calcium contents of sediments, especially pelites, as a function of age and degree of metamorphism. *Chem. Geol.,* 12: 1–37.

Varentsov, I.M., 1964. *Sedimentary Manganese Ores.* Elsevier, Amsterdam, 119 pp.

Veizer, J., 1971. *Chemical and Strontium Isotopic Evolution of Sedimentary Carbonate Rocks in Geologic History.* Thesis, Australian National University, Canberra, 151 pp. (unpublished).

Veizer, J., 1973. Sedimentation in geologic history: recycling vs. evolution or recycling with evolution. *Contrib. Mineral. Petrol.,* 38: 261–278.

Veizer, J. and Compston, W., 1973. $^{87}Sr/^{86}Sr$ evolution of sea water in geologic history (abstr.). *Fortschr. Mineral.,* 50: 133.

Veizer, J. and Compston, W., 1974. $^{87}Sr/^{86}Sr$ composition of sea water during the Phanerozoic. *Geochim. Cosmochim. Acta,* 38: 1461–1488.

Viljoen, M.J. and Viljoen, R.P., 1969. A reappraisal of the granite-greenstone terrains of shield areas based on the Barberton model. *Geol. Soc. S. Afr. Spec. Publ.,* 2: 245–274.

Vine, J.D. and Tourtelot, E.B., 1973. Copper deposits in sedimentary rocks. *Geol. Soc. Am. Abstr.,* 5/7: 848–849.

Vinogradov, A.P., Ronov, A.B. and Ratynskii, V.Y., 1952. Evolution of the chemical composition of carbonate rocks. *Izv. Akad. Nauk U.S.S.R., Ser. Geol.,* 1: 33–60 (in Russian).

Volkov, V.N., 1968. Coal and bituminous shales. In: A.I. Semenov and A.D. Shcheglov (Editors), *Geological Framework of the U.S.S.R., 4.* Nedra, Moscow, pp. 458–471 (in Russian).

Wangersky, P.J., 1972. The control of sea water pH by ion pairing. *Limnol. Oceanogr.,* 17: 1–6.

Weaver, C.E.J., 1967. Potassium, illite and the ocean. *Geochim. Cosmochim. Acta,* 31: 2181–2196.

Wedepohl, K.H., 1969. Die Zusammensetzung der Erdkruste. *Fortschr. Mineral.,* 46: 145–174.

Wedepohl, K.H., 1971. "Kupferschiefer" as a prototype of syngenetic sedimentary ore deposits. *Soc. Mining Geol. Japan,* Spec. Issue, 3: 268–273.

Wedow, H. Jr., Kiilsgaard, T.H., Heyl, A.V. and Hall, R.B., 1973. Zinc. In: A.A. Brobst and W.P. Pratt (Editors), *United States Mineral Resources – U.S. Geol. Surv. Prof. Pap.,* 820: 697–711.

Winkler, H.G.F., 1967. *Petrogenesis of Metamorphic Rocks.* Springer, New York, N.Y., 237 pp.

Wyllie, P.J., 1971. *The Dynamic Earth.* Wiley, New York, N.Y., 416 pp.

Chapter 2

ORE DEPOSITS IN THE LIGHT OF PALAEOCLIMATOLOGY

HELMUT WOPFNER and MARTIN SCHWARZBACH

INTRODUCTION

Recent developments in mineral exploration, together with data derived from the study of modern metalliferous sedimentary environments, have led to much increased (and still increasing) awareness and appreciation of the important role played by sedimentary processes in the formation of ore bodies. Many an ore deposit, the formation of which was thought to have depended upon some imaginary igneous intrusion, hidden in the depth of the earth, can now be interpreted in terms of comparatively common depositional and geochemical factors. The role of certain rock properties, such as porosity and permeability and the importance of rock or basin geometry to form a "trap" or host, concepts long familiar in petroleum geology, are now increasingly being recognized as important factors in the formation of ore deposits; so too are biogenic processes.

Considering the governing role of climate on a large range of depositional environments and on such aspects as type of erosion, products of weathering[1] and mode of transport, it will be apparent that the same controlling forces must surely also influence the formation of syngenetic ore deposits. These processes have been discussed extensively by Schwarzbach (1963, 1974), who also pointed out the importance of certain ore bodies (e.g., bauxites) as palaeoclimatic indicators.

The influence of climate on the formation of syngenetic ore bodies lies, firstly, in the mobilisation, both by chemical and physical events, of metal-ions and, secondly, in the controlled deposition and enrichment of the "mobilised" ions in a certain sedimentary environment. Needless to say, such climatic influences may be dominant controlling factors, as for instance in the formation of bauxites, or they may be accessory, just contributing one of many factors which combine in the accumulation of a deposit, leaving only a subtle imprint, often no longer decipherable with certainty from the geological record. Climatic control may also manifest itself as a secondary factor by enriching the mineral concentration of low-grade deposits to a grade at which they are

---

[1] Editor's note: The reader is referred to a related chapter on pedogenesis, weathering and supergene ores by Lelong et al. (Chapter 3, this Vol.).

economically exploitable. This latter type of control may be operational in syngenetic as well as in deposits of igneous origin. And, thirdly, we would mention certain climatically induced features which can be utilized as aids in mineral exploration.

As Strakhov (1967) has pointed out, the most favourable environment for the formation of syngenetic ore deposits exists under warm, humid climates. It is here that both the chemical and mechanical "mobility" of such metal-ions as Fe, Al, Mn, Cu and Zn, is greatest. Thus, large amounts of mineral matter can be liberated from the compound rock-phase and transported, either in solutions or in colloidal or particulate form, into adjacent basins, thus providing the potential material for mineral concentrations. On land it may lead to the formation of deposits *in situ*, either by primary concentration through vertical ion-exchange or through lateral removal of certain materials, resulting in the concentration of residual elements or minerals.

Arid zones, although less active in the provision of "source material", nevertheless provide receptacles for material derived from neighbouring zones of high rainfall and may then be destined to become foci of metal concentrations.

The least favourable environment, but for a few exceptions, appears to exist under glacial environments. When all available water exists in solid form for all or most of the year, chemical weathering reaches a minimal activity, and erosion and transport are restricted to mechanical processes. It should be stressed, however, that these conditions apply only to hyperglacial regions and are no longer valid in the periglacial environments, where mechanical concentration may form important placer deposits.

Climate also plays a decisive role in the formation of specific sedimentary facies, such as red-beds or reef complexes which may become host rocks for metal concentrations.

Besides the regional climatic influences which governed the existence of particular depositional environments within specific climatic zones, there are also factors to be considered which resulted from climatic variations of global magnitude. Eustatic changes of sea level during glacial periods, for instance, although not necessarily primary causes for mineral concentrations, have led to the development of important placer deposits by creating optimal erosional gradients or sorting conditions within coastal regions. On the other hand, there were no doubt periods in our planet's history when its polar regions were not covered by large ice caps. Absence of polar ice caps would have resulted not only in a latitudinal spread of climatic belts, but it would have also eliminated the source for cold, oxygenated bottom waters, which profoundly influence the dynamics of our hydrosphere and the deposition of oceanic sediments, as observed today. Borchert (1960) suggested that lack of these oxygen-rich bottom currents in the past could have led to much higher concentrations of organic material (black-shale facies) in oceanic regions. This has been regarded with some reservations by Schwarzbach (1974, p. 126). Lack of bottom currents would have led further to considerable environmental changes along continental margins, where the depositional conditions are influenced by the upwelling of such cold bottom currents. Further determining factors resulted from variations of the atmospheric composition itself. The most significant of these was undoubtedly the change from the anoxygenous atmosphere of the Early Precambrian to the oxygen/nitrogen atmosphere of the Phanerozoic Era (Holland, 1962 and many other authors).

The picture emerging from these considerations thus demonstrates that climatic influences, although variable in magnitude and complexely interwoven with other factors, may be found in almost any sedimentary ore deposit.

Naturally, we recognise the fact that the frequency of occurrence of ore deposits is often substantially increased within specific geological provinces and hence we realise the importance of "endogenic" tectonic forces, controlling such aspects as basin geometry, internal heat-flow and erosional gradients. However, we wish to emphasize the many instances where only the appropriate interplay between tectonic forces and the factors created within our earth's atmosphere fulfils the condition required for the formation of an ore deposit.

We make no claim that this chapter is a complete treatise on all ore deposits, syngenetic or otherwise, the formation or economic viability of which was due to climatic influence. Rather we have restricted our discussion to some typical metallogenic environments upon which climatic influences can be demonstrated with a great amount of certainty.

CLIMATIC CONTROL OF Fe AND Mn DEPOSITS

Banded iron-formations[1]

The oldest ore deposits for which a governing climatic control can be postulated are the Precambrian banded iron-formations. They are also referred to as jaspilites or itabirites, although the latter term should, in accordance with its first usage in Brazil, perhaps be applied more specifically to the metamorphosed variety of this type of ore.

Banded iron-formations and itabirites have gained a significant economic role as the major suppliers of the worlds iron-ore demand and deposits of this type provide by far the largest reserves of iron ore, usually in the magnitude of several billions of tonnes. For instance, the deposits on the Ukrainian Shield and the adjacent platform (Krivoj Rog and Kursk) are alone thought to contain about $60 \cdot 10^9$ tons of iron ore (U.N. Survey quoted by Walther and Zitzmann, 1973). Similarly, the mineable reserves of the Hamersley Basin in northwestern Australia are estimated to be in the vicinity of $20 \cdot 10^9$ tons (Campana, 1966). The sizes of many other deposits of banded iron-formations are of comparable magnitude.

Substantial parts of banded iron-formations are mined in their unenriched, primary form. These low-grade iron ores, containing between 15 and 30% Fe are called taconite. In many instances, however, the banded iron-formations have been the "raw material" from which, by subsequent processes of enrichment, high-grade iron ores were produced.

---

[1] Editor's note: see, in Vol. 7, Eichler's Chapter 4.

In this part of our contribution we shall consider only those climatic aspects which may have influenced the formation of the primary banded iron-formations.

Climatic processes leading to subsequent enrichment of banded iron-formations, as for instance the formation of the high-grade iron ores of the Hamersley Ranges in Western Australia, will be discussed in the part dealing with laterites and bauxites.

Banded iron-formations are known from most of the great Precambrian shield areas of the world. Large deposits have been investigated on the Canadian Shield (Gross, 1965) and the banded iron-formations in the region of Lake Superior have contributed substantially to the iron-ore production of the United States (Bayley and James, 1973; Klemic et al., 1973b). The Russian deposits, briefly mentioned before, have been discussed most recently by Alexandrov (1973). Banded iron-formations are further known from Brazil (Dorr, 1973), from India (Wadia, 1966), South Africa (Cullen, 1963; Beukes, 1973) and Australia (Miles, 1955; Campana et al., 1964; Trendall, 1968 and 1972).

Although the distribution of banded iron-formations is worldwide, their occurrence appears to fall into a comparatively narrow time-slot of the Middle Proterozoic. Although none of these formations have been dated by direct methods, radiometric dates of underlying or associated igneous rocks seem to indicate a maximum development around $2 \cdot 10^9$ years before present. Since this comparatively narrow time limit (in terms of Precambrian time) appears to have some genetic significance, it may be important to qoute some examples.

The age of the Gun Flint Iron-Formation in Ontario is placed between 1.6 and $2.0 \cdot 10^9$ years (Schopf, 1970) and that of the Biwabik Iron-Formation in Minnesota, thought to be a lateral equivalent of the Gun Flint Iron-Formation, is stated to be in excess of $1.7 \cdot 10^9$ years (Cloud and Licari, 1968). The Hamersley Group, containing the banded iron-formations in Western Australia has an age between 1.8 and $2.0 \cdot 10^9$ years (Trendall, 1968), whereas the Pretoria Series, the main source of iron in South Africa, is about $1.9 \cdot 10^9$ years old (Cloud and Licari, op. cit.).

The deposits in Russia and Fennoscandia are placed between 2.1 and $1.7 \cdot 10^9$ years (Walther and Zitzmann, 1973). The oldest known banded iron-formation, according to Cloud and Licari (op. cit.) is the Soudan Iron-Formation, for which an age in excess of $2.7 \cdot 10^9$ years is indicated. This latter formation, like most older banded iron-formations is of the Algoma-type, whereas those with ages around $2.0 \cdot 10^9$ years generally belong to the Superior-type.

Although these ages can only be regarded as general indications, they would seem to demonstrate a temporal restriction for the formation of banded iron-formations to within the Early and Middle Proterozoic. No true banded iron-formations are known from the Late Precambrian or the Phanerozoic.

The main characteristic of the banded iron-formations, as implied by the name, is a regular, often rhythmical interbedding of light and dark coloured "bands". The thickness of the individual band varies between 5 and 30 mm and reflects primary differences in composition. The light bands are composed of micro- to cryptocrystalline silica or "chert", whereas the darker bands are iron-rich, containing hematite, magnetite, iron silicates or, to a lesser degree, siderite or pyrite. Depending on the assemblages of Fe-minerals, a silicate, oxide or carbonate facies may be distinguished.

Microlamination in the range of fractions of a millimetre are reported from the Griqua-town Series in South Africa (Cullen, 1963) and from the Hamersley Group in Western Australia (Trendall, 1968) (see Fig. 1). These microlaminations are grouped by the next-order cycle in the centimetre range. Trendall (1972) distinguishes three such "rhythms" in the Brookman Iron-Formation of the Hamersley Group, which he termed "micro-bands", "mesobands" and "macrobands", each of which he interprets in terms of climatic variations. Fig. 1, showing a geological map of the Hamersley Basin in Western Australia and some sections of the basin fill (reproduced from Trendall, 1968), may serve as an example of the uniform and large areal extent of banded iron-formations.

Microbanding, although less widespread than in South Africa and Australia, also occurs in some of the North American iron-formations.

Not uncommonly, algal stromatolites are associated with banded iron-formations as, for instance, in the Gunflint and Biwabik Iron-Formations of North America (Cloud and

TABLE I

Chemical composition (wt. %) of banded iron-formation from three continents

	(1)	(2)	(3)
SiO_2	46.40	46.86	38.9
Al_2O_3	0.90	0.43	0.55
TiO_2	0.04	0.05	0.01
Fe_2O_3	18.70	24.68	9.7
FeO	19.71	17.19	27.6
MnO	0.63	n.dtm.	0.01
MgO	2.98	2.58	0.8
CaO	1.60	1.49	3.1
Na_2O	0.04	0.16	0.22
K_2O	0.13	0.10	0.01
P_2O_5	0.08	0.25	0.08
CO_2	6.90	5.81	17.9
C	0.17	n.dtm.	n.dtm.
H_2O^+	1.60	0.57	0.1
H_2O^-	0.32	0.08	
	100.20	100.25	98.98
Ignition	1.72	n.dtm.	n.dtm.

(1) North America; Biwabik Iron Formation (calculated average composition of entire formation after Bayley and James, 1973, p.942)

(2) Australia; Brockman Iron Formation, Dales Gorge Member (after Trendall and Blockley, 1970, p.134)

(3) South America; Mina Grande Lode, Nova Lima Group, Minas Gerais (after Dorr 1973, p.1008).

Licari, 1968). The same formations also contain a varied assemblage of phytomicro-fossils (Barghoorn and Tyler, 1965; Cloud and Licari, 1968) which, according to Schopf (1969) "appear to be procaryotic in affinity". Eucaryotic organisms apparently did not develop until the beginning of the Late Precambrian, $1.3 \cdot 10^9$ years ago (Schopf, 1970). The palaeobiologic evidence thus indicates that organisms, capable of photosynthesis were an integral part of the environment in which the banded iron-formations were deposited.

The chemical composition of the banded iron-formations is characterized by the dominance of silica (SiO_2) and iron (Fe_2O_3 and FeO), constituting at least 75%, but usually more than 80%, of the whole rock composition. A further typical feature is the very low participation of Al_2O_3 and TiO_2 as well as Na_2O and K_2O. This composition, although varying in degree, may be taken as typical and, as indicated by the analyses in Table I, can be observed on a worldwide scale. The global similarity in chemical composition of banded iron-formations, together with the great lithological similarities, are difficult to explain unless identical modes of deposition are envisaged, requiring a primary control by supra-regional factors.

Although volcanic and magmatogenic processes have been invoked for the formation of the banded iron-formations in the past, it is now generally accepted that these deposits are of sedimentary origin. However, there is still divergence between the specific depositional models proposed, particularly with respect to the origin of the iron and the climatic factors which influenced both the formation of the source and deposition. Even those hypotheses proposing a direct climatic influence are at variance as to the palaeo-temperature, precipitation and the composition of the palaeo-atmosphere. Therefore, in the textbooks "Climates of the Past" (Schwarzbach, 1963) and "Das Klima der Vorzeit" (Schwarzbach, 1974), no attempt was made to discuss the paleoclimatic relationship of the banded iron-formations. We shall attempt here to summarise the salient points which, to us, appear to have the greatest palaeo-climatic significance.

Sakamoto (1950), Alexandrov (1955) and Hough (1958), amongst others, suggested that the iron was formed on senile land surfaces by chemical weathering, under warm humid climatic conditions and was subsequently removed by erosion and deposited in large adjoining closed basins.

Cullen (1963) accepted the same mode of origin for the source material, but postulated a specific tectonic model for the deposition of the South African banded iron-formations. His concept envisaged chemical precipitation of the iron-formation on a large shelf, synchronous to the deposition of clastic material in an adjoining geosynclinal trough.

An interesting explanation for the formation of banded iron-formations as presented by Khodyush (1969) is reported by Alexandrov (1973). Khodyush suggested, like others before him, that the iron accumulated in sea water at a time when the earth's atmosphere lacked oxygen. He suggested further that the banding within the iron-formation reflects climatic variations characterised by changes from warm humid periods to dry cold pe-

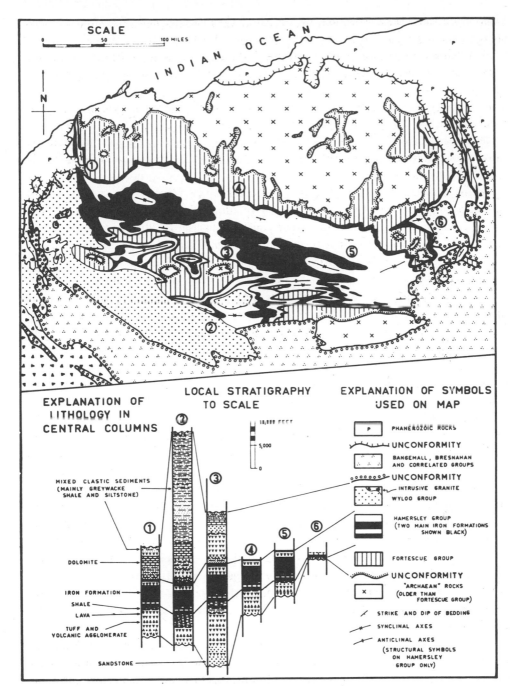

Fig. 1. Geological map of the Precambrian Hamersley Basin in Western Australia, showing distribution of banded iron-formations. The stratigraphic sequence and the position of the iron-formations are shown in the columns *1–6*. (Reproduced from Trendall, 1968; courtesy Geol. Soc. Am.)

riods. Flourishing organic life during the warm periods would have created, through photosynthesis, an increase in oxygen content and pH, thus leading to the oxidation of iron bicarbonate in sea water. The cold periods would have caused a decrease in organic activity and oxygen content leading to increased acidity and greater precipitation of silica.

Possible biogenic influence through oxygen production by photosynthesis has also been pointed out by American palaeo-biologists. Cloud and Licari (1968), for instance, suggested that banded iron-formations were formed as a chemical precipitate under an anoxygenous atmosphere, whereby the ferrous iron served as an acceptor of biologically produced oxygen.

Chemical precipitation had already been postulated by Miles (1955) as the mode of deposition of some Australian banded iron-formations. Trendall (1968, 1972) also suggested chemical precipitation as the depositional mechanism for the banded iron-formations of the Hamersley Basin, but furthermore presented strong and cogent evidence for both seasonal and supra-seasonal control of deposition. Trendall (op. cit.) and Trendall and Blockley (1970) demonstrated that the microbanding clearly consisted of couples, each couple representing an annual increment of precipitate. Like varves, they can be correlated over tens and even hundreds of kilometres. From an analysis of these seasonal deposits, Trendall and Blockley arrived at a depositional time span of roughly 6 m.y. Such regular deposition of an evenly fine-grained sequence over the entire area of the Hamersley Basin (see Fig. 1) requires a model which allows for a uniform and simultaneous basin-wide deposition of alternating layers of iron- and silica-rich sediment. The basin-wide lithological and chemical uniformity of the banded iron-formations would seemingly exclude mechanical transport as a significant depositional factor. We conclude, therefore, that chemical precipitation is the only mechanism which provides a satisfactory explanation for all the sedimentary and chemical features exhibited by the banded iron-formations. The general absence of clastic and pyroclastic components lends further support to that conclusion. Occasional layers of physically transported granular iron and clastic components, as reported by Mengel (1973) from iron formations near Lake Superior, do not negate the principal concept of an essentially precipitating environment, as some clastic intake and reworking of prior deposited layers would be expected along the basin margin or during periods of increased gradient and run off. A marginal, at times even sub-aerial, environment is indicated by Mengel's (op. cit.) observation of possible imprints of rain drops. Trendall (1968) observed that granular iron-formations are common in the Lake Superior Ranges, but absent or extremely rare elsewhere in North American deposits.

Therefore, if one accepts the banded iron-formations as a chemically precipitated sediment, and the evidence in favour of this concept outweighs all others, an arid climate would appear most likely. Such climatic conditions, together with a stable, low, surrounding relief, would also account for the paucity of clastic components within banded iron-formations, as the much reduced run off in an arid climate would have kept the clastic intake to a minimum. Trendall (1972), who also emphasized the similarities between laminated evaporites and banded iron-formations, suggested (p. 297) a palaeo-latitude for the iron-formations of the Hamersley Group "close to one of the tropics, whatever the latitude of these was at the time".

More difficult to decipher is the palaeo-temperature. However, if the compositional differences reflect seasonal and supra-seasonal climatic variations, then their cause was most likely a change in concentration of salinity, brought about by different rates of evaporation, favouring a seasonally hot climate. Chert pods, described by Beukes (1973) from South African banded iron-formations, are surprisingly similar to shallow-water chert associations in certain red-bed environments. Eugster and I-Ming Chou (1973) have pointed out the similarity between chert nodules after magadiite and chert pods from Precambrian iron-formations. Such conditions would also be compatible with the biologic record and indeed would seemingly combine climatic and biogenic factors for the ultimate control of deposition. This interrelationship between climate and biologic population seems important to us in explaining the restriction of the banded iron-formations to the time between about 2.7 and $1.6 \cdot 10^9$ years.

Such a model would contradict a continental source for the large amounts of iron accumulated. A seasonally hot, arid climate would not only be unfavourable for the chemical weathering processes required to mobilise the iron, it would also lack the necessary volume of transport medium to transfer the iron from the land into the adjacent basin. In view of the anoxygenous atmosphere and the lack of vegetative cover of the Precambrian land surfaces, it is generally questionable if lateritisation, as we know it from the Phanerozoic, could have existed. The extremely low content of the banded iron-formations in Al_2O_3 and TiO_2 (see Table I) is also difficult to reconcile with a lateritic source area. Cogent arguments against a large-scale inflow of continental waters and a continental source of iron, although based on different considerations, have been put forward by Holland (1973).

Oolitic[1] iron ores[2]

Deposits of oolitic iron ores are as much a Phanerozoic phenomenon as the banded iron-formations are typically a feature of the Precambrian (for exceptions, see Gross, 1965). Although some iron oolites are also known from the Proterozoicum, their major developments are restricted to the Phanerozoicum. In addition to the obvious physical differences between the two types of iron-formations, there are also marked geochemical differences which were discussed by Taylor (1969), who drew particular attention to the different Ca/Mg and Fe/Al ratios.

The main minerals which participate in the formation of oolitic iron deposits are

---

[1] We follow the recommendation of Teichert (1970, p. 1748) to use the term "oolite" (= "oolith") for the rock and "ooid" for the individual particle.

[2] Editor's note: see, in Vol. 7, Dimroth's Chapter 5, for one example.

goethite, limonite, hematite, siderite, chamosite and, to a lesser degree, pyrite. Chamosite
is the dominant constituent of many Palaeozoic iron oolites, as, for instance, in the
Ordovician chamosite deposits of Thuringia and northwestern France or the Silurian
Clinton Formation of eastern U.S.A. Pyritic oolites occur, together with chamosite, in the
Ordovician iron-ore deposit of Wabana in New Foundland (Hayes, quoted by Taylor,
1969), and oolitic iron ores consisting of siderite, magnetite, chlorite and limonite were
described from the Senonian and Early Tertiary of Nigeria (Jones, 1965). The ooids
composing the famous Jurassic Minette ores of Europe and the U.K. are composed in
order of dominance, of goethite, limonite, hematite, siderite and chamosite. Mixed-ore
types, consisting of mechanically concentrated granular limonite as well as oolites
("Trümmer Erze") occur at the base of the transgressive Neocomian along the southern
and southeastern margins of the North German Basin.

Oolitic iron ores, and particularly the Jurassic Minette ores, have contributed substantially to the foundation of the central European and English iron industries. These deposits are still of some economic significance, providing about 15% of the iron-ore reserves of the European continent (Walther and Zitzmann, 1973).

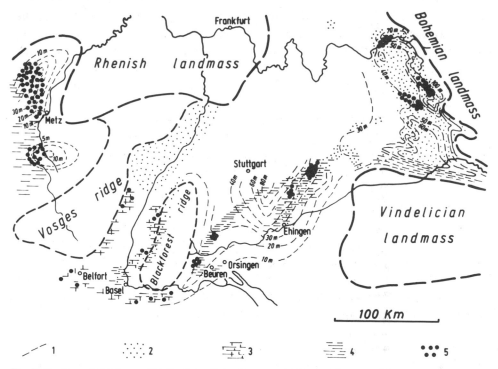

Fig. 2. Facies and thickness distribution of the Dogger β in Central Europe. In the region of Lorraine
the Lias is included with Dogger β. Explanation of symbols: *1* = isopachous lines (interval: 10 m); *2* =
sand facies; *3* = calcareous facies; *4* = argillaceous facies; *5* = iron oolite facies. (After Karrenberg,
1942; courtesy *Arch. Lagerstättenforsch., Berlin*.)

The occurrences of iron oolites are not evenly distributed over the Phanerozoic, but are concentrated at specific times. Periods of particularly extensive formation of iron oolites were the Ordovician and Silurian (Clinton, Wabana, Normandy and Bretagne, Thuringia and Bohemia, Sardinia, Wales, Western Morocco) and the Jurassic and Early Cretaceous (southern and northern Germany, Lorraine, England, Chile). Within the Jurassic the most prolific development of iron oolites occurred in the Dogger, reflected also in German terminology which refers to that time-interval as "brauner Jura". It was in this period in which the famous Minette ore deposits were formed (see Fig. 2). Late Cretaceous and Tertiary deposits are known from the Ukraine and from North Africa (Algeria, Tunesia, Egypt) and Nigeria.

Other periods, like the Late Palaeozoic and the Triassic have produced either no or hardly any iron oolites. This temporal distribution of iron oolites may be taken as indicative that their formation was dependent on specific supra-regional influences, one of which would be the prevailing climate.

Iron oolites formed preferentially in epicratonic marine basins. Geosynclinal regions appear to have been generally unfavourable areas for their formation, although some very thin and locally much restricted iron oolites occur in the Carnian (Raibler Beds) of the "Northern Limestone Alps".

By far the most important oolitic iron ores are the Jurassic occurrences of Central Europe, of which the Minette ores of Luxembourg and Lorraine, with $11.3 \cdot 10^9$ tons constitute the largest single deposit of its kind. The ore-bearing sequence extends from the Upper Liassic into the Dogger and contains between 6 and 10 lodes, with an average iron content of between 25 and 30%. The thickness varies from comparatively small lenticular bodies in the lower part of the sequence to an average of 3—4 m in the so-called "grey lode" (Berg and Karrenberg, 1942). In many horizons, configurations of ore bodies and distribution of iron concentrations show a northeasterly trend.

The ooids composing the individual lodes consist of goethite, limonite, siderite and chamosite. In certain areas pyrite may be present. Although there are many variations, the mineral sequence listed above also constitutes a distinctive facies-sequence, which is controlled by water depth and the related physicochemical conditions of the depositional medium. According to Borchert (1960) hydroxides and oxides form in shallow, aerated waters, followed, with increasing water depth, by siderite, chamosite and, ultimately, by pyrite. This facies differentiation corresponds to a vertical zonation of the sea water, comprising, in descending order an O_2, CO_2 and H_2S zone. In Borchert's model (op. cit.), which was also followed by Braun (1962), the iron is not derived directly from continental weathering solutions but from within the marine environment, whereby iron is mobilized from the sediments by waters with increased CO_2 concentration in the CO_2-zone.

The model proposed by Borchert (op. cit.) is most convincing in the case of Black Sea-type conditions. Such conditions are thought to have been prevalent during ice-free periods when, due to the absence of polar ice caps and cold, oxygenated bottom currents, a predominance of black-shale facies was developed on the ocean floors. Ice-free periods were, therefore, according to Borchert (op. cit.), preferred periods for the formation of sedimentary iron ores. A certain contradiction to this concept seems to arise, however, from palaeoclimatic evidence for Ordovician and Early Silurian times with the now well-established Sahara-Pakhuis glaciations (see discussion in Schwarzbach, 1974) and yet, during the same periods, an abundance of iron oolites was formed second only to that of the Jurassic.

Based on studies of the mineralogical and chemical composition of minette and other oolitic iron ores and their associated matrix material, Bubenicek (1968) proposed a model which combines controlled pedologic processes with a complex cycle of regression and transgression. Discussing the climatic requirements, he suggested conditions similar to those prevailing in Southeast Asia of today, where rivers entering the basin of Tonle Sap, Cambodia, contain more than 1% of iron.

The cradle for the Jurassic iron oolites was an epicontinental marine basin which, during the Jurassic, covered large parts of Central Europe, southeast England and most of the North Sea. The basin was subdivided by landmasses like the Rhenish Massif, and by peninsulas and shoals as indicated in Fig. 2. Oolitic iron ores formed in proximity to these land areas, whereby embayments and re-entrants appear to have been preferrential loci for their formation.

Calcareous oolites require a well-aerated and agitated shallow-water environment (Illing, 1954), although some also form at lower energy levels (Freeman, 1962). Very similar, if not identical, conditions are generally accepted for the formation of iron oolites. Observations on recent iron oolites confirm this concept (see below).

According to the classical concept (e.g., Berg and Karrenberg, 1942; Correns, 1942), which is still adhered to by many authors, iron was derived from lateritised land areas. From these, iron was eroded and transported by rivers (and/or groundwaters) partly as weathering solutions, but primarily in colloidal and adsorbed form, into the basin area and deposited in the shallow regions of the sea. However, the formation of iron oolites did not take place in the immediate coastal region, as could be expected, if iron entered an oxygenated marine environment, but at some distance from the coast. The distribution of the oolitic iron ores shown in Fig. 2 suggests that the foci of oolitic iron deposition approximately followed the boundary between the sand and mud facies. As normal oxygenated sea water can only carry trace amounts of iron (Wattenberg, 1942) a delay-mechanism for the deposition of the iron has to be envisaged. Correns (1942) suggested that the retardation of iron flocculation was due partly to the low concentration of iron colloids and further to the presence of humus and humic acids which would have acted as dispersants.

According to Borchert's (op. cit.) model, iron, presumably in adsorbed and chemically bound form derived from land areas, is transported into deeper basin areas, whence it is mobilised in the presence of CO_2-bearing waters. Marine currents carry the mobilised iron into shallow areas where, upon contact with the oxygenated surface waters, iron oolites form. This mechanism is no doubt valid in some instances.

Iron originating from the erosion of lateritic terrain is transported not so much as true solution, but primarily in particulate, colloidal and adsorbed form. It is difficult to envisage that these fractions, upon entering the marine environment, should manage to pass the zone of agitated shallow waters without being flocculated and deposited.

An intake of land-derived, particulate and presumably also colloidal iron, penecontemporaneous with the formation of iron oolites is demonstrated by the Neocomian "Trümmer Erze" of the Cretaceous basin of northern Germany. According to Kolbe (1968), this process took place in a tropical climate. Participation of terrestrially derived (lateritic) iron is also suggested by the generally high content of Al_2O_3 and the small, but systematic presence of TiO_2 in the Minette ores (Bubenicek, 1968).

The models discussed so far would all suggest warm, or even hot, climatic conditions. Considering the participation of iron derived from laterites (and a penecontemporaneous process of erosion and formation of laterite), a seasonally humid climate, proximal to seasonally arid, is indicated.

Such an interpretation is corroborated by investigations of recent iron oolites forming in Lake Chad (Lemoalle and Dupont, 1973). Although Lake Chad is substantially a fresh-water lake, its salinity increases considerably due to evaporation during the dry season, so that certain comparisons with the shallow-marine and brackish-water environment can be made. Of particular significance is the observation of these authors, that the maximum reactive iron concentration occurs at the beginning of the seasonal flood, the iron being transported either in particulate form or as adsorbed oxides or hydroxides. The concentration of dissolved iron is very low and, according to Lemoalle and Dupont (op. cit.), plays no significant part in the formation of the iron oolites. The oolites form in proximity to the major river-inlets in the southeast of the lake and in water depths between 1 and 3 m. It appears significant that their distribution lies on the distal side of the sand—mud line.

Taylor (1969) reports that all the known occurrences of chamosite in recent marine shelf sediments are within 10° of latitude on either side of the equator. In his discussion it is pointed out that investigations of recent marine sediments by the Royal Dutch Shell Group indicate that "chamosite develops on a large scale in shallow waters where the temperature exceeds 20°C". These observations lend further support to the postulation of a warm, seasonally humid climate for the formation of oolitic iron ores.

This is also evident from distribution maps published by Brockamp (1942) for the Northern Hemisphere and more recently by Strakhov (1967). Brockamp shows the distribution of iron oolites for the Silurian, Devonian, Jurassic, Cretaceous and Tertiary Periods in relation to the respective evaporite belts. The latter, which are easy to reconstruct, are a good palaeoclimatic reference datum, reflecting the position of the subtropical high-pressure belt. The maps show that all occurrences of oolitic iron ores, but for two in the Cretaceous, fall close to or south of the respective evaporite belts.

Today, however, the Palaeozoic and Mesozoic deposits are situated in relatively northerly positions. This indicates a migration of the warm climatic belts and, therefore, also a migration of the equator.

Manganese deposits[1]

The geochemical behaviour of manganese is very similar to that of iron. Within the weathering and precipitation cycle, however, manganese is considerably more mobile, being dissolved more easily by CO_2-laden water than iron but, under equal conditions, precipitating later than iron. Furthermore, $Mn(HO)_4$ which derives from the weathering process is electro-negative, in contrast to the electro-positive $Fe(OH)_3$. Hence $Mn(HO)_4$ is capable of absorbing cations which may lead to some enrichment of Ni, Co, Ba, Li and other metals in manganese deposits.

As the main separation between Fe and Mn occurs during the weathering process, it is easily understandable that by far the major portion of manganese deposits of economic dimensions is furnished by either sedimentary deposits or residual (lateritic) accumulations. According to Varentsov (1964), more than 78% of the world reserves of manganese ore are provided by sedimentary manganese deposits of Palaeogene age, and over 6% by residual deposits of lateritic derivation. Manganiferous jaspilites of Precambrian age provide about 6% of the reserves, whereas about 3% occur in red-bed–carbonate associations. The remaining 7% are provided by gondite/queluzite-type deposits of Precambrian age, by occurrences of volcanogenic associations and by some of uncertain or controversial genetic affinities.

Climatic factors have probably influenced the formation of most, if not all, of the sedimentary manganese deposits but, apart from the residual, lateritic deposits in which the climatic control is self-evident, they are best decipherable in the marine, oolitic manganese deposits and in the accumulations of the red-bed–carbonate association.

It is apparent from the preceding figures on ore-reserves, that marine manganese deposits, similar to the oolitic iron formations, are not distributed evenly through geologic time. Apart from the manganese occurrences in the Precambrian (gondite/queluzite and jaspilite deposits), the Cambro–Ordovician, the Carboniferous and, above all, the Early Tertiary were periods of abundant manganese concentrations. Although one may again suspect a strong, if not governing climatic influence, our knowledge of past environmental factors is still too sketchy to allow an exact answer as to just what combinations of conditions were responsible for the preferred concentration of specific elements at specific geologic times.

Marine, oolitic manganese deposits show considerable similarities to the oolitic iron-formations, particularly with regard to their geometry and facies development. As this type of deposit is exemplified by the large manganese concentrations of the southern Ukraine, they are referred to collectively as the Nikopol Typus (Varentsov, 1964).

---

[1] Editor's note: see, in Vol. 7, various chapters on Mn-deposits.

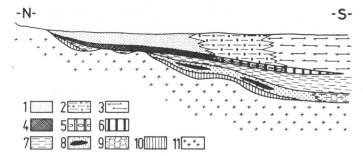

Fig. 3. Schematic profile through the Nikopol manganese deposit, showing the transgressive nature of the sedimentary sequence onto weathered, crystalline basement of the Ukrainean Platform. *1, 2*, and *3*: Kharkov Stage, Oligocene (*1* = sand; *2* = siltstones, sands and clays; *3* = clays and marls). *4, 5*, and *6*: Manganese ore zone (*4* = oxide ores; *5* = mixed oxide–carbonate ores; *6* = carbonate ores). *7* = Kiev Stage (clays, marls and siltstones); *8* and *9*: Buchaksk Stage (sands, clays and coal); *10* = weathering-crust on basement; *11* = crystalline basement (granites, gneisses and schists). (After Varentsov, 1964.)

The Nikopol deposit, which may be regarded as the *locus typicus*, is of Early Oligocene age and of the Lower Kharkov Stage. Within the marginal areas of the basin, these sediments rest unconformably on the crystalline basement of the Ukrainean Shield or on a weathering crust developed on the basement. Further south, in a basinwards direction, older deposits (Eocene) are intercalated between the basement and the sediments of the Kharkov Stage. The pre-Oligocene sedimentary sequence is composed of sandstones, silt-stones and shales with occasional lenticular coal seams near the base (Varentsov, op. cit.).

The Lower Oligocene (Kharkov Stage) overlaps the older sediments and transgresses onto the ancient crystalline basement. The manganese deposits occur near the base of this transgressive sequence (Fig. 3). Mature quartz sands and glauconitic clays, varying in thickness from a few centimetres to 30 m, occur between the actual transgression surface and the manganese deposit. According to Varentsov (op. cit.), this quartz sand–glauconite association of the underlying sediments is characteristic for this type of manganese deposit. The ore bed, the maximum thickness of which is 4.5 m, comprises an oxide facies (pyrolite and psilomelane) proximal to the basin margin, a facies of mixed ores (psilomelane and manganite) and a distal carbonate facies (rhodochrosite and mangano-calcite). The latter facies thins out basinward and interdigitates with the normal basin sediments (Fig. 3). The ore is composed partly of oolitic, partly of concretionary, irregularly shaped forms enclosed in a clayey sand matrix. Towards the north, in the direction of the basin margin, the manganese deposits wedge out and their stratigraphic position is taken up by carbonaceous sands and clays. Coal-bearing beds are also known.

Apart from other manganese deposits on the southern Ukrainean Shield, Varentsov also includes in the same type the Oligocene manganese deposits of Georgia (e.g., Chiatura), northern Caucasus, and the Paleocene occurrences of the North Ural. They are characterized by their marginal position in cratonic basins and their similar sedimentary sequences and the geometry of the ore bodies and distribution of ore facies.

Relative enrichments of Ni, Co, Cr, V, Cu and Mo are considered typical of these manganese ores, but enrichments of Ba and Sr also occur. The manganese is thought to have been derived from chemical weathering of land areas whence it was transported into littoral areas of adjoining epicratonic, marine basins. Varentsov (op. cit.) suggested a humid, presumably warm humid, climate for the formation of the Ukrainean and Georgian deposits, an environment bordering an arid zone for the north Caucasian deposits and a temperate climate for those of the northern Ural. Unfortunately, he does not elaborate on the reasons for the rather wide climatic spectrum assigned to these deposits. In the case of the north Caucasian deposits, gypsiferous red-beds occur at the base of the Oligocene sequence, but they are separated from the ore zone by about 50 m of grey marine clays and marls.

A distinctly different type of manganese deposits, characterised by red sandstone and dolomite associations is exemplified by the manganese occurrences of Morocco. These were discussed in some detail by Bouladon and Jouravsky (1952, 1956) and Vincienne (1956); only the most salient points are repeated here.

Geotectonically, the Moroccan deposits are also products of epicratonic depositional areas. However, the periods in which they were formed are widely dispersed throughout the Mesozoic, ranging in age from Permo-Triassic to Early Jurassic and Late Cretaceous.

Although the various sequences vary in their age and detailed lithologies, they are characterised overall by red, arkosic sandstones at the base, and overlain by primary dolomites, followed by red, gypsiferous shales. The latter may grade laterally into cherty dolomites. Apart from the Permo–Triassic deposit at Narguechoum, where the manganese occurs within the red-bed sequence, the manganese is found at the base of the dolomite or interlayered with it (e.g., Imini and Tasdremt). The deposits are generally persistent along the strike over tens of kilometers. Basinward they pinch out into pure dolomite, whereas in the direction of the ancient shoreline, the facies changes to manganese-impregnated sandstones and, finally, into sterile sandstones (Bouladon and Jouravsky, op. cit.).

The ores are composed mainly of polianite and psilomelane. Of special significance appears to be their high content of Ba (up to 6% $BaSO_4$), Pb, Zn and Cu.

All main features of the Moroccan manganese deposits of Mesozoic age show mildly evaporitic affinities. This is attested to by the red-bed–dolomite association, by the presence of gypsum and the exceptionally high content of barium (cf. Fig. 8). An arid climate is thus indicated for their formation. In the chart, Fig. 12, they are placed together with the accumulations of the red-bed suite. Supergene (lateritic) manganese deposits are discussed under "laterites".

In both the oolitic and the Moroccan type of manganese deposits, a terrestrial source is envisaged. As mentioned above, manganese is more readily soluble than iron. However, this feature alone does not explain the low iron content generally found in manganese deposits. The geochemical factors involved in the separation of Fe and Mn have been discussed by Krauskopf (1957) and Strakhov (1969) and it will suffice here to state that Mn precipitates at a higher pH than iron. Thus, a differentiation between the two elements can already occur in the source area, but may be expected also with the transition from low-pH terrestrial waters to the higher-pH marine environment. Varentsov (op. cit.), for instance, has demonstrated, that the Mn/Fe ratios in the Ursinsk deposit rise systematically from an average value of 1 at the southern, shoreward side, to about 4 in the northern, distal parts of the manganese deposit. This he interpreted as evidence for the terrestrial provenance of the manganese.

A similar, differential mechanism may be envisaged in the case of the Moroccan deposits, where iron may have remained bound or was precipitated within the terrestrial environment, whereas manganese was deposited within the higher-pH environment along the periphery of mildly evaporitic basins.

ORE DEPOSITS OF WARM AND HOT CLIMATES

Laterites

Some of the clearest evidence for climatic control on the formation of economic ore bodies is presented by those deposits which owe their existence to lateritic processes. Perhaps it would be more appropriate to refer to these occurrences as ore or mineral "concentrations" rather than "deposits", as this type of mineral accumulation results primarily from *in situ* redistribution of elements and not, as might be inferred, from depositional or sedimentary factors. This is not to say that we do not recognise secondary laterites which owe their formation to reworking, akin, for instance, to the K-cycle model, suggested by Butler (1967), or other catena-like processes.

Lateritic profiles are essentially fossil soil profiles (regoliths), which resulted from prolonged chemical weathering on stable, senile land surfaces or peneplains, where erosional energy was at a low level. A seasonally humid and warm (tropical–subtropical) climatic regime not only provided for a certain vegetative cover, but also for optimal variation of the ground-water table.

In the application of laterites as palaeoclimatic indicators, care should be taken to stay within close limits of the original characterisation of laterites and lateritic soils, as less critical application of the term may result in serious misinterpretation of the fossil climate. Identification of any ferricrete as a "laterite" without the support of additional evidence, may be misleading and result in erroneous palaeoclimatic models. An example are the "bauxites" of the Vogelsberg Basalts (Germany), which are often quoted as tropical, but more probably are the product of edaphically favoured red weathering in a subtropical or even only warm-temperate climate (Wirtz, 1972; Schwarzbach, 1968).

The basic feature of laterites and their associated deep weathering profiles are well known through the classical description of Buchanan (reviewed by Fox, 1936), Walther (1915) and many others. More recently, the subject has been dealt with by Prescott and

Pendleton (1952), Strakhov (1967) and Maignien (1966). Of the many papers dealing with regional occurrences, perhaps those of Eyles (1952), Brückner (1957), Mulcahy (1967), Kužvart and Konta (1968) and Valeton (1972) should be mentioned.

The classic and typical laterite profile consists, in ascending order, of a ferruginous, often pisolitic layer, a mottled zone and a bleached zone. The latter, also referred to as the pallid zone, grades down into unweathered rock (Walther, 1915; Prescott and Pendleton, op. cit.). Lenses, veins or irregular bodies of opaline or cryptocrystalline silica may form locally on specific parent rocks, but are not typical for laterite profiles as a whole. The view expressed by Whitehouse (1940), that the extensive silcrete layers in western Queensland and northern South Australia are part of a lateritic profile, is not supported by more recent field evidence (Wopfner and Twidal, 1967; Wopfner, 1974). No continuous siliceous layers are known from the bauxite deposits of Northern Australia (Grubb, 1973) and silcrete layers are also absent from the vast, lateritic terrains of Western Australia.

Depending on the prevalence of Fe or Al in the upper part of the profile, a basic subdivision into ferruginous and bauxitic laterites is sometimes made. Further subdivisions according to geomorphic criteria, nature of parent rock and presence of specific, resistant minerals are also common. The last-mentioned laterites often lead to the formation of residual mineral concentrations (see below). Whilst the physical properties and mineralogical and chemical constituents of lateritic profiles are well established and characterised on a worldwide basis, there is still some discussion on the complex geochemical processes involved in the formation of laterites. Both experimental and field data on chemical weathering, recently reviewed by Valeton (1972) in relation to bauxites, still leave some uncertainty as to the control of selective solubility and the complex interplay between pH, Eh, solution concentration and many other factors, which decide whether a lateritic, a bauxitic or a siliceous–kaolinitic profile will ultimately be formed.

Very much generalised, the process of lateritisation involves the transformation of silicates into Fe- and Al-oxides, the concentration of these oxides within the upper part of the profile and the lateral removal of silica. To achieve this transformation, which takes place mainly via ionic solutions, a high, but seasonally fluctuating groundwater table is required. This demands above all else a seasonal humid climate, but also high climatic temperatures to facilitate the necessary "ionic mobility". A further prerequisite is a stable landscape. Only under stable geomorphic conditions is it possible for deep chemical weathering profiles to form without their being destroyed by mechanical erosion. A multiple-phase, climate-dependent process for the formation of laterites in West Africa was suggested by Brückner (1957, p. 255), who envisaged "a climatic cycle involving three stages — arid, humid, semi-arid — of which the last two are essential", as the prime factor in the formation of laterites and bauxite. He thought that differences in the composition of the parent rock were the deciding factor as to whether laterite or bauxite would form. According to the data presented by Valeton (1972) and Plumb and Gostin (1973), the development of primary bauxites requires more humid conditions than are necessary for the formation of ferruginous laterites. As indicated above, for the formation of laterites we envisage a warm, seasonally humid climate, possibly in the outer zones of the monsoonal belts, with a moderate, perhaps savannah-like vegetation cover.

Although the existence of laterites has been suggested from Early and Middle Proterozoic sequences (Alexandrov, 1955; see also discussion on banded iron-formations), the evidence for such occurrences is by no means conclusive. To the best of our knowledge no reliably identified lateritic profiles are known from the Precambrian. This is not at all surprising, since the anoxic or poorly oxygenated atmosphere, at least of the earlier Precambrian and the lack of vegetative cover would have created different environmental conditions to those envisaged for the formation of laterites. Some authors (e.g. Kužvart and Konta, 1968) have argued that vegetative cover is not essential for the formation of laterites and that lateritisation may be effected solely by CO_2 absorbed from the atmosphere. It appears doubtful, however, whether such processes would result in the same lateritic products such as we know them from the post-Devonian periods of earth history. The existence of lateritic profiles in the Phanerozoic, on the other hand, is well documented, particularly since the Silurian. By far the greatest proportion of known laterites are of Cainozoic age, but this is at least partly related to the fact that laterites are easily destroyed by mechanical erosion.

In the following a few typical examples of economic ore deposits formed by lateritic processes are discussed. They include supergene enrichment of low-grade iron-ore and manganese deposits, nickeliferous laterites and some miscellaneous concentrations by lateritic processes.

Supergene enrichment of low-grade iron ore and manganese. In many instances lateritic processes can, through lateral removal of silica, enrich iron deposits of uneconomic grade to high-grade iron ores. This natural benefication is of particular significance to banded iron-formations, as it reduces the undesirable silica content and increases the grade of the

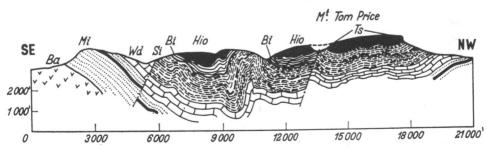

Fig. 4. Section through the Precambrian sequence at Mt. Tom Price in the Hamersley Ranges, Western Australia, showing the structural position of the high-grade iron-ore bodies (*Hio*). *Ba* = Basal volcanics; *Mi* = Marra Mamba Iron-Formation; *Wd* = Wittenoom Dolomite; *Si* = Mt. Sylvia Iron-Formation; *Bi* = Brockman Iron-Formation, from which the high-grade ore developed; *Ts* = Tertiary land surface. Lateritic processes on the Tertiary land surface removed silica from the Brockman Iron-Formation, thus forming the high-grade ore. (Reproduced from Campana, 1966; courtesy *Miner. Deposita*.)

ore by a factor of 2 to 3. An excellent example of this process are the recently discovered large iron-ore deposits in the Hamersley Ranges in Western Australia, (Fig. 1). The sequence of the Hamersley Basin contains three banded iron-formations with a total thickness of up to 900 m (Fig. 1). Of these, only the Brookman Iron-Formation, which reaches a thickness of about 600 m, is of economic significance at present. The average grade of the Brookman Iron-Formation ranges, according to Campana (1966), between 20 and 35% Fe.

Contrasting with these values are the iron contents of the high-grade iron ores of the economically exploited deposits. The massive, dark-blue hematite ore of Mt. Tom Price has an average iron content of 64–66% (Campana, 1966), and MacLeod (1964) reports grades for similar ores in the Mt. Newman area of between 64 and 69%. Of somewhat lower grades (between 58 and 62% Fe) are the hematite-goethite ores, whereas pure goethite concentrations range between 50 and 58% in iron content (Campana, op. cit.). Some of the ores show a high manganese content, suggesting that manganese was retained in the system together with the iron.

Most of the high-grade ore zones have a thickness between 20 and 30 m, but local thicknesses in excess of 100 m are reported (MacLeod, 1964). All the ore bodies are exposed on the surface, requiring no removal of overburden (Fig. 4).

The secondary enrichment of iron which led to the formation of these high-grade ore bodies is clearly related to lateritic weathering (MacLeod, 1964; Campana, 1966). Thus, the climatic influence must be regarded as one of the major controlling agencies, although other factors, like stratigraphy and structural configuration in relation to the existing land surface were also of importance.

Prolonged lateritic weathering was facilitated by the existence of a stable low topography, which prevailed not only in Western Australia but was the dominant landform over large parts of the Australian continent in Tertiary times. The formation of a lateritic profile on the banded iron-formation of the Hamersley Basin resulted in the removal of silica, thus concentrating the iron in the residual lateritic crust. Shallow plunging synclines were particularly favourable loci for high-grade ore concentrations (Fig. 4) as the structurally controlled run-off of groundwater optimised the selective removal of silica (Campana, 1966). In addition of this "peneplain"-type of lateritisation, Campana also reports the formation of laterites in valleys and as talus deposits.

The age of lateritisation, and thereby the time during which a warm and seasonally humid climate prevailed in the region of the Hamersley Basin, cannot be deduced with certainty from the local evidence of that area. Johnstone et al. (1973) report from the Carnarvon Basin that lateritisation has affected rocks of Late Eocene and older age, but not strata younger than Middle Miocene, indicating an Early Miocene–Late Oligocene period of lateritisation. A similar age was also deduced by Wopfner (1974) for the period of lateritisation in northern South Australia.

It may be of interest to consider possible reasons for the establishment of a warm humid climate in Australia at that particular time. One explanation is offered by the drifting away of Australia from Antarctica in Early Tertiary times. This not only brought Australia into lower latitudes, but also allowed for better entrance of moisture-laden marine air onto the continent. As the separation between Western Australia and India had already commenced in the Neocomian (Johnstone et al., 1973), deep weathering processes in Western and Northern Australia may have started earlier than in the southern and southeastern parts of the continent.

Lateritic enrichment of banded iron-formations is not restricted to the Australian occurrences. Concentration of iron by similar supergene processes is reported from South American deposits (Dorr, 1973), from India (Wadia, 1966), from Lake Superior, and another example is the iron deposit of Bilbao in Spain, just to mention a few. One may indeed speculate with Campana (op. cit.) on the many cases in which a simple genesis has been obscured by subsequent tectonic events.

Lateritic processes, or deep-weathering processes akin to them, may also lead to the formation of high-grade manganese deposits. Although rarely of comparable size to the oolitic deposits, they nevertheless form important accumulations of manganese. Large deposits of this type occur in South Africa (Postmasburg) (De Villiers, 1956; Kupferburger et al., 1956) and in Gabon (Moanda − 200 million tons of ore; Varentsov, 1964; Dorr et al., 1973). Secondary enrichment by supergene processes is also suggested for the vast Kalahari deposits (De Villiers, 1956). Important deposits of manganese oxide also formed on the Precambrian Dharwas Slates in Singhbum, Jabalpur and Bellary, India (Wadia, 1966). Other occurrences are in Brazil, Ghana, and Zaire (Dorr et al., op. cit.).

Considering the higher mobility and greater solubility of manganese than those of iron, the question arises why manganese was retained within the weathering profile. In most of the examples mentioned above, the lateritic profiles developed on low-grade, manganiferous limestones and dolomites. As these sediments are inherently low in iron and silica, manganese takes the position of iron within the laterite profile, whereas carbonate is removed in solution. The higher pH which presumably prevailed in these profiles would have been an additional factor in manganese retention.

Of some interest from a genetic point of view are the deposits of manganese oxide in the Pilbara area in Western Australia. These are thought to have formed through deposition of manganiferous solutions along fossil drainage channels on a Tertiary land surface. The highest grades of manganese ores in that region were formed where manganese was precipitated from solutions into cavernous limestones (Pratt, 1971). The source for the solutions is provided by slightly manganiferous rocks of Precambrian age.

Nickeliferous laterites. Wherever lateritic profiles formed on ultrabasic igneous rocks like peridotite, serpentinite, dunite or pyroxenite, the resultant laterites are invariably rich in nickel and commonly also cobalt. Depending on the composition of the ore, one distinguishes between nickeliferous iron laterites and nickel−silicate laterites. Nickeliferous iron laterites are characterised by high iron contents (45−55%) and nickel grades between 0.9 and 1.5%, whereby the nickel is mainly bound to goethite and limonite (Fig. 5). In the nickel—silicate laterites, iron rarely exceeds 30% and the nickel, averaging about 1.5%, occurs either as garnierite or is bound to clay minerals. Differences of the parent material are thought to be responsible for the different types of nickel laterite; nickel—iron laterites being formed preferentially on serpentinite and the nickel—silicate type through the weathering of peridotite, dunite and pyroxenite (Cornwall, 1973). However, morphological differences may also lead to the development of different types of nickeliferous laterites as demonstrated by Lersch (1973) in some Brazilian deposits. Additional influences through slight climatic variations may be possible, but have as yet not been established. Like other laterite profiles, nickeliferous laterites show a distinct vertical differentiation.

Trescases (1973) has described profiles from New Caledonia which, in descending order consist of an iron crust (0.3 m), granular and plastic laterite (7 m), fine-grained saprolite (21 m) and coarse-grained saprolite (1.7 m). The latter forms the transition to the underlying, fresh peridotite. Profiles observed in other nickel-laterite regions (Philippines, Australia, Oregon, and California, Cuba, etc.) follow substantially the same scheme (Philippines Rep., 1958; De Vletter, 1955). The presence of microcrystalline and porcelanitic quartz in the form of veins and silica boxwork in some nickeliferous laterites in Oregon (Hotz, 1964), may be of genetic significance. Siliceous layers have been described also from South American profiles (Lersch, 1973) and lenses of chrysoprase are also common in Australian deposits. The retention of silica within the profile (see also Fig. 5) may be symptomatic for restricted subterranean drainage and a fluctuating groundwater table.

The element distribution within nickeliferous laterite profiles, as demonstrated by Fig. 5, shows an iron concentration in the top of the profile and maximum nickel enrichment in the lower part.

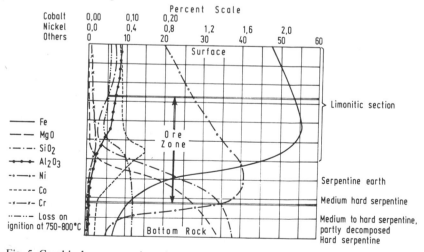

Fig. 5. Graphical representation of mineral contents and vertical zonation of Ocujal lateritic nickel ore, Cuba. (After De Vletter, 1955; courtesy *Eng. Min. J.*)

The depth of weathering is largely dependent on the type of land surface, the deepest profiles being observed on stable peneplains. In many cases subsequent erosional dissection of the peneplains may lead to modifications of the original profile and the formation of piedmont and valley deposits (e.g., Lersch, op. cit.; Trescases, op. cit.). On the average, the thickness of a profile may vary between 10 and 30 m.

Most of the formation of nickeliferous-laterite profiles appears to have taken place in Cainozoic times. A Miocene age is indicated for the Australian deposits, whereas a Miocene—Pliocene age is suggested for the main formational period of the New Caledonian nickel laterites (Trescases, op. cit.).

The largest laterite—nickel deposits occur in the region of the Southwest Pacific (Philippines, Indonesia, New Caledonia, Australia and Solomon Islands) and in Cuba. Smaller deposits occur in northern South America, the United States, U.S.S.R. and in the Balkans. Nickeliferous laterites provide about 35% of the total identified nickel resources (Cornwall, 1973).

Miscellaneous laterite deposits. A considerable number of elements other than those discussed so far are obtained from lateritic sources, either as byproducts of the deposits already dealt with, or in separate operations. Gallium is recovered from bauxites in Arkansas and Hungary, and vanadium from certain French bauxites.

Lateritic weathering may also lead to concentrations of titanium oxide, mainly in the form of leucoxene. These deposits are not as yet of commercial value. According to Klemic et al. (1973b) they are widespread in Hawaii and other tropical countries. The same authors also report a major occurrence in ferruginous bauxites from the Columbia River Basalt in Oregon.

Concentration by lateritic processes was responsible for the formation of the rare-earth ore deposits on Morro de Ferro in the Pocos de Caldas district of Brazil (Wedow, 1967). Similar lateritic enrichment and subsequent formation of eluvial placers are thought to have been the forming agent of the beddeleyite and zircon deposits in the same region (Klemic et al., 1973c).

Bauxite

Bauxites provide by far the most important source for the world's rapidly rising demand for aluminium. Like laterites, the formation of bauxites results primarily from deep chemical weathering processes on stable landforms. However, bauxite deposits are frequently multicyclic, leading to the formation of complex catenae from upland bauxites or to reworking and redeposition as may be the case when peneplain bauxites were uplifted and dissected by subsequent erosion (Valeton, 1972, 1973). Consequently, a principal distinction between autochthonous and allochthonous bauxite deposits can be made, although the criteria observed in the field are not always sufficient to allow a unique answer. Hence, there are still many bauxite deposits the origin of which is contro versial (see, for instance, the discussion on Australian bauxites by Plumb and Gostin, 1973). Further subdivisions based on regional aspects, geomorphological affinities or chemical characteristics have been attempted. Russian authors for instance (e.g., Strakhov, 1968; see also Plumb and Gostin, 1973) make a distinction between bauxite deposits developed in geosynclinal regions and those formed on platform areas. Valeton (1972), in discussing the merits of bauxite classification, has stressed the importance of texture and mineral composition.

Of considerable importance for the formation of bauxites is the geomorphological framework in providing the proper balance between chemical weathering and ground-water movement which is necessary for the removal of silica and the neoformation of bauxite mineral. Thus, Grubb (1973) distinguishes between upland bauxites and low-level peneplain bauxites. The two types are also characterized by different pH-values, those for upland bauxites ranging between 4.8 and 5.5 and those for north Australian peneplain deposits between 6 and 7. Typical profiles for two north Australian Tertiary peneplain deposits, Weipa and Gove Peninsula, are shown in Fig. 6.

Differences in temperature are also thought to be responsible for variations in the mineral composition of bauxites; boehmite—diaspore bauxites supposedly having formed at higher temperature than those composed primarily of gibbsite. This assumption, how-ever, is not fully supported by other evidence, as was pointed out by Grubb (1973).

A rather special case is the karst-bauxites which may be classed separately from the above types. They include the oldest known bauxite deposits (China, Ural) and provide

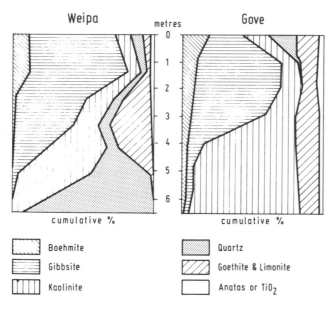

Fig. 6. Composition of bauxite ores of the bauxite deposits at Weipa (Cape York Peninsula) and Gove Peninsula in northern Australia (After Grubb, 1973; courtesy *Miner. Deposita*.)

the majority of European deposits, ranging from Devonian to Mid-Miocene. They are also known from Fiji, Jamaica, the U.S.A. and India (Valeton, 1972). The origin of karst-bauxites is still very much under discussion, but their sometimes bedded nature and the occasional presence of marine fossils make a secondary detrital origin in a coastal-plain environment most likely (Nicolas, 1968).

Clearly allochthonous are the stratified bauxite deposits, as exemplified by the baux-ites in the basal Wilcox Group at the Saline Dome in Arkansas. These resulted from erosion of autochthonous bauxites and redeposition of the material in colluvial and stratified deposits (Gordon et al., 1958, quoted by Valeton, 1972). Deposits of this nature frequently show a distinctive facies differentiation. In the case of the Saline Dome occurrences, the bauxite layers are associated with lignites which either overlie or are developed basinwards from the bauxite beds. Lateral facies changes from bauxite to coal were also described by Schüller (1957) from the Honan province of China.

Although specific parent rocks, rich in alumina, were once thought necessary for the formation of autochthonous bauxites, more recent evidence has shown that bauxite may form on almost any silicate-bearing parent rock which is low in iron, providing that the right climatic and groundwater conditions prevail over sufficiently long periods of time. The huge bauxite deposits of Weipa in northeasternmost Australia even formed on a kaolinitic quartz-sandstone, composed of more than 90% of SiO_2 (Evans, 1959; see also Fig. 6).

There is general agreement amongst most authors dealing with bauxites that, in order to remove the silica, good drainage plays an essential role in the formation of bauxites. Thus, an evenly high moisture level would seemingly facilitate optimal conditions, and this would best be provided by a humid climate. The capacity for moisture retention would be further enhanced by a dense vegetative cover as was pointed out by Grubb (1970). This would further decrease mechanical erosion and through addition of humic acids would provide an optimal environment for the solution and removal of silica. That the formation of bauxites requires higher annual precipitation than that of laterites is also indicated by the distribution of bauxites and laterites in Australia. They are all fossil occurrences, and although not all of them can be dated accurately, a Mid-Tertiary age appears to be the most likely. Ferruginous laterites occur throughout the centre of the continent, whereas bauxites are situated in coastal regions, where a more humid climate prevailed than in the interior. Since neither laterites nor bauxites form in Australia today, this suggests that during the Mid-Tertiary the climate over the whole of Australia was more humid than it is at present.

According to Ségalen (in Valeton, 1972, p. 58), "ferralitic soils with aluminium-hydroxide" form in more humid and somewhat cooler climates (mean annual tempera-ture between 18 and 25°C, and 1—2.5 m annual precipitation) than "tropical ferruginous soils" (mean annual temperature between 25 and 30°C and annual rainfall around 0.5—0.8 m). Tenyakow (quoted by Plumb and Gostin, 1973) suggested a climate with 3000—4000 mm annual rainfall and temperatures between 28 and 32°C for the formation of bauxites. This would place bauxite development within the hot and humid climatic zone. Considering the present distribution of bauxite, this would require considerable variations of the climatic belts of the past. The map on Fig. 7 compiled from Strakhov (1967), shows in a generalised form the worldwide distribution of "bauxites" (in the broad sense of the meaning) since the Devonian. Also shown on the map is the distribution of the recent red soils which, forming a belt on both sides of the equator, clearly demonstrate the dependence of this particular soil-type on the present climatic zonation.

In contrast to the present red-soil belt, the fossil occurrences of "bauxites" show a marked southerly shift with decreasing age. This shift in "bauxite" (and also laterite) distribution thus indicates variations in the distribution of Phanerozoic climatic zones. Today, the development of this trend is explained by most geologists in terms of continental drift and plate tectonics, but variations of the lateral width of climatic zones, depending on the absence or presence of polar-ice caps must also be considered.

As indicated above, bauxites are known since Ordovician times, but Mesozoic and particularly Tertiary deposits account for the majority of economical deposits. Views expressed against the formation of Precambrian laterites apply equally or even more so in the case of bauxites.

World bauxite reserves are at present estimated to be in the region of $12-15 \cdot 10^9$ tons (Patterson and Dyni, 1973), but large tonnages are still likely to be discovered and

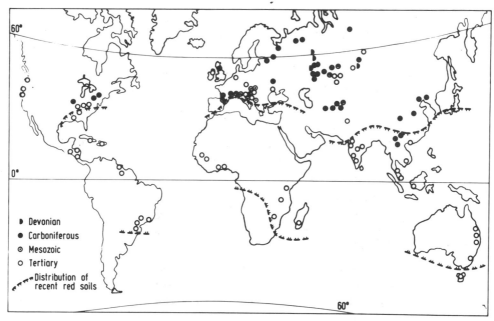

Fig. 7. Distribution of "bauxite" deposits from Devonian to Tertiary times. Compiled from several maps of Strakhov (1967). Only Strakhov's data were used, wherefore some deposits (e.g., Weipa, Gove Peninsula) are not shown, and the stratigraphic position of others may be subject to review. Distribution of red soils after Van Houten (1961).

developed in those regions where extensive peneplains were exposed to a warm and humid climate, particularly during the Tertiary as for instance in northwestern Australia (Sofoulis, 1965).

Supergene uranium deposits

Concentrations of uranium minerals (mainly uraninite, carnotite and coffinite) in sedimentary rocks form one of the principal sources of uranium (see Chapter 3, Vol. 7). Under the term "supergene" deposits we group those uranium occurrences in sedimentary rocks which, according to the general consensus of opinion, were formed by the percolation of uranium-bearing groundwaters through a suitable host rock with a reducing environment. The reducing environment may either be provided by lignitic material (fossil wood) or may be self-generating. We do not intend to discuss here the mostly Precambrian uranium deposits associated with carbonaceous sediments of apparently alluvial affinities, as, for instance, those of the Witwatersrand Basin in South Africa (see Chapter 1, Vol. 7). Uranium occurrences in red-bed associations are discussed briefly under "arid concentration deposits". This seemingly arbitrary separation appears justified in view of the different genetic models.

Based on the form of North American occurrences, two principal types of uranium deposits of groundwater percolation origin are recognised; the Wyoming "roll"-type and the Colorado Plateau "peneconcordant"-type (Fischer, 1970).[1]

The Wyoming roll-type, so named for the semi-cylindrical cross-section of its solution front, is found characteristically in loosely consolidated or uncemented arkosic sandstones, commonly containing carbonised fossil wood. The sediments are usually of fluviatile origin and not infrequently infillings of broad channels. The highest concentration of uranium occurs immediately behind the solution front, which sharply separates ore grades from barren rock at the convex side of the roll. Roll-type deposits generally occur at shallow depth below the surface, and below the groundwater table. Details of this type of deposits have been discussed amongst others, by King and Austin (1966), Fischer (1970), Harshman (1970), Rubin (1970) and need not be further elaborated on by us.

The Colorado Plateau-type is found in well-lithified, usually quartzose sandstones. They are often calcareous and also contain fossil wood, or lignitic or asphaltic material. Commonly, lenses of vari-coloured tuffaceous shale are intercalated within the sandstones. A fluviatile origin of the host-rock sediments is indicated. Uranium deposits of the Colorado Plateau-type form tabular, discrete ore bodies which lie more or less parallel to the bedding. Uranium occurs as cement or as replacement. Vanadium and copper are commonly associated with the uranium (Laverty and Gross, 1956; Fischer, 1970).

A somewhat special type of stratiform uranium deposits appears to be represented by

---

[1] Editor's note: see arguments presented by Rackley in Chapter 3, Vol. 7, against this two-fold subdivision.

the radioactive lignites of Paleocene age in Dakota, where uranium is concentrated in those parts of the lignites which are in unconformable contact with overlying volcanic rocks of Oligocene and Miocene age (Denison and Gill, 1956).

The character of the host sediment was, no doubt, partly shaped by climatic factors. The presence of carbonised fossil wood and lignite and the fluviatile nature, often with large channel-developments, indicate a humid or seasonally humid climate. However, we feel that the major climatic control imposed on the formation of these uranium deposits, is to be found in the climatic influence on the formation of the uranium solution. The nature of these solutions was not only influenced by the climate at their place of origin (mode of weathering), but also along their path of transport.

Today, the generally accepted model envisages solution of uranium either from granitic and volcanic or from sedimentary rocks by meteoric waters, transport by groundwaters and precipitation in a reducing environment. The reductant is thought to be provided by the organic material and/or bacterial action (King and Austin, 1966; Fischer, 1970; Rubin, 1970). In many instances, volcanic material in the immediate vicinity of or associated with the host rocks are thought to provide the necessary source of uranium, whereas in other cases more distant sources have to be accepted. Hostetler and Garrels (1962) suggested slightly alkaline and moderately reducing groundwaters as the most suitable carrier for the dissolved uranium in the case of the Colorado Plateau-type deposits. Fischer (1970) thought that the alterations of the host rock, associated with Wyoming roll-type deposits favoured an oxidizing solution. He suggested further (p. 782) that uranium can be transported as a carbonate complex in a neutral or alkaline solution that is either oxidising or reducing.

Some authors envisage elevated groundwater temperatures, and climatic models range from "warm climates with periods of elevated groundwater temperatures" (Adler, 1970, p. 331) to semi-arid or arid climates (Robertson, 1970). The uranium vein deposits of the Central Massif in France were thought by Moreau et al. (1966) to have formed by supergene concentrations along fractures and shear zones under a tropical to subtropical Triassic paleoclimate. These examples show that there is general agreement on a warm climate, but fairly divergent views on the precipitation-levels of the prevailing climatic regime.

Gabelman (1970) has cast doubt on the solution and mineralisation potential of normal meteoric water, maintaining that present fresh groundwater in the U.S.A. does not mineralise aquifers. Further he drew attention to the fact that the uranium content in normal groundwaters is very low, even in those emanating from uraniferous regions, whereas brines may contain much higher concentrations not only of uranium, but also of other elements. However, he found it difficult to envisage the formation of brines in fresh-water basins and he implied that the presence of "ions not so easily leached such as fluorine and barium" (p. 317), often associated with sedimentary uranium deposits, is hard to explain by supergene processes. This led him to suggest a formation of the uraniferous solutions by mixing of meteoric with hydrothermal waters.

However, concentrated brines and highly reducing environments with vigorous bacterial action are present today in endorheic or partly closed arid basins, as, for instance, in the Lake Eyre Basin in central Australia. There, normal meteoric water entering shallow groundwater horizons to the south of the Macdonnell Ranges reach the stage of saturated brines some 20 km north of Lake Eyre, over a distance of about 350 km (Wopfner and Twidale, 1967). Similar concentrations are experienced along the lower courses of Cooper's Creek and the Diamantina River. The waters of these strictly seasonal rivers, originating in the semi-humid monsoonal regions of northeastern Queensland, become highly concentrated brines by the time they reach the vicinity of Lake Eyre. The groundwater below the normally dry playa surface of Lake Eyre is strongly reducing and the recent to sub-recent lake sediments show considerable concentrations of barium and strontium (Fig. 8). Some groundwaters in the region also show abnormally high fluorine contents. Similar examples may be quoted from other hot and arid regions as for instance from North Africa and Iran (e.g., Krinsley, 1970).

Authors dealing with sedimentary uranium deposits have suggested an intramontane basin framework for their formation (particularly for the Wyoming roll-type, Fischer, 1970; Harshman, 1970) and the spatial restriction of these deposits within the basin areas is well known. Thus, a climate akin to modern arid basins, with drainage areas reaching into seasonally humid regions would fulfil most, if not all requirements, viz. elevated groundwater-temperatures, leaching capacity by corrosive brines and strongly reducing groundwaters within the focal points of endorheic systems.

We suggest, therefore, a hot and arid to seasonally arid climate as the most conducive to the formation of supergene uranium deposits.

Arid concentration deposits

Under this heading we include those deposits, which owe their existence to mineral concentrations in hot and arid climatic zones, to subsequent transport of metalliferous solutions by brines and final precipitation of the metals in dominantly reducing environments, either on land or in nearby marine basins. These processes may take place contemporaneously with sedimentation or during early diagenesis. In this category belong continental red-beds and their nearshore-facies equivalents.

Arid basins are characterised by specific hydrologic regimes. Many of them are endorheic and, as evaporation exceeds precipitation, ascendant groundwater flow is quite common. Wherever primary ore bodies occur within the basement of the basin floor, ascending, metalliferous solutions will precipitate new minerals on or near the surface and may thus form rich mineral concentrations. Metalliferous solutions enriched through weathering may also be derived from the basin margin, whereby the latter are often situated in much more humid climatic zones (see discussion of U-deposits). With increasing aridity, increasing amounts of calcium, magnesium and sodium, normally removed by meteoric water, remain within the system, leading to precipitation or intergranular crystallisation of dolomite, gypsum and halite. Minor elements, like Sr, Ba, F, Br, Li and Cs also become progressively concentrated (Strakhov, 1967). In closed, endorheic systems enrichment of metals brought in solution from distant source areas, may occur. This is evidenced for instance by the concentration of zinc (0.2%) in evaporitic sediments of Mio—Pliocene age beneath Lake Eyre North, situated in the southwestern part of the Great Artesian Basin in Australia (Jones, 1965) (Fig. 8).

Where hot and arid continental regions are open to the sea, complex hydrologic systems ensue. Heavy, terrestrially derived, hyposaline brines may either enter the marine system directly and interact with bottom sediments, particularly if reducing conditions prevail, or they may migrate into porous, nearshore sediments, and react with the interstitial pore fluid to form mineral concentrations. A further mechanism is the sabkha-process, whereby marine waters migrate into terrestrial sediments, becoming progressively more saline with increasing immigration path (Shearman, 1966; Renfro, 1974). Thus, in the arid region there does not exist a really sharp genetic demarcation between deposits in terrestrial red-beds, accumulations in tidal flats and lagoons and concentrations which formed by infusion during early diagenesis.

According to Schneiderhöhn (1962), the ores of arid concentration deposits occur as stratiform impregnations or in diffuse distributions in individual beds, of usually thick, terrestrial continental deposits of arid regions, ranging from graded fanglomerates to arkoses, sandstones and argillaceous sandstones. Particularly frequent are copper ores (chalcocite, bornite, covellite, among others), in other places pure Ag-ores, Ag-halogenides and argentite, or vanadium—uranium ores (carnotite), or galena, pyromorphyte, anglesite, cerrusite and sphalerite. Ore-impregnated plant remnants are common as carbonaceous materials act as a reducing agent. The process of mineral concentration within the sediment may take place either syndepositionally or later, during early stages of diagenesis. Vertical transitions to weathering zones of primary ore bodies may be observed as well as lateral transitions to bedded marine ore-deposits of the Kupferschiefer- or Kupfermergel-type.

Although most geologists favour a syngenetic formation of stratiform or diffuse ore concentrations in red-bed type sediments, there is by no means unanimity of opinion. This applies equally to small occurrences like the lead deposits of Mechernich—Maubach in the Buntsandstein of the northern Eifel, Germany, as well as to the large copper deposits of the copperbelt in North Rhodesia, Katanga (Congo) and Zambia (see below and Chapter 9, Vol. 6).

The lead deposits of Mechernich—Maubach in the Early Triassic Buntsandstein may serve as a typical example of a red-bed type ore deposit. The mineralised zones are contained within a red, slightly kaolinitic and in places conglomeratic facies of the Buntsandstein. The terrestrial nature of these beds is evidenced by amphibian tracks and remnants of *Cyclotosaurus* (Jux and Pflug, 1958). The mineral suite comprises Pb, Zn and Cu-minerals. For many years the deposits were regarded as endogenetic; hydrothermal solutions, derived from a hypothetical plutonic body, were thought responsible for the mineralization. Schneiderhöhn (1962), in his latest view on Mechernich, still envisaged a certain participation of secondary hydrothermal processes. In addition to already existing exogenetic concentrations, he assumed sterile thermal water becoming enriched by leach-

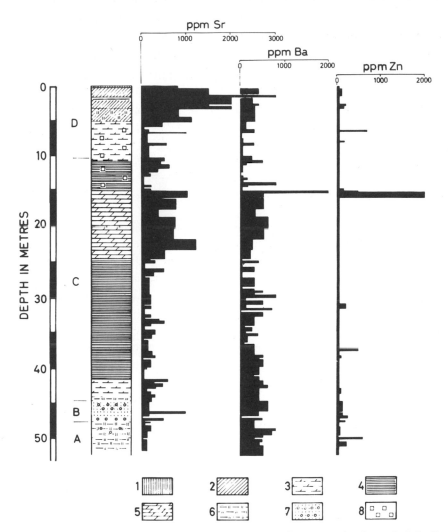

Fig. 8. Lithological and geochemical log of Cainozoic–Recent section beneath Lake Eyre North in South Australia (Lake Eyre bore 20). Lake Eyre is the "sink" for an endorheic system, which drains an area of 1,280,000 km^2. Periods of aridity are characterized by high concentrations of Sr and elevated values of Ba and Zn.
Stratigraphy: A = Eyre Formation (Paleocene–Eocene) (only uppermost part of formation is shown); B = Ferrugineous conglomerate with Fe-ooides, (?) equivalent of Doonbara Formation (Oligo–Miocene); C = Etadunna Formation (Miocene–(?) Pliocene); D = Unnamed Pleistocene–Recent.
Lithology: 1 = halite; 2 = gypsum; 3 = dolomitic shale; 4 = shale and mudstone; 5 = dolomite; 6 = siltstone; 7 = conglomerate; 8 = pyrite. Geochemical log compiled from data presented by Johns (1963).

ing of ore shoots in the basement and subsequent migration of the solution to the surface. Jux and Pflug (1958), on the other hand, suggested lateral intake of metal solutions, derived from the peneplained Rhenish Massif which formed the southern margin of the basin, thus implying climatic control. Samama (1973), in a study of Lower Triassic ore deposits in France and Western Germany, has found an interrelationship between the intensity of reddening of the Buntsandstein and the elemental composition of the individual occurrences. In the red and kaolinitic facies, copper is always present and in some deposits also uranium, whereas in the white or pale-coloured sandstones, mineralisation is restricted to lead and zinc with only minute traces of copper. From this evidence and other studies, Samama (op. cit.) suggested a controlling influence of element distribution by pre-selective concentration of certain elements by specific weathering processes.[1] Depending on the type of weathering, and thus on the prevailing climate, certain elements may be retained within the weathering profile, leading to an enrichment of these elements within the weathering crust, whereas others may be expelled. In subsequent erosional episodes, pre-concentration of a specific element or a group of elements within the soil profile will thus provide a substantially enriched source material.

For the lead deposits in the Buntsandstein at Largentière, situated on the eastern slope of the Central Massif, Samama (1968) suggested that mineralisation occurred along the interface between continental sulphate waters and reducing waters under hot, arid conditions.

Examples of other arid concentration deposits of the "red-bed type" are manifold in the Rotliegendes and the Buntsandstein of Europe as well as in their equivalents in Asia, and the U.S.S.R., for example, the Cu-deposits of the Perm district in the Urals (Nechaev, 1961). Also mentioned should be the Se, Mo, Pb concentrations of the Colorado Plateau and the Cu—As deposits of Central Peru. (See Chapter 8 by Smith, in Vol. 6.)

Similar modes of origin are envisaged for the uranium occurrences in the Permo—Triassic red beds of the Alps (Schulz and Lukas, 1970; Brondi et al., 1973). The mineralisation of the Permo—Triassic alpine "Buntsandstein" comprises Fe—Cu—U minerals and occurs typically in grey interbeds, sandwiched between thick red sandstones. Highly carbonised plant fossils and pyritised bacteria are thought to have provided the reducing environment for the ore formation out of aqueous weathering solutions (Schulz and Lukas, 1970; Schulz, 1972). A similar mode of formation is also envisaged for the Pb—Zn mineralisations in the Permian Grödner Sandstone in South Tyrol. Again, mineralisation is bound to grey sandstones with lenses of carbonaceous material, which are interbedded in predominantly red arenites. In some areas, small amounts of uranium and copper are associated with the Pb—Zn mineralisation. The mineralisation is regarded as being of synsedimentary origin, the metals having been derived from weathered, Permian volcanic rocks (Mostler, 1966; Brondi et al., 1973).

---

[1] Editor's note: see, in this Volume, the chapter by Lelong et al.

A position ranging from the red beds to the marginal marine facies is represented by the deposits of the African Copperbelt. Garlick (1961) and Schneiderhöhn (1962) considered the ore deposits of the Copperbelt as purely exogenic, more or less syngenetic sedimentary concentration deposits (see Vol. 6, Chapter 6 by Fleischer et al.). Although there are still some voices defending a hydrothermal origin, the evidence for a sedimentary or early diagenetic origin far outweighs that supporting the opposite point of view. Pelletier (1964) opined that the majority of geologists working on the Copperbelt favoured the sedimentary theory of the ore genesis. The mineralised sediments belong to the Roan Supergroup and consist mainly of conglomerates, sandstones, dolostones and carbonaceous shale. The sandstones frequently show large-scale cross-bedding, varyingly interpreted as eolian or nearshore marine (? antidune) (Bartholomé et al., 1973; Van Eden, 1974). The rocks were not affected, or only very slightly, by metamorphism and mineralisation took place before deformation (see Bartholomé, p. 23). Their depositional environment ranges from terrestrial to shallow marine. Specific depositional models were suggested recently for the Mufulira deposit in Zambia by Van Eden (1974) and for the Kamoto deposit in Katanga by Bartholomé et al. (1973). The latter authors suggested a tidal-flat environment, evidenced by the presence of dolomite and magnesite as well as laminated shale and stromatolites. They envisaged an environment similar to that prevailing today along the Trucial Coast and, hénce, an arid and hot climate. The ore solutions were thought to have consisted of hypersaline brines, derived by evaporation at the surface. From a palaeoclimatic point of view, their model comes very close to that proposed by Renfro (1974), who suggested a sabkha process for the formation of stratiform ore-deposits. It appears that an early diagenetic mode of formation is favoured at the moment, rather than a strictly synsedimentary origin. There is general unanimity that the formation of the Copperbelt ore bodies required a hot and arid climate.

Similar genetic processes to those proposed by Bartholomé et al. (1973) for the Katanga deposits were suggested by Brown (1971) and White (1971) for the copper mineralisation of the Nonesuch Shale at White Pine in northern Michigan.

Metals in evaporitic basins

Evaporite basins are invariably surrounded by hot, arid regions and are, therefore, often equivalent with red-bed sequences. Alternatively, negative crustal movements may lead to transgressive overlap by evaporites or penesaline deposits across older red-bed sequences as in the case of the Rotliegendes—Zechstein Basin in Europe. Evaporite basins are characterised by an excess of evaporation, and by limited run-off from the surrounding continental regions, resulting in limited intake of clastic material. Richter-Bernburg (1968) suggested that the actual precipitation of evaporite minerals is preceded by long periods of time during which the salt concentration of the sea water is raised slowly until saturation point is reached. During this penesaline period, euxinic conditions are prevalent (Sloss, 1953), providing conducive conditions for the formation of sulfide minerals through the reducing action of $H_2 S$.

One of the best known examples of mineralisation in such a pre-evaporitic environ-ment is the Mansfeld Kupferschiefer of the Zechstein in Central Germany (cf. Chapter 9, Vol. 6), which has supported mining operations since the year 1200 and is still being mined. The Kupferschiefer transgressively overlaps the red-bed sequence of the Rotlie-gendes and occurs near the base of the thick evaporite pile of the Zechstein. Lithological-ly, the Kupferschiefer is a black to dark brownish-grey, marly saprolitic shale with an average carbonate content of 30%. In the nearshore facies, dolomite is dominant, whereas in the distal deposits calcite is more prevalent. The mineralisation comprises primarily sulfides of Cu, Pb and Zn with minor concentrations of Ag, V, Cr, Ni, Mo and Co. Mineral contents are greatest in the southeastern part of the Zechstein Basin, i.e., that part which was proximal to land. Towards the distal northwestern regions of the basin mineral concentrations diminish (Wedepohl, 1964). To the far southeast, in Lower Silesia, equiva-lents of the Kupferschiefer, consisting of interlaminated layers of bituminous shales and carbonates, contain primarily galena (Harańczyk, 1970).

Wedepohl estimated that the maximum water depth of the Zechstein Basin did not exceed 300 m and from considerations of the element balance he concluded that the normal Cu-contents of sea water were completely inadequate to account for the amounts concentrated in the Kupferschiefer. He rejected endogenetic sources for reasons of the even regional distribution of metal contents and the strict interdependence of the miner-alisation and lithology and envisaged, therefore, a lateral intake of terrestrially derived metal solutions. He pointed out that mineralisation in bituminous sediments apparently occurs only where they transgress over red beds or arkoses. From this observation he deduced that the elements were derived laterally from the now underlying Rotliegend sandstones, mainly during the transgression of the Zechstein Sea. Bleached horizons in the upper Rotliegende (the Weissliegende und Grauliegende) were interpreted by him as a kind of regolith whence the mineral solutions for the concentration in the Kupferschiefer were derived. This interpretation approaches somewhat the concept of Samama (1973) of a pre-concentration of certain elements in soil horizons, or in this case, in older sediments.

The stratigraphic position of the Kupferschiefer, sandwiched as it is between the red-bed facies of the Rotliegendes and the thick evaporite sequence of the Zeichstein allows us to place it with great certainty in a hot, arid climate. Transport of mineral solution by more or less saturated brines, not envisaged by Wedepohl (1964), could have been the means of effecting the element transfer from the terrestrial into the marine environment. A hot, arid environment is also suggested by Harańczyk (op. cit.) for the lead-bearing equivalents of the Kupferschiefer in Silesia. He proposed that the lead was precipitated in a hypersaline lagoon, separated from the euxinic Zechstein Basin by a barrier. Normal Kupferschiefer mineralisation took place in the adjacent marginal parts of the basin.

Another model, requiring a hot and arid climate not only for the Kupferschiefer, but also for many other sedimentary copper deposits, was put forward by Renfro (1974). He observed that many stratiform metal deposits were underlain by continental red beds and overlain by evaporitic sequences. As mentioned above, Wedepohl (op. cit.) had already drawn attention to the interrelationship between basal red beds or arkoses, overlain by bituminous shale, and mineralisation. Renfro proposed that these stratiform deposits were mineralised by a sabkha process, whereby, very generalised, saline, metal-bearing groundwaters react interstitially with hydrogen sulfide in a low-pH environment generated by bacterial decay of the buried algal mat. Renfro (op. cit., p. 34) lists 15 metal deposits, including the Kupferschiefer of Germany, the African Copperbelt and the Nonesuch of Michigan, which fulfill the criteria of his rather elegant model. This early-diagenetic process of mineralisation requires a hot and arid climate with a substantial debit of evaporation.

It will be apparent from the foregoing discussions of arid concentration-deposits, evaporite-associated metal deposits and also uranium deposits, that the hot and arid climatic zones are most favourable habitats for synsedimentary or early-diagenetic mineralisation. The influence of the arid climate may, however, reach beyond the realm of exogenetic processes. For instance, the recent concentrations of heavy-metal brines in the Red Sea (see Chapter 4, Vol. 4), although related to excess circulating water, heat-flow and volcanic phenomena associated with the rift mechanism (Bäcker and Richter, 1973), are not devoid of climatic influence, in particular as connate fluids are required. Such metal concentrations can only form as long as they are not diluted by a large intake of detrital material. In the case of the Red Sea, the arid climate with its limited run-off thus provides an important additional factor on the heavy-metal concentration. Later, during diagenesis, such brines may conceivably migrate and find their way into suitable host rocks, such as reef carbonates, which again owe their formation to specific climatic conditions.

Ores in reef complexes and carbonate shelves

Reef complexes and carbonate shelves are well known as favourable loci for ore accumulations. As the formation of reefs is controlled to a large degree by specific climatic conditions, ore bodies formed within the reef complex must also be considered as climate-dependant, no matter if the mineralisation occurred syngenetically or if the reef functioned solely as a preferential host facies (or "reservoir rock") for endogenetically derived mineral solutions.

Present-day coral reefs occupy a comparatively narrow belt on either side of the equator. Although smaller coral banks are known from middle latitudes, the belt of major development of typical reef complexes does not extend far beyond the tropics of Capricorn and Cancer. Prolific coral growth requires a minimum water temperature of 21 °C. Facies associations of fossil reefs and the known solution equilibria of $CaCO_3$ in sea water in relation to water temperature indicate that reef formation in the geologic past depended on similar climatic factors (Teichert, 1958; Strakhov, 1969). Reef complexes may be found from hot arid regions to warm, seasonally humid climatic zones. Thus, they may be climatic facies-equivalents of red beds and evaporites, as well as of laterites.

Typical reef complexes are composed of several facies units, comprising in a seaward direction the lagoonal and back reef facies, the main reef body and the fore-reef facies. Principle depositional models of reef complexes and carbonate shelfs were discussed recently by Wilson (1974). Monseur and Pel (1972, 1973) suggested and demonstrated on examples, that in each facies unit, certain mineral associations are selectively concentrated. In post-Cambrian reef complexes the suggested sequence from shelf to reef is, in simplified form, Mn–Pb, Fe–Pb, Fe–Mn, Pb–Zn, Zn–Pb. In Cambrian bioherms, Cu is formed on the shelf side and Pb–Zn in the biohermal carbonates. The latter sequence is to a certain degree indicated by geochemical studies of a number of Lower Cambrian bioherms of the Adelaide Geosyncline in South Africa, where the bioherms show the highest Pb–Zn values, whereas Cu tends to be higher on the shelf (Thomson, 1965). A causal connection between facies and Cu, Pb, Zn mineralisation was also suggested by Bogdanov and Kutyrev (1973).

Although there is by no means agreement as to the nature of ore formations in many of these deposits, there can be little doubt that at least some of them are of syngenetic or early diagenetic origin. Schulz (1968), for instance, described and pictured interbedding between layers of carbonate and thin bands of sphalerite from the Ladinian Pb–Zn deposit of Bleiberg in Austria. This, together with other clear sedimentary fabrics convincingly demonstrates the synsedimentary nature of this mineralisation.

Also very much discussed is the origin of metal solutions. Three possible sources have been considered: hydrothermal solutions (for epigenetic deposits); submarine volcanic exhalations and lastly, weathering solutions, derived from the nearby land under warm, arid to semi-humid climates.

Whatever the source or sources may have been in any specific case, the fact remains that without the existence of the climatically controlled formation of the host rock, no mineralisation could have taken place. The reason for the selective concentration of certain elements within specific parts of the facies spectrum is suggestive of direct environmental control (different energy levels, organic content, pH, Eh and early diagenetic hydrodynamics). Monseur and Pel (op. cit.) report preferential mineralisation of dolomitized coral-rudistic reef bodies from Reocin (Spain) which "were sufficiently rich in organic matter to permit the precipitation of metals from solutions". Enrichment of metals, particularly Zn, through metabolic processes of certain reef dwelling organisms may also play a role.

The most important of the strata-bound Pb–Zn deposits are the Mississippi Valley-type deposits in the U.S.A. (see pertaining chapters in Vol. 7). Those in southeastern Missouri show a dominance in Pb, whereas sphalerite–galena associations are present in the Upper Mississippi Valley and Tennessee. American geologists seem to favour a diagenetic–epigenetic origin for these deposits (see Morris et al., 1973) whereas the Amstutz-school prefers a syngenetic-diagenetic genesis. Under the same category falls, according to Bogdanov and Kutyrev (op. cit.) the Mirgalinsai deposit in the U.S.S.R., whereas the Central Asian deposits are separately placed by these authors under the "coastal marine type". Of largely syngenetic (and in places two-cyclic) origin are the Triassic Pb—Zn deposits of the Alps and the Carpathians. A synsedimentary origin is also likely for the Pb—Zn mineralisation of the completely undeformed and magnificently exposed Upper Devonian reef complex of the northern margin of the Canning Basin in Western Australia, although a later mineralisation through circulating groundwaters was also suggested (Playford and Lowry, 1966).

Examples of mineralised Cambrian bioherms are the Pb—Zn deposits at Ediacara and the willemite deposits south of Copley, all in Lower Cambrian *Archaeocyathid* limestones of the Arrowie Basin of the Adelaide Geosyncline, South Australia. At least two cycles of mineralisation may be assumed for the willemite deposits: an original syngenetic enrichment of Zn within certain dolomitized zones of the bioherm and subsequent concentration of willemite (and baryte) by circulating supergene solutions along major fault zones.

ORE DEPOSITS OF GLACIAL AND COOL-TEMPERATE CLIMATES

Influence of glacial climate on ore formation and ore prospection

In glacial climates, weathering is almost completely restricted to mechanical factors such as frost action and glacial abrasion. Glaciers may transport blocks or boulders of ores and occasionally sufficient ore may become accumulated within the moraine to form an economic deposit. Such deposits have been referred to as "glacial placer" deposits; however, this term is somewhat unfortunate, as placer deposits characteristically form through gravity separation, a process which is not effective in the case of transport by glacier ice. This is not to say that sorting through water action cannot take place subglacially in wet-bottom glaciers and naturally in the fluvio-glacial environment.

An example of a workable moraine accumulation is known from the slope below the Kennecott copper deposit on Prince William Sound, Alaska. The primary ore body, which occurs at 2000 m above sea level is affected only by mechanical weathering. The talus, including many boulders and blocks of chalcocite, accumulates at the foot of the precipitous rock walls and forms an ice-cemented glacial deposit.

Much more abundant, however, are occurrences of individual ice-transported ore erratics and these can be of considerable importance in the discovery of primary orebodies. Mapping of such ore erratics results in a typical fan-shaped distribution pattern, the confluence of the fan indicating their place of origin. The Swedish explorer Daniel Tilas applied this method in Finland as early as 1739 and described it in a publication (see Borgström, 1942, p. 417, and Laitakari, 1942, pp. 438—439). It was unknown at the time, of course, that these distribution patterns had resulted from glacial transport; the concept of the Pleistocene glaciation was not established until 100 years later. Tilas, however, observed correctly that in Finland the primary orebodies were always situated to the north or northwest of the loose "drift rocks", according to the current direction of the "*Diluvio magno*". The small copper deposit of Tilasinvuori was discovered in this manner by a farmer (and subsequently named after Tilas).

The most significant orebody of this kind in Finland is the copper deposit of Outokumpu. The first loose ore boulder was discovered during the construction of a canal in east Finland. A sample of this erratic, first mistaken for a meteorite, was forwarded to the Geological Survey in Helsinki, where O. Trüstedt, recognising the real nature of the sample, immediately initiated a systematic survey. In 1910, after 2 years of investigations, the ore body was located 50 km to the northwest from the place where the first ore boulder had been found. Drill holes were sunk and the third hole struck the orebody at a depth of 90 m below the surface.

Numerous finds of ore erratics were reported to the Geological Survey of Finland by the local population after the importance of such occurrences was published in schools and through articles in the newspapers. Monetary rewards provided additional incentives. This method, however, did not always lead to the discovery of ore deposits; often the number of ore erratics was quite small, resulting in insufficient control of the fan pattern and the primary rocks, not exposed on the surface, but covered by extensive Quaternary drift, had to be located and investigated by drilling.

The method of locating orebodies by means of glacially transported ore boulders has been used also in Sweden and Canada, but it is gradually being replaced by more sophisticated geochemical and geophysical methods.

Lake and bog iron ores of cool-temperate climates

Cool climates provide optimal conditions for the formation of humus and as certain iron-rich minerals are particularly susceptible to weathering and attack by humic acids, groundwaters become enriched with humates and bicarbonates of iron. When such iron-rich groundwater comes in contact with the oxygen-saturated and usually less acid waters of a lake, iron precipitates as iron-hydrogel. At a few metres depth extensive layers of

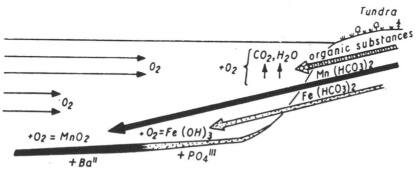

Fig. 9. Schematic geochemical system of iron and manganese, leading to the formation of lake- and bog-iron ores in cold, temperature climates (after Borchert, 1960; courtesy *Trans. Inst. Min. Metall.*)

iron ochre or solid limonite concretions may thus form lake ore deposits. In marshy or swampy terrain similar processes lead to the formation of bog iron ore (or the German equivalent: "Wiesenerz" and "Rasenerz"). A model of the chemical conditions leading to the formation of iron deposits within swamp and lake environments of cold-temperate climates, as presented by Borchert (1960), is shown in Fig. 9.

These iron ores usually form only thin deposits, but they are sometimes comparatively rich in manganese. Organic and mineral impurities are common. Nevertheless, in the past centuries, they have been mined and smelted in many places, particularly in northern Europe.

WEATHERING OF PRIMARY ORE BODIES

Weathering may profoundly modify and redistribute the metal content and the mineral composition of the upper, near-surface part of an orebody, creating a distinct vertical zonation. This modified portion is termed the weathering zone. In the upper part of the orebody, situated at or close to the surface, where oxygen is readily available, oxidation of the original mineral takes place, thus forming the oxidation zone. Within this zone, sulfides are altered to oxides and carbonates and iron concentration on the surface may form a distinctive cap, the so-called gossan. With increasing depth, but particularly within the reducing environment of the groundwater level, certain minerals are precipitated from descending solutions. This neomineralisation leads to the development of the secondary sulfide zone, sometimes also referred to as the "cementation zone".

As supergene processes are largely dependent on climate factors, the developments of oxidation zones and secondary sulfide zones also show climate-dependent differences.

According to Schneiderhöhn (1962), the relationships between climate and the weathering of primary orebodies are as follows: The most enriched oxidation and secondary sulfide zones occur in normal-arid and tropic-arid regions, fairly frequently also in temperate-arid, semi-arid and temperate-humid regions, depending on the character of the country rock. Under favourable conditions, rich oxidation zones may also form in hyper-arid climates, although in highly permeable rocks the oxidation zones may be completely lacking. The secondary sulfide zones in hyper-arid climates is invariably missing or only poorly developed. The oxidation zone is never enriched in tropical-humid climates and becomes more impoverished with increasing precipitation; often it is completely leached and free of metals. Enrichment in the secondary sulfide zone is also reduced. In glacial regions, where precipitation occurs in the solid form and where the groundwater is mostly frozen, the weathering zone is usually absent and the fresh, primary ore minerals are exposed on the surface. The same also applies in high mountain regions and glaciated areas, although here the high rates of mechanical erosion are additional factors.

Fig. 10 shows, in a generalised and schematic form, weathering profiles of different climatic regions.

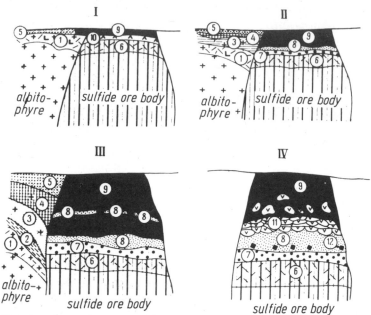

Fig. 10. Schematic representation of climatic influence on the weathering zone of a sulfide ore body. (After I.I. Ginsburg, in V.I. Smirnow, 1970). I. cold climate. II. temperate-humid climate. III. warm-humid climate. IV. arid, semidesert climate. Zones of the general weathering profile: *1* = disintegration zone; *2* = sericite and hydromica; *3* = hydromica; *4* = kaolinite; *5* = ortstein in I and II, laterite in III. Subzones of the weathered ore body: *6* = cementation zone; *7* = granular sulfide; *8* = granular baryte; *9* = limonite; *10* = sulfates; *11* = jarosites; *12* = halogenides.

PLACER DEPOSITS[1]

Placer deposits owe their formation primarily to the greater specific weight of most ore minerals which permits the lighter material to be winnowed out by water or air currents, or by wave action. This winnowing process may act at the place of origin (residual deposits) or somewhere along the path of transportation (accumulation deposits). Intermediate to these are the eluvial or slope deposits. As a fourth category, marine placers may be added. Greater resistance of certain ore minerals to mechanical or chemical weathering processes may also influence the formation of mineral concentrations. A further factor may be imposed by additional concentration through colloidal or solution migration in certain soil profiles, both prior to or after the formation of the placer deposits (e.g., cassiterite-coatings in tin placers as described by Aleva et al., 1973). The occurrence of giant gold nuggets like the "Welcome Stranger", weighing more than 70 kg, is also difficult to explain by purely mechanical transport.

Broadly speaking, the formation of every terrestrial placer deposit is somehow in-

---

[1] Editor's note: for a general summary, see, in this volume, Chapter 5 by Hails.

fluenced by climatic conditions, since mode of weathering and the temporal distribution of energy-levels (e.g., water volumes and current velocities) are largely dependent on the prevailing climate, among other things. Today, the maximum efficiency of water erosion occurs in semi-arid climates, but as Schumm (1968) has pointed out, erosional efficiency is largely controlled by the capacity for water retention of the vegetation cover, and that was certainly not always the same throughout the earth's history. Thus, the above factors are often difficult to decipher from the record of the geologic past.

Accumulation deposits seem to form without any specific preference as to temperature requirements, providing the right transport energy is available to effect gravity separation. Periglacial to late glacial conditions were responsible for the formation of alluvial gold deposits on the Yukon River in Alaska and on the south island of New Zealand (e.g., Karawau Gorge and Landis River), whereas a seasonally hot and arid climate is indicated for the auriferous alluvial gravels at the base of the Late Jurassic Algebuckina Sandstone along the western and southern margin of the Great Artesian Basin of Australia (Wopfner et al., 1970).

A periglacial environment was also thought by Wiebols (1955) to be responsible for the formation of the Witwatersrand gold deposits (cf. Chapter 1, Vol. 7), but this view has been challenged by other South African workers. Some geologists even advocate a hydrothermal origin of the gold. Haughton (1969) has argued for a seasonally humid climate, which in his opinion was necessary to provide for the fluviatile transport of the large mass of clastic material brought into the Witwatersrand Basin.

Gold concentrations of between 30 and 100 ppb are reported over a large area of the floor of the northern Bering Sea (Simons and Prinz, 1973), but it is as yet unknown if these concentrations are affected by climatic factors or result solely from specific source or sea-floor conditions.

Multiple climatic influences have been demonstrated by Aleva et al. (1973) for the formation of most of the tin placer deposits of the Southeast Asian tin belt. Although the participation and importance of each single factor varies widely in the formation of any specific placer deposit, the above authors suggest three essential factors: (1) deep chemical weathering; (2) selective removal of light-weight material; and (3) the presence of catchment areas. Whereas the first factor can be related directly to most tropical climatic conditions, the latter two may have been influenced at least in part by changing sea-level stands during the Pleistocene. The "Kaksa" deposits of the two main tin islands of Indonesia, Billiton and Bangka may here be cited as a typical example for the multiple climatic control of cassiterite placer deposits. According to Krol (1960) and Aleva et al. (op. cit.) this type of deposit comprises "the normal weathering residue" of a deeply weathered near-peneplain within the humid tropics. These lag deposits are composed of "coarse grains and angular pieces of vein quartz from the original bedrock and all insoluble heavy minerals contained in that bedrock". The presence of a once extensive laterite cover is evidenced by the occasional occurrence of rounded pebbles of bauxite, whereas sporadic finds of tectites (Billitonites) indicate a maximum age of 500,000 years. The kaksa deposits occur in shallow, but well-defined valleys, incised onto the flat, very low landscape (Fig. 11). Aleva et al. (op. cit.) suggest that the periods of increased erosion were related to eustatic sea-level changes during the later Pleistocene.

From the evidence presented by these authors the following sequence of events may be deduced: (1) Formation of a lateritic profile on stanniferous granites, whereby the associated deep chemical weathering not only "liberated" the cassiterite from the enclosing rock fabric, but also effected a certain pre-concentration. It may be assumed that this process took place in the Late Tertiary, perhaps extending into the Early Pleistocene. (2) Eustatic lowering of the sea level increased the erosional gradient and thus led to the dissection of the lateritised near-peneplain surface. (3) When the sea level commenced to rise, erosional force was decreased sufficiently to leave the heavier tin minerals behind, but was still strong enough to remove the specifically lighter mineral fractions. With further rise, normal valley fill accumulated above the placer. This process may have been repeated several times during the Pleistocene, although the main formational period may have been during the minimal sea-level stand in the later Pleistocene.

Varying erosional gradients caused by eustatic sea-level changes during the Pleistocene may have influenced a larger number of other terrestrial placer deposits in which these influences have not been recognised.

During periods of low sea-level stands, alluvial tin-deposits formed in areas now again covered by the sea. Fossil stream beds containing commercial grades of cassiterite concentrations are known from Southeast Asia, where they are exploited by means of seagoing dredges. Sainsbury and Reed (1973) report that some placers in Rondonia (Brazil) were covered by the sea during certain interglacial periods.

An intermediate position between terrestrial and marine placers is taken by deposits of those areas which, due to Pleistocene climatic changes, have fluctuated between land and shallow marine cover. According to Sainsbury and Reed (op. cit.), some tin deposits in Cornwall and in Tasmania fall into that category. Similar deposits have been described by Aleva et al. (op. cit.) from offshore Billiton and West Thailand.

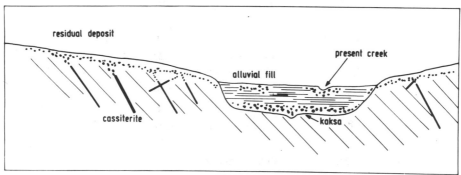

Fig. 11. Cross-section through valley containing tin-placer deposit on Billiton Island, Indonesia. The section demonstrates the relationship between the placer deposit, valley fill and lateritic residual deposits. (After Aleva et al., 1973; coutesy Bur. Miner. Res., Canberra.)

The influence of eustatic sea-level changes was particularly significant for the formation of marine placer deposits. In this category fall all those deposits, the development of which was dependent on eustatically controlled positions of Pleistocene shorelines, either above or below the present level. Some of the best examples of this type are the heavy-mineral sands along the eastern coast of Australia (New South Wales and Queensland), which were largely accumulated during and immediately after a high sea-level stand in the Late Pleistocene. These sands, containing rutile, ilmenite, zircon, monazite and garnet, are the world's major supplier of titanium and zirconium.

Concentrations of ore minerals by wind action are quite common along certain coastal areas and in desert regions, but they are generally thin and thus of no commercial significance.

CONCLUSIONS

In this brief review it was only possible to outline those climatic factors which had an important or governing influence on the formation of sedimentary ore deposits. Many occurrences could only be touched upon, whereas many more deposits with proven or suspected climatic influence may not have been mentioned. But then, every sedimentary ore deposit, by virtue of being of sedimentary origin, contains some measure of climatic control. Painting with a rather broad brush, we hope to have been able to indicate the vast range of influence of climatic factors on ore genesis. However, whereas the megaclimate is largely controlled by extraterrestrial forces, local climate itself is influenced by tectonism, such as the formation of mountain chains or the global distribution of land and sea, resulting in an intricate interplay of cause and effect.

Climatic influences affect physical, chemical and biologic processes in the source area, along the transport path and finally within the depositional environment. Within the source area climate controls the mode of erosion and physicochemical aspects of weathering and hence the geochemical behaviour of metals. Pedogenic processes determining variations in mineral solubility, selective mineral concentration and selective leaching are primarily governed by the prevailing climatic regime. Precipitation levels, temperatures and evaporation/precipitation ratios have a decisive influence on the type of hydraulic system and the nature of the transport medium along the path between the source area and the place of deposition. Within the latter, specific climatic environments for the reduction of metal solutions, precipitation of colloids or the formation of favourable facies associations which may act as host rocks for metals of both supergene or hydrothermal provenance.

In Fig. 12 we have attempted to summarise the major types of ore deposits discussed in this review in a simple, binary climatic diagram. The most conducive conditions for the formation of sedimentary ore deposits exist in hot, arid to seasonally arid climates. Within this zone has to be placed the important facies sequence red-beds—sabkha—euxinic

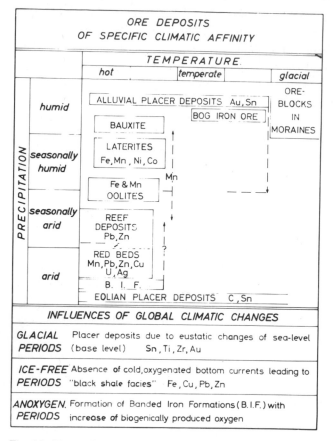

Fig. 12. Binary diagram showing distribution of sedimentary ore deposits in relation to temperature and precipitation. Dashed vertical lines indicate possible range of ore formation in relation to precipitation levels. Some, partly hypothetical influences of global climatic variations, which could not be accommodated in the binary diagram are listed separately below the diagram.

shale facies—reef complex, containing some of the most significant copper, uranium and lead—zinc deposits. The same climatic zone probably produced also the banded iron-formations, but in an atmosphere deficient in oxygen.

Also within hot—warm climatic regions, but at slightly increased levels of precipitation, we have placed the oolitic iron and manganese deposits, although the latter may extend into both more humid and more arid climates.

A somewhat wider climatic band on either side of the equator was responsible for the formation of laterites and bauxites, although still within the hot and warm regions of the globe. The most conducive conditions for the formation of laterites are in the seasonally humid zones. Enrichment of low-grade deposits of iron and manganese to economic grades and concentrations of nickel, cobalt, titanium and tin are important resources of these metals, created by lateritic processes. Bauxites, requiring higher precipitation levels for their formation than laterites, are placed within the humid zone. The latter provide almost the total of the world's reserves of aluminium ore.

Temperate and glacial climates play a lesser role in the formation of ore deposits, but Quaternary eustatic sea-level changes were of importance in the formation of placer deposits.

Considering the decisive role played by climate in the formation of sedimentary ore deposits, regional palaeoclimatic studies are thus an important tool in the prediction of favourable habitats for metalliferous concentrations.

REFERENCES

Adler, H.H., 1970. Interpretation of colour relations in sandstone as a guide to uranium exploration and ore genesis. *Proc. Int. At. Energy Agency, Vienna, Uranium Explor., Geol.*, pp. 331–344.

Aleva, G.J.J., Fick, L.J. and Krol, G.L., 1973. Some remarks on the environmental influence on secondary tin deposits. In: N.H. Fisher (Editor), *Metallogenic Provinces and Mineral Deposits in the Southwestern Pacific — Bur. Miner. Resour., Canberra, Bull.*, 141.

Alexandrov, E.A., 1955. Contribution to studies of origin of Precambrian banded iron ores. *Econ. Geol.*, 36: 465–489.

Alexandrov, E.A., 1973. The Precambrian banded iron formations of the Soviet Union. *Econ. Geol.*, 68: 1035–1062.

Allen, V.T., 1948. Formation of bauxite from basaltic rocks of Oregon. *Econ. Geol.*, 43: 619–626.

Bäcker, H. and Richter, H., 1973. Die rezente hydrothermal–sedimentäre Lagerstätte Atlantis-II-Tief im Roten Meer. *Geol. Rundsch.*, 62: 697–741.

Barghoorn, E.S. and Tyler, S.A., 1965. Microorganisms from the Gunflint chert. *Science*, 147: 536–577.

Bartholomé, P., Evrard, P., Katekesha, F., Lopez-Ruiz, J. and Ngongo, M., 1973. Diagenetic ore-forming processes at Kamoto, Katanga, Republic of the Congo In: G.C. Amstutz and A.J. Bernard (Editors), *Ores in Sediments*. Springer, Berlin, pp. 21–41.

Bayley, R.W. and James, H.L., 1973. Precambrian iron formations of the United States. *Econ. Geol.*, 68: 934–959.

Berg, G. and Karrenberg, H., 1942. Die oolithischen Eisenerze Lothringens. *Arch. Lagerstättenforsch., Berlin*, 75: 79–84.

Beukes, N.J., 1973. Precambrian iron formations of southern Africa. *Econ. Geol.*, 68: 960–1004.

Bogdanov, Y.V. and Kutyrev, E.I., 1973. Classification of stratified copper and lead–zinc deposits and the regularities of their distribution. In: G.C. Amstutz and A.J. Bernard (Editors), *Ores in Sediments*. Springer, Berlin, pp. 59–63.

Borchert, H., 1960. Genesis of marine sedimentary iron ores. *Trans. Inst. Min. Metall.*, 69: 261–279.

Borgström, L.H., 1942. Geschichte der Geologie in Finnland. *Geol. Rundsch.*, 32: 415–434.

Bouladon, J. and Jouravsky, G., 1952. Manganèse, géologie des gîtes minéraux marocains. *Int. Geol. Congr., 19th, Monogr. Rég., 3e Sér., Maroc. No. 1, Rabat.*

Bouladon, J. and Jouravsky, G., 1956. Les gîtes de manganèse du Maroc (suivi d'un description des gisements du Précambrien III). *Symp. del Manganeso, 20th Congr. Geol. Int., Mexico*, 2: 217–248.

Braun, H., 1962. Zur Entstehung der marin-sedimentären Eisenerze. *Z. Erzbergbau Metallhüttenwes.*, 15: 613–623.

Brockamp, B., 1942. Die paläogeographische Stellung der Eisen-ablagerungen. *Arch. Lagerstättenforsch., Berlin*, 75: 181–186.

Brondi, A., Carrara, C. and Polizzano, C., 1973. Uranium and heavy metals in Permian sandstones near Bolzano (Northern Italy). In: G.C. Amstutz and A.J. Bernard (Editors), *Ores in Sediments*. Springer, Berlin, pp. 65–77.

Brown, A.C., 1971. Zoning in the White Pine copper deposit, Ontonagon County. Michigan. *Econ. Geol.*, 66: 543–573.

Brückner, W.D., 1957. Laterite and bauxite profiles of West Africa as an index of rhythmical climatic variations in the tropical belt. *Eclogae Geol. Helv.*, 50: 239–256.

Bubenicek, L., 1968. Géologie des minerais de fer oolithiques. *Miner. Deposita*, 3: 89–108.

Butler, B.E., 1967. Soil periodicity in relation to landform development in Southeastern Australia. In: J.N. Jennings and J.A. Mabbutt (Editors), *Landform Studies from Australia and New Guinea*. Aust. Nat. Univ. Press, Canberra, pp. 231–255.

Campana, B., 1966. Stratigraphic–structural–paleoclimatic controls of the newly discovered iron ore deposits of Western Australia. *Miner. Deposita*, 1: 53–59.

Campana, B., Hughes, F.H., Burnes, W.G., Whitcher, I.G. and Muceniekas, E., 1964. Discovery of the Hamersley iron deposit. *Aust. Inst. Min. Metall. Proc.*, 210: 1–30.

Cissarz, A., 1965. *Einführung in die Allgemeine und Systematische Lagerstättenlehre*. Schweizerbart, Stuttgart, 2e Aufl., 228 pp.

Cloud Jr., P.E. and Licari, G.R., 1968. Microbiotas of the banded iron formations. *Proc. Nat. Acad. Sci., U.S.A.*, 61: 779–786.

Cornwall, H.R., 1973. Nickel. In: D.A. Brobst and W.P. Pratt (Editors), *United States Mineral Resources. U.S. Geol. Surv., Prof. Pap.*, 820: 437–442.

Correns, C.W., 1942. Der Eisengehalt der marinen Sedimente und seine Entstehung. *Arch. Lagerstättenforsch.*, 75: 47–57.

Cullen, D.J., 1963. Tectonic implications of banded ironstone formations. *J. Sediment. Petrol.*, 33: 387–392.

Denison, N.M. and Gill, J.R., 1956. Uranium-bearing lignite and its relation to volcanic tuffs in eastern Montana and North and South Dakota. *U.S. Geol. Surv., Prof. Pap.*, 300: 413–418.

De Villiers, J., 1956. The manganese deposits of the Union of South Africa. *Symp. del Manganeso, 20th Congr. Geol. Int., Mexico*, 2: 39–72.

De Vletter, D.R., 1955. How Cuban nickel ore was formed – a lesson in laterite genesis. *Eng. Min. J.*, 156: 84–87 and 178.

Dorr II, J.V.N., 1973. Iron formation in South America. *Econ. Geol.*, 68: 1005–1022.

Dorr II, J.V.N., Crittenden, M.D. and Worl, R.G., 1973. Manganese. In: D.A. Brobst and W.P. Pratt (Editors), *United States Mineral Resources. U.S. Geol. Surv., Prof. Pap.*, 820: 385–399.

Eugster, H.P. and I-Ming Chou, 1973. The depositional environments of Precambrian banded iron formations. *Econ. Geol.*, 68: 1144–1168.

Evans, H.J., 1959. The geology and exploration of the Cape York Peninsula bauxite deposits in northern Queensland. *Chem. Eng. Min. Rev.*, 51 (11): 48–56.

Eyles, V.A., 1952. The composition and origin of the Antrim laterites and bauxites. *Mem. Geol. Surv. North. Irel., Belfast*, 90 pp.

Fischer, R.P., 1970. Similarities, differences, and some genetic problems of the Wyoming and Colorado Plateau types of uranium deposits in sandstone. *Econ. Geol.*, 65: 778–784.

Fox, C.S., 1936. Buchanan's laterite of Malabar and Kanara. *Rec. Geol. Surv., India*, 69: 389–422.

Freeman, T., 1962. Quiet-water oolites from Laguna Madre, Texas. *J. Sediment. Petrol.*, 32: 475–483.

Gabelman, J.W., 1970. Speculations on the uranium ore fluid. *Proc. Int. At. Energy Agency Vienna, Uranium Explor., Geol.*, pp. 315–330.

Garlick, W.G., 1961. The syngenetic theory. In: F. Mendelsohn (Editor), *The Geology of the Northern Rhodesian Copper Belt*. Mc Donald, London, pp. 146–165.

Gross, G.A., 1965. Geology of iron of Canada, 1. *Geol. Surv. Can., Econ. Geol. Rep.*, 22: 181 pp.

Grubb, P.L.C., 1970. Mineralogy, geochemistry and genesis of the bauxite deposits on the Gove and Mitchell Plateaux, Northern Australia. *Min. Deposita*, 5: 248–272.

Grubb, P.L.C., 1973. High-level and low-level bauxitization: a criterion for classification. *Miner. Sci. Eng., Johannesb.*, 5 (3): 219–231.

Harańczyk, C., 1970. Zechstein lead-bearing shales in the fore-sudetian monocline in Poland. *Econ. Geol.*, 65: 481–495.

Harshman, E.N., 1970. Uranium ore rolls in the United States. *Proc. Int. At. Energy Agency, Vienna, Uranium Explor., Geol.*, pp. 219–232.

Haughton, S.H., 1969. *Geological History of South Africa.* Geol. Soc. S. Afr., Cape Town, 535 pp.

Holland, H.D., 1962. Model for the evolution of the earth's atmosphere. In: A.E. Engel, H.L. James and B.F. Leonard (Editors), *Petrologic Studies: A Volume in Honor of A.F. Buddington.* Geol. Soc. Am., 447–477.

Holland, H.D., 1973. The oceans: A possible source of iron in iron formations. *Econ. Geol.*, 68: 1169–1172.

Hostetler, P.B. and Garrels, R.M., 1962. Transportation and precipitation of uranium and vanadium at low temperatures, with special reference to sandstone-type uranium deposits. *Econ. Geol.*, 57: 137–167.

Hotz, P.E., 1964. Nickeliferous laterites in southwestern Oregon and northwestern California. *Econ. Geol.* 59: 355–396.

Hough, J.L., 1958. Fresh-water environment of deposition of Precambrian banded iron formations. *J. Sediment. Petrol.*, 28: 414–430.

Illing, L.V., 1954. Bahaman calcareous sands. *Bull. Am. Assoc. Pet. Geol.*, 38: 1–95.

Johns, R.K., 1963. Investigation of Lake Eyre, Part I. *Geol. Surv. South Aust. Rept. Invest.*, 24: 1–69.

Johnstone, M.H., Lowry, D.C. and Quilty, P.G., 1973. The geology of southwestern Australia – a review. *J. R. Soc. West Aust.*, 56 (1): 5–15.

Jones, H.A., 1965. Ferruginous oolites and pisolites. *J. Sediment. Petrol.*, 35 (4): 838–845.

Jux, U. and Pflug, H.D., 1958. Alter und Entstehung der Trias-ablagerungen und ihrer Erzvorkommen am Rheinischen Schiefergebirge. *Abh. Hess. Landesamt Bodenforsch.*, 27: 1–50. .

Karrenberg, H., 1942. Paläogeographische Übersicht über die Ablagerungen der Dogger-β-Zeit in West- und Südwestdeutschland. *Arch. Lagerstättenforsch., Berlin*, 75: 78–79.

Khodyush, L.Y., 1969. Banding of ferruginous quartzite and its origin. In: *Problems of Formation of Precambrian Ferruginous Rock.* Acad. Nauk. Ukrain SSR, Inst. Geol. Nauk, Kiev, pp. 242–258.

King, J.W. and Austin, S.R., 1966. Some characteristics of roll-type uranium deposits at Gas Hill, Wyoming. *Min. Eng.*, 18 (5): 73–80.

Klemic, H., Gottfried, D., Cooper, M. and Marsh, S.P., 1973a. Zirconium and hafnium. In: D.A. Brobst and W.P. Pratt (Editors), *United States Mineral Resources. U.S. Geol. Surv., Prof. Pap.*, 820. 713–722.

Klemic, H., James, H.L. and Eberlein, G.D., 1973b. Iron. In: D.A. Brobst and W.P. Pratt (Editors), *United States Mineral Resources. U.S. Geol. Surv., Prof. Pap.*, 820: 291–306.

Klemic, H., Marsh, S.P. and Cooper, M., 1973c. Titanium. In: D.A. Brobst and W.P. Pratt (Editors), *United States Mineral Resources. U.S. Geol. Surv., Prof. Pap.*, 820: 653–665.

Kolbe, R.K.H., 1968. Influence of salt tectonics on formation and conservation of sedimentary iron ores. In: *UNESCO, 1972, Geology of Saline Deposits. Proc. Hannover Symp., 1968*, pp. 243–245.

Krauskopf, K.B., 1957. Separation of manganese from iron in sedimentary processes. *Geochim. Cosmochim. Acta*, 12 (1-2).

Krinsley, D.B., 1970. A geomorphological and paleoclimatological study of the playas of Iran. *U.S. Geol. Surv. Rep.*, AFCRL–70–0503: 486 pp.

Krol, G.L., 1960. Theories on the genesis of the kaksa. *Geol. Mijnbouw*, 10: 437–443.

Kupferburger, W., Boardman, L.G. and Bosch, P.R., 1956. New considerations concerning the manganese ore deposits on the Postmasburg and Kuruman areas, northern Cape Province, Union of South Africa. *Symp. del Manganeso, 20th Congr. Geol. Int., Mexico*, 2: 73–87.

Kužvart, M. and Konta, J., 1968. Kaolin and laterite weathering crusts in Europe. *Acta Universitatis Carolinae – Geologica*, 12: 1–19.

Laitakari, A., 1942. Hauptzüge der Erzforschung in Finnland und ihre Ergebnisse. *Geol. Rundsch.*, 32: 435–451.

Laverty, R.A. and Gross, E.B., 1956. Paragenetic studies of uranium deposits of the Colorado Plateau. *U.S. Geol. Surv., Prof. Pap.*, 300: 195–201.

Lemoalle, J. and Dupont, B., 1973. Iron-bearing oolites and the present condition in Lake Chad (Africa). In: G.C. Amstutz and A.J. Bernard (Editors), *Ores in Sediments*. Springer, Berlin, pp. 167–178.

Lersch, J., 1973. Prospektion und geologische Untersuchung lateritischer Nickellagerstätten am Beispiel Barro Alto, Brasilien. *Z. Dtsch. Geol. Ges.*, 124: 135–148.

MacLeod, W.N., 1964. Iron ore deposits in the eastern section of the Hamersley iron province. *Geol. Surv. West. Aust., Annu. Rep. – 1963*: 34–38.

Maignien, R., 1966. Review of research on laterites. *UNESCO Nat. Resour. Res., IV.*

Mengel, J.T., 1973. Physical sedimentation in Precambrian cherty iron formations of Lake Superior-type. In: G.C. Amstutz and A.J. Bernard (Editors), *Ores in Sediments*. Springer, Berlin, pp. 179–193.

Miles, K.R., 1955. The geology and iron ore resources of the Middleback Range area. *Geol. Surv. South. Aust., Bull.*, 33: 247 pp.

Monseur, G. and Pel, J., 1972. Reef facies, dolomitization and stratified mineralization. *Miner. Deposita*, 7: 89–99.

Monseur, G. and Pel, J., 1973. Reef environment and stratiform ore deposits. In: G.C. Amstutz and A.J. Bernard (Editors), *Ores in Sediments*. Springer, Berlin, pp. 195–207.

Moreau, M., Poughon, A., Puibaraud, Y. and Sanselme, H., 1966. L'uranium et les granites. *Chron. Min. Rech. Minières*, 350: 47–51.

Morris, H.T., Heyl, A.V. and Hall, R.B., 1973. Lead. In: D.A. Brobst and W.P. Pratt (Editors), *United States Mineral Resources. U.S. Geol. Surv., Prof. Pap.*, 820: 313–332.

Mostler, H., 1966. Sedimentäre Blei–Zink–Vererzung in den mittel-permischen "Schichten von Tregiovo". *Miner Deposita*, 1: 89–103.

Mulcahy, M.J., 1967. Landscapes, laterites, and soils in southwestern Australia. In: J.N. Jennings and J.A. Mabbutt (Editors), *Landform Studies from Australia and New Guinea*. Aust. Nat. Univ. Press, Canberra, pp. 211–230.

Nechaev, Y.A., 1961. Lead and zinc in copper-bearing sandstones of the Perm district. *Geochemistry*, 5: 478–487.

Nicolas, J., 1968. Nouvelles données sur la genèse des bauxites à mur karstique du sud-est de la France. *Miner. Deposita*, 3: 18–33.

Patterson, S.H. and Dyni, J.R., 1973. Aluminium and Bauxites. In: D.A. Brobst and W.P. Pratt (Editors), *United States Mineral Resources. U.S. Geol. Surv., Prof. Pap.*, 820: 5–43.

Pelletier, R.A., 1964. *Mineral Resources of South-Central Africa*. Oxford Univ. Press, Cape Town, 277 pp.

Philippines Republic, 1958. Iron, nickel. *Special Project Series, Bur. Min., Publ. 17.*

Playford, P.E. and Lowry, D.C., 1966. Devonian reef complexes of the Canning Basin, Western Australia. *Geol. Surv. West. Aust., Bull.*, 118: 1–150.

Plumb, K.A. and Gostin, V.A., 1973. Origin of Australian bauxite deposits. *Bur. Miner. Resour., Rec.*, 1973/156, 71 pp.

Porath, H., 1967. Palaeomagnetism and the age of Australian ore bodies. *Earth Planet. Sci., Lett.*, 2: 409–414.

Pratt, R., 1971. Manganese – the supply/demand position. *Bur. Miner. Resour., Canberra, Q. Rev.*, 23: 78–95.

Prescott, J.A. and Pendleton, R.L., 1952. Laterite and lateritic soils. *Commonw. Agric. Bur., Bucks, England*, 51 pp.

Renfro, A.R., 1974. Genesis of evaporite-associated stratiform metalliferous deposits – a sabkha process. *Econ. Geol.*, 69: 33–45.

Richter-Bernburg, G., 1968. Salzlagerstätten. In: A. Bentz and H.J. Martini (Editors), *Lehrbuch der Angewandten Geologie, Vol. II, part 1*. Enke, Stuttgart, pp. 918–1061.

Robertson, D.S., 1970. Uranium, its geological occurrence as a guide to exploration.

Rubin, B., 1970. Uranium roll front zonation in the southern Powder River Basin Wyoming. *WGA Earth Sci. Bull.*, Dec. 1970: 5–12.

Sainsbury, C.L. and Reed, B.L., 1973. Tin. In: D.A. Brobst and W.P. Pratt (Editors), *United States Mineral Resources. U.S. Geol. Surv., Prof. Pap.*, 820: 637–651.

Sakamoto, T., 1950. The origin of the Precambrian banded iron ores. *Am. J. Sci.*, 248: 449–474.

Samama, J.C., 1968. Contrôle et modèle génétique de minéralisations en galène de type "Red-Bed" – Gisement de Largentière – Ardèche, France. *Miner. Deposita*, 3: 261–271.

Samama, J.C., 1973. Ore deposits and continental weathering: A contribution to the problem of geochemical inheritance of heavy metal contents of basement areas and of sedimentary basins. In: G.C. Amstutz and A.J. Bernard (Editors), *Ores in Sediments*. Springer, Berlin, pp. 247–265.

Schneiderhöhn, G., 1962. *Erzlagerstätten, Kurzvorlesungen*. Fischer, Stuttgart, 4th ed., 371 pp.

Schopf, J.W., 1969. Recent advances in Precambrian palaeobiology. *Grana Palynologica*, 9: 147–168.

Schopf, J.W., 1970. Precambrian micro-organisms and evolutionary events prior to the origin of vascular plants. *Biol. Rev.*, 45: 319–352.

Schüller, A., 1957. Mineralogie und Petrographie neuartiger Bauxite aus dem Gun Distrikt, Honan-Provinz (China). *Geologie*, 6: 379–399.

Schulz, O., 1968. Die synsedimentäre Mineralparagenese im oberen Wettersteinkalk der Pb–Zn-Lagerstätte Bleiberg–Kreuth (Kärnten). *Tschermaks Mineral. Petrogr. Mitt.*, 12: 230–289.

Schulz, O., 1972. Neuergebnisse über die Entstehung paläozoischer Erzlagerstätten am Beispiel der Nordtiroler Grauwackenzone. *Int. Symp. Miner. Deposits Alps, 2nd, Geol. Trans. Rep.*, Ljubljana, 15: 125–140.

Schulz, O. and Lukas, W., 1970. Eine Uranerzlagerstätte in permotriadischen Sedimenten Tirols. *Tschermaks Mineral. Petrogr. Mitt.*, 14: 213–231.

Schumm, S.A., 1968. Speculations concerning paleohydrologic controls of terrestrial sedimentation. *Geol. Soc. Am. Bull.*, 79: 1573–1588.

Schwarzbach, M., 1963. *Climates of the Past: An introduction to Paleoclimatology*. Van Nostrand, London, 328 pp.

Schwarzbach, M., 1968. Das Klima des rheinischen Tertiärs. *Z. Dtsch. Geol. Ges.*, 118: 33–68.

Schwarzbach, M., 1974. *Das Klima der Vorzeit*. Enke, Stuttgard 3rd ed., 380 pp.

Shearman, D.J., 1966. Origin of marine evaporites by diagenesis. *Trans. Sect. B, Inst. Min. Metall.*, 75: 207–215.

Simons, F.S. and Prinz, W.C., 1973. Gold. In: D.A. Brobst and W.P. Pratt (Editors), *United States Mineral Resources. U.S. Geol. Surv., Prof. Pap.*, 820: 263–275.

Sloss, L.L., 1953. The significance of evaporites. *J. Sediment. Petrol.*, 23: 143–161.

Smirnow, V.I., 1970. *Geologie der Lagerstätten Mineralischer Rohstoffe*. Dtsch. Verlag Grundstoff-ind., Leipzig, 563 pp.

Sofoulis, J., 1965. Bauxite deposits of the north-Kimverley region, Western Australia. *Geol. Surv. West Aust., Ann. Rep.*, pp. 69–70.

Strakhov, N.M., 1967, 1969. *Principles of Lithogenesis*. Oliver and Boyd, Edinburgh, I: 245 pp, II: 609 pp.

Taylor, J.H., 1969. Sedimentary ores of iron and manganese and their origin. *Proc. Inter-Univ. Geol. Congr., 15th, 1967*, pp. 171–186.

Teichert, C., 1958. Cold- and deep-water coral banks. *Bull. Am. Assoc. Pet. Geol.*, 42: 1064–1082.

Teichert, C., 1970. Oolite, oolith, ooid: Discussion. *Bull. Am. Assoc. Pet. Geol.*, 54: 1748–1749.

Thomson, B.P., 1965. Source and distribution of heavy metals in Cambrian and Marinoan shelf sediments in South Australia. *Commonw. Min. Metall. Congr., Aust. N.Z., 8th.*

Trendall, A.F., 1968. Three great basins of Precambrian banded iron formation deposition: A systematic comparison. *Geol. Soc. Am. Bull.*, 79: 1527–1544.

Trendall, A.F., 1972. Revolution in earth history. *J. Geol. Soc. Aust.*, 19(3): 287–311.

Trendall, A.F. and Blockley, J.G., 1970. The iron formations of the Precambrian Hamersley Group, Western Australia. *West. Aust. Geol. Surv. Bull.*, 119, 366 pp.

Trescases, J.J., 1973. Weathering and geochemical behaviour of the elements of ultramafic rocks in New Caledonia. In: N.H. Fisher (Editor), *Metallogenic Provinces and Mineral Deposits in the Southwestern Pacific. Bur. Miner. Resour., Canberra, Bull.*, 141: 149–161.

Valeton, I., 1972. *Bauxites — Developmental in Soil Science, 1.* Elsevier, Amsterdam, 226 pp.

Valeton, I., 1973. Laterite als Leithorizonte zur Rekonstruktion tektonischer Vorgänge auf den Festländern. *Geol. Rundsch.,* 62: 153—161.

Van Eden, J.G., 1974. Depositional and diagenetic environment related to sulfide mineralization, Mufulira, Zambia. *Econ. Geol.,* 69: 59—70.

Van Houten, F.B., 1961. Climatic significance of red beds. In: A.E.M. Nairn (Editor), *Descriptive Paleoclimatology.* Interscience, New York, N.Y., pp. 89—139.

Varentsov, I.M., 1964. *Sedimentary Manganese Ores.* Elsevier, Amsterdam, 119 pp.

Vincienne, H., 1956. Observation géologiques sur quelques gîtes marocains de manganèse syngenetique. *Symp. del Manganeso, 20th Congr. Int., Mexico,* 2: 249—268.

Wadia, D.N., 1966. *Geology of India.* Macmillan, London, 536 pp.

Walther, H.W. and Zitzmann, A., 1973. Die Lagerstätten des Eisens in Europa. *Z. Dtsch. Geol. Ges.,* 124: 61—72.

Walther, J., 1915. Laterit in West Australien. *Z. Dtsch. Geol. Ges.,* 67 (B): 113—140.

Wattenberg, H., 1942. Das Vorkommen des Eisens im Meere. *Arch. Lagerstättenforsch.,* 75: 36—47.

Wedepohl, K.H., 1964. Untersuchungen am Kupferschiefer in Nordwestdeutschland. *Geochim. Cosmochim. Acta,* 28: 305—364.

Wedow, H., 1967. The Morro de Ferro thorium and rate-earth ore deposit, Brazil. *U.S. Geol. Surv. Bull.,* 1185-D: 1—34.

White, W.S., 1971. A paleohydrologic model for mineralization of the White Pine copper deposit, northern Michigan. *Econ. Geol.,* 66: 1—13.

Whitehouse, F.W., 1940. Studies in the late geological history of Queensland. *Univ. Queensl. Pap.,* 2 (1): 1—74

Wiebols, J.H., 1955. A suggested glacial origin for the Witwatersrand conglomerates. *Trans. Geol. Soc. S. Afr.,* 58.

Wilson, J.L., 1974. Characteristics of carbonate-platform margins. *Bull. Am. Assoc. Pet. Geol.,* 58: 810—824.

Wirtz, R., 1972. Beitrag zur Kenntnis der Paläosole im Vogelsberg. *Abh. Hess. Geol. Landesamt,* 61: 1—158.

Wopfner, H., 1974. Post-Eocene history and stratigraphy of northeastern South Australia. *Trans. R. Soc. South Aust.,* 98(1): 1—12.

Wopfner, H. and Twidal, C.R., 1967. Geomorphological history of the Lake Eyre Basin. In: J.N. Jennings and J.A. Mabbutt (Editors), *Landform Studies from Australia and New Guinea.* Aust. Natl. Univ. Press, Canberra, pp. 118—143.

Wopfner, H., Freytag, I.B. and Heath, G.R., 1970. Basal Jurassic—Cretaceous rocks of western Great Artesian Basin, South Australia: Stratigraphy and environment. *Bull. Am. Assoc. Pet. Geol.,* 54 (3): 383—416.

Chapter 3

PEDOGENESIS, CHEMICAL WEATHERING AND PROCESSES OF FORMATION OF SOME SUPERGENE ORE DEPOSITS

F. LELONG, Y. TARDY, G. GRANDIN, J.J. TRESCASES and B. BOULANGE

INTRODUCTION

In metallogeny, the term supergene is applied to ores or ore minerals which have been formed by generally descending meteoric waters, as opposed to hypogene, which corresponds to ascending waters (Routhier, 1963). But this term is also used by surface or subsurface geologists in a more general sense, to designate all processes originating at or near the earth's surface, such as weathering and denudation, and the continental deposits and sediments formed through such processes. In this general sense, the term supergene is nearly synonymous with the term epigene (see "Geochemistry of epigenesis", Perel'man, 1967), but it is more restrictive than the term exogene, in which the processes of marine sedimentation and diagenesis are included.

The main processes corresponding to supergene evolution may be defined as follows:

(1) Pedogenesis: formation of soil profiles, with more or less vertically differentiated horizons, which are derived from the in-situ evolution of preexisting parent rocks, through the action of meteoric and biological agents (waters, carbonic and humic acids, living organisms, etc.).

(2) Weathering: any alterations, either mechanical (e.g., fragmentation) or chemical (e.g., decaying) of materials of the earth's crust under the influence of atmospheric agents (heat, frost, surface and subsurface waters, etc.). The weathering acts not only in soil profiles, but also in deeper cracks, veins, faults and any porous layers, which are connected to the atmosphere. The resulting formations are termed regoliths, saprolites or alterites; they may have been differentiated either strictly in-situ (residual, sedentary alterites) or at a certain distance from their present location (reworked, migratory, or transported alterites). Thus, the concept of weathering is more general but not as well defined and delineated as is the concept of pedogenesis.

(3) Erosion: removal and displacement of materials on the land surface, which are produced either by the mechanical work of meteoric forces (rainfall, running water, stream water, winds, ice, etc.) or by the chemical dissolving action of waters, combined with the force of the earth's gravitational attraction. The mechanical and chemical erosion provokes the lowering of the land surface (denudation) and the flattening of the

landform, as the transported substances (solid or dissolved) tend to be displaced from higher to lower land surfaces.

(4) Continental sedimentation: local accumulation of debris (detrital sedimentation) or reprecipitated matters (chemical sedimentation) which have been moved and transported by erosional forces, into propitious continental sites, low lands, depressions and basins. These accumulations are often transitory deposits, since they are reworked by the continuing erosion and ultimately carried from the continents to the ocean basins. "Burial in the sea is the ultimate fate of most sediments" (Ordway, 1972).

The geochemical differentiations occurring during these supergene processes, are produced by two opposite mechanisms: (a) mechanisms of subtraction which proceed through leaching of the more mobile constituents and result in relative residual accumulation of the less mobile ones; and (b) mechanisms of addition, which proceed through migration of the more mobile constituents and result in absolute, lateral accumulation of these constituents. The former are considered to prevail during the weathering processes, whereas the latter are considered to prevail during the sedimentation processes (Millot, 1964). In fact, however, the supergene evolution often shows composite differentiations, proceeding through both mechanisms: for example, the B horizons in illuviated soils result simultaneously from relative and absolute accumulation of clay minerals (formation in-situ of these minerals through leaching of vulnerable primary silicates, translocation of clay minerals coming from the A leached horizon). In some cases, the two mechanisms may act successively, for example, in the genesis of placer deposits, which result firstly from the residual accumulation of heavy, stable minerals during weathering and secondly from their absolute accumulation during sedimentation of alluviums or alluvial fans.

In the present contribution, the role of the chemical weathering processes in metallogeny will be principally considered. This role is indeed very important, being at one and the same time direct and indirect. Weathering processes govern directly the genesis of economically considerable residual ore deposits, such as bauxite, laterites and associated mineral concentrations. They also indirectly govern the formation of continental or even marine mineral concentrations, since the lithologic differentiations encountered in sedimentary basins are really correlative formations of the weathering differentiations which occur in the neighbouring landscapes (Erhart, 1956). In the first section of this contribution, the problem of the geochemical differentiation produced through pedogenetic processes will be developed. The following sections are devoted to the problem of the genesis of strictly residual (or so considered) ore deposits, such as bauxites, laterites, and associated concentrations (Mn, Ni). In the last section, we shall attempt a more general approach to the supergene behaviour of chemical elements, in order to understand other aspects of supergene metallogeny, such as mineral concentrations related to continental, detrital or chemical sedimentation.

Even so delineated, the subject remains large and must be further reduced for the present treatment:

(1) The limits of the weathering zone, beneath the land surface, are not sharply fixed and the weathering processes are limited in deeper zones by cementation and hydrothermal processes. Schematic relationships between the different subsurface hydrodynamic zones and the resulting supergene differentiations, soil profiles, weathered crust and cementation zone, etc., are depicted in Fig. 1. The zone of weathering is assumed to correspond to the zones of percolation and of active circulation of ground waters, which are located above the hydrostatic level. This zone is characterized by the more or less intense renewal of meteoric waters, by the presence of oxygen and the influence of biological factors. The underlying zone of catagenesis is characterized, on the contrary, by the slowness or stagnation of the waters, by the lack or scarcity in oxygen and by the existence of highly mineralized water (Perel'man, 1967). The processes occurring in this latter zone will not be considered here.

(2) Another limitation of the subject consists in excluding the study of paleo-landsurfaces which have been more or less weathered during past geological time, then buried and now covered unconformably or disconformably by sediments. Varied and sometimes complex metallic concentrations are often found along these paleosurfaces. However, it is often difficult to establish to what degree weathering processes or subsequent epigenetical, diagenetical or even hydrothermal processes were responsible for these concentrations. We do not try to resolve this problem.

(3) The last limitation we introduce concerns the peculiar weathering processes which develop in ore-bearing parent rocks or hydrothermal vein ore deposits; these processes give varied superficial oxidized products termed iron caps ("chapeaux de fer", Eisener Hut, gossan). They are greatly complicated by the role played by some unusual constituents, anionic or cationic, which may react together to give quite complex mineral paragenesis. In spite of their interest, these peculiar weathering products are actually very localized and we now consider only the metallogenic aspects of the weathering differentiation of ubiquitous or common parent rocks.

The experimental study of the supergene geochemical processes has progressed far these last years, through alteration and synthesis experiments under laboratory conditions (pressure and temperature), near the land surface conditions. Inasmuch as these processes may be considered as resulting from equilibrium reactions between primary minerals,

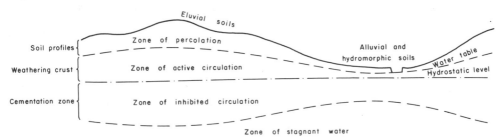

Fig. 1. Schematic relationships between the principal supergene processes, the hydrodynamic zones and the resulting formations: soil profiles, weathered crust and cementation zone.

aqueous solutions and secondary minerals, the thermodynamical method may be successfully applied; this method permits to calculate and to predict the fields of stability and
mobility of minerals and mineral products, in relation with a few physico–chemical
variables (ionic concentration or activity, pH, Eh, temperature, etc.) Some examples of
such calculations will be given incidentally in this chapter (see p. 118 and p. 127), but we
do not attempt a general presentation of the thermodynamical approach to the supergene
processes, because the available thermodynamical data concerning weathering minerals
and chemical trace elements are scarce, absent or still subject to discussion.

GEOCHEMICAL DIFFERENTIATION DURING CHEMICAL WEATHERING AND PEDOGENETIC PROCESSES

The weathering and pedogenetic processes are the subject of a great number of works
which deal mainly with: (a) morphological characteristics of soil profiles, from fresh or
little weathered parent rock (D or C horizon) to superficial eluvial horizons (A horizons),
with occasionally an intermediate weathered or illuvial horizon (B horizon); (b) granulometric, mineralogical and chemical characteristics from bottom to top; (c) origin and
evolution of clays and secondary minerals; (d) evolution of the organic matter and its
interaction with the mineral matter; (e) weathering balances.

Since all these aspects cannot possibly be listed here, we recommend to the reader
some recent comprehensive works, which treat the subjects with an either geological
(Millot, 1964; Loughnan, 1969), pedological (Duchaufour, 1968), or geochemical emphasis (Perel'man, 1967).

Soviet authors make a distinction between the weathering products: the soil profiles,
the formation of which is directly influenced by the living matter (biogenetic accumulations, action of the roots of vegetables, etc.) and the weathered crusts. The latter designate the different products accumulated residually: fragmented rocks, weathered or not,
and the more or less hydrolysed deep layers in the zone of active circulation of water;
they generally exhibit the relict structure of the parent rocks, and their thickness, which
varies strongly according to the climatic conditions as well as the lithological properties of
the rock, is often much greater than the thickness of the soil profiles. Here, the term
weathered crust will not be used and the concept of soil profile will be considered with its
widest meaning: all the residual products and deep layers of weathering will be included
in the more or less weathered bottom C horizon.

Pedogenesis and chemical weathering types

The problem with which we are concerned, considered from a metallogenic point of
view, is to state precisely what are the geochemical differentiations due to superficial
chemical weathering and pedogenesis. This problem raises two questions: (a) Are the

types of differentiations as numerous as the types of pedogenesis? (b) How do the pedogeneses have to be classified with respect to their geochemical effects?

Major zonal pedogenesis. To the first question, it may be answered that geochemical differentiations are actually very numerous when referring to the variations in main factors of soil formation: climate, vegetation, parent rock, topography and time (Jenny, 1941). But for parent rocks of common composition, e.g., granitic rocks, which evolve superficially under normal topographic conditions (good drainage and limited erosion), the type of soil depends mainly on bioclimatic conditions (climate, vegetation). For the sake of simplicity, it may be said that five major pedogeneses related to the five main climatic zones occur (Erhart, 1935; Millar et al., 1958; Ganssen, 1965 and 1968; Pédro, 1968; Kovda et al., 1969):

(1) Humid temperate—cold regions, where the typical vegetation is moorland and conifers, give podzolic soils (spodosols).

(2) Mild temperate regions, where the vegetation is hardwood plants and meadows, give brown soils, leached brown soils and acid brown soils (inceptisols or alfisols).

(3) Either cold or hot continental but always more or less dry regions, where the vegetation is steppe, give isohumic soils (mollisols or aridisols).

(4) Subtropical regions with a more or less xerophytic vegetation of dry savanna, give fersiallitic and tropical ferruginous soils (alfisols or ultisols).

(5) Equatorial or wet regions with a dense forest vegetation, give ferrallitic soils (oxisols).

The soil nomenclature used above, is that of the French classification; the American classification terms are noted in parentheses.

Soils related to these pedogeneses occur in roughly parallel belts extending from the poles to the equator, but the steppe isohumic soils develop not only in temperate or even cold continental zones, but also in arid regions with hot climates. From a climatic point of view, the five zones are characterized by a decreasing mean annual temperature and increasing annual range of temperatures from the equator to the poles, and by a rainfall presenting a maximum in the equatorial zone and a minimum in subtropical continental regions.

Under very extreme climatic conditions (arctic or subarctic zones, or desert regions), there is practically no geochemical differentiation, since very low temperatures or the lack of rain reduce the weathering processes to nearly zero.

"Weathering balance". A good understanding of geochemical differentiations due to each pedogenesis calls for a "weathering balance" giving the mineralogical and chemical absolute variations from bottom to top of the soil profiles and allowing the losses and gains of matter to be defined in the different horizons. Therefore, a reference constituent must be selected.

Up to now, the more commonly selected reference element was Al, Ti or Fe. These elements are generally fairly stable in superficial conditions, if considered on the scale of broad areas, but in certain soil profiles they undergo vertical or lateral migrations, which result in a relatively significant removal from some horizons and accumulations in some other ones. The isovolumic method (Millot, 1964) is a more accurate means for calculating geochemical balances and has been successfully employed by many authors (Tardy, 1969; Novikoff, 1974), but this method can only be used when the structures of the parent rock are conserved, which is not always the case.

For illustrating here the complete geochemical trends, up to the destructured superficial horizon, the isoquartz method has been chosen (Lelong, 1969; Souchier, 1971): in soil profiles, even in intensively weathered ones, the quartz is indeed the only constituent which shows a progressive increase from bottom to top, except in illuvial horizons; this fact confirms its relative stability.

A suitable reference constituent having been selected, the balances still have to meet the following conditions:

(1) They have to deal with sedentary (not reworked) soil profiles derived from homogeneous rocks.

(2) They must be established from numerous samples in order to get mean results and well-defined standard deviations.

(3) Parent rocks of soil profiles have to be of the same nature, since the pedological evolution depends largely on the nature of the parent rock.

(4) Polycyclic soil profiles resulting from several different evolutions must be excluded.

"Weathering balance" of zonal pedogeneses

The balances already published are very few and deal especially with the pedogeneses of temperate regions (Dejou, 1959 and 1967; Tardy, 1969; Lelong and Souchier, 1970) or of equatorial regions (Harrison, 1933; Bonifas, 1959; Leneuf, 1959; Lelong, 1969; Tardy, 1969; Novikoff, 1974). Moreover, the method used (isoalumina method, isoquartz method, isovolume method, etc.) differs from one author to another so that the results cannot be easily compared.

Examples of pedogeneses in equatorial and temperate regions. We will give only two examples of balances which are established in a quite similar way ("bilan isoquartz"), and which correspond respectively to soils under a wet temperate climate and soils under an equatorial climate (Lelong and Souchier, 1972). Their results are summarized in Table I and Fig. 2. In both cases, the conditions of location are: relatively high topographic position, low gradient slope, dense forest cover. Consequently, the soils are well drained and little affected by erosion. They are mature soils of Quaternary (Post-Würm) age for the temperate soils, and older soils evolving perhaps from the final Tertiary or Early Quaternary Period for the equatorial soils.

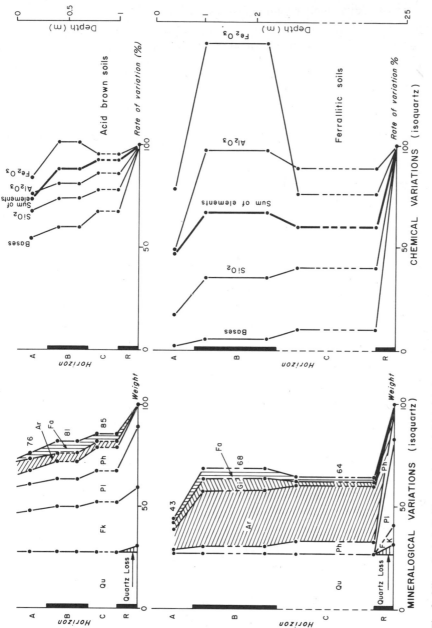

Fig. 2. Mineralogical (left) and chemical (right) balances in soil profiles of temperate and equatorial soils, on granitic rocks, referring to a constant weight of quartz (after Lelong and Souchier, 1972). The mineralogical abbreviations are similar to those of Table I.

TABLE I

Mean weathering balances referring to constant weight of quartz (isoquartz balances) in profiles of "acid brown soils" (temperate climate) and of ferrallitic soils (equatorial climate)[1,2]

Horizons	Mineralogical balance[3]							
	Primary minerals				Secondary mineral fraction			Total
	Qu	FK	Pl	Ph	Ar	Gi	Fa	
Acid brown soils								
A	27.5	20	12.5	7	6	–	3	76
B	27.5	22	13.5	8	5	–	5	81
C	27.5	24	15	12	2.5	–	4	85
granite	30	29	29	12				100
Ferrallitic soils								
A	27			1	10	3	2	43
B	27			3	27	6	5	68
C	27			6	27	1	3	64
granite	32	9	42	17				100

[1] After Lelong and Souchier (1972).

[2] The quartz losses (7.5% in temperate soils and 15% in equatorial soils) are corrected.

The soils occurring under a wet temperate climate belong to the group of acid brown—podzolic soils, developed on Hercynian granitic rocks in the Massif des Vosges (France) at a relatively high altitude (600—950 m) under a vegetation of fir trees and piceas. Climatic conditions are: annual rainfall of 1200—1800 mm and mean annual temperature of 8°C. The soils found under an equatorial climate are ferrallitic soils, developed in French Guiana on granitic rocks of the Precambrian basement, under large dense forest. The climatic characteristics vary with each particular station: from 2500 to 4000 mm rainfall/year and a mean annual temperature of 26—27°C.

The results for the balances along the soil profiles for each climatic zone (Table I and Fig. 2) are *mean* results, obtained from at least ten samples/horizon in soils of the same type. The distinct study of the "bilans" made from separate profiles reveals notable differences within each climatic zone, which are due to the particularities of each location. Despite this, the respective fields of global variations corresponding to the temperate soils and to the equatorial soils are clearly distinct and this fact justifies a mean balance type for each climatic zone, as represented in Fig. 1.

The mineralogical balances represent the aggregate variations of mineralogical constituents from the bottom to the top of the profiles: the mineral-matter losses produced during the evolutionary process appear in the right-angle of the diagrams. The chemical balances give the variations of each element expressed as a variation rate (% of each element compared with the amount originally present in the parent rock); the four bases

Chemical balance[4]					
SiO_2		Al_2O_3	Total iron	Bases	Total mineral matter
(quartz)	(combined)				
92.5	67	74	84	54	76
92.5	74	88	101	60	81
92.5	78	92.5	95	68	85
100	100	100	100	100	100
85	17	49	79	1	43
85	35	97	157	5	68
85	40	89	76	10	64
100	100	100	100	100	100

[3] Results in weights, referring to a constant weight of quartz. The abbreviations used are: Qu = quartz; FK = potassic feldspars; Pl = plagioclases; Ph = phyllites; Ar = clay minerals; Fa = inorganic amorphous matter (mainly oxides and hydroxides of Al and Fe); Gi = gibbsite.
[4] Result in variation rates (% of each element compared with the initial quantity in the fresh granite).

Na_2O, K_2O, MgO, CaO, are grouped together for simplification. Losses are figured by variation rates lower than 100% and gains by rates higher than 100%.

The results call for some remarks:

(1) A very marked contrast exists between the moderate and progressive evolution of temperate soils and the much deeper and more intense chemical weathering which affects even the bottom part of the equatorial soils. Thus, in temperate soils, destruction of the primary vulnerable minerals is only partial (40–50%). The chemically weathered fraction, termed "complexe d'altération" (secondary minerals such as clay minerals and inorganic amorphous products), is about 10%; the losses, which are not very high, do not exceed 20–25% of the initial amount present in the parent rock. On the other hand, in the equatorial soils, the destruction of vulnerable minerals is almost total and affects even the bottom part. The chemically weathered fraction is very great (30–40%) and the losses are considerable (40–50% of the initial amount).

(2) The disparities in chemical weathering intensity are accompanied by differences in the nature of the secondary minerals: temperate soils contain 2/1 clay minerals especially (illites, vermiculites and montmorillonites) which are due to the progressive transformation of pre-existing phyllite minerals and to neoformations from elements released by destruction of unstable primary minerals, the plagioclasese. Equatorial soils contain 1/1 clay

minerals predominantly (kaolinite, halloysite) or even Al-hydroxides (gibbsite), which are neoformed from elements released by the destruction of the most primary minerals.

(3) Despite these disparities, a certain geochemical similarity exists between the two pedogeneses. Elements seem to be released and removed in the same order: bases > combined silicon > Al > Fe. Indeed, both pedogeneses are acid and produce an important removal of mineral matter: the major part of the released bases is leached (mainly Na and Ca which do not enter into any secondary minerals) and only a small part of them is retained by the adsorbing complex (clay and humus). The combined silicon is partly leached and partly recombined in the clay minerals formed in the soil. The Fe and Al metallic elements are mainly fixed in an oxide or hydroxide state, or enter into clay minerals; but they are partially leached or translocated from the A to B horizon (illuviation) or even migrate further away. In both soil types, the Fe accumulates more intensely than the Al. The intensity of these accumulations is maximum in the most weathered soil profiles, which explains the formation of very high supergene concentrations of Al and especially of Fe in equatorial soils.

The other zonal pedogeneses. We have no similar balance for the other major zonal pedogeneses. However, the mineralogical and chemical data available allow us to extrapolate the preceding results, with consideration to the different bioclimatic conditions. These pedogeneses will be classified in accordance with the zonal order from the high-latitude regions where the climate is cold and wet, downward to the low latitudes where the climate is hot, whether dry or not:

(1) Under cold wet climates, the decay of vegetal debris is very slow and an accumulation horizon of raw organic matter occurs at the surface. It is formed of vegetal debris which is slowly degrading and which releases soluble organic compounds, liable to complex the metallic cations[1] and to carry them toward deeper layers. Therefore, the soil profiles (podzolic soils and podzols) are very well differentiated (Ponomareva, 1966; Duchaufour, 1970; Targulian, 1971): underlying the $A_0 A_1$ organic horizon, is an A_2 horizon of leached mineral matter, followed by a B illuvial horizon with accumulations of organic compounds and metallic oxides which are translocated from the surface horizons. The clay fraction is constituted of 2/1 and 2/2 clay minerals which are mixed with amorphous matter in the illuvial horizon and which are more or less degraded in the superficial leached horizons. Because of the low-temperature conditions, the hydrolytic processes are rather limited, but under the influence of very acid organic compounds, all the mineral matter released near the surface is leached downward and eventually accumulated in the B horizons (redistribution).

(2) Under the contrasting continental climatic conditions of the medium latitudes, the organic matter undergoes an intense humification[2]; it is incorporated deep inside the

---

[1] See Chapter 5 by Saxby in Vol. 2.

[2] Synthesis of dense and stable complex organic molecules (humic acids, humine), liable to associate intimately with the clay fractions of the soil.

profile and gives a dark and thick grumelous horizon (isohumism). The chemical weathering of parent-rock minerals is low, Fe and Al are not released and the clay minerals formed are generally swelling clays (montmorillonites). The only elements to be released in rather large amounts are the bases which may be redistributed to form a carbonated accumulation horizon generally more or less diffuse (chernozems and chestnut steppe soils). The vertisols and eutrophic brown soils described by Paquet (1969), Bocquier (1971) and Boulet (1974) in large areas of Sahelian Africa, exhibit many similar characteristics, namely the richness in swelling montmorillonites. Most of these characteristics are also met within the steppe soils (brown and reddish brown soils) which are developed in dry and subdesert areas, but the organic matter is much less abundant and the massive and often thick carbonate accumulations form a calcareous crust or caliche (Ganssen, 1965; Duchaufour, 1968; Loughnan, 1969). Thus, in these pedogeneses, the leaching and hydrolysis processes are very limited and only the more mobile elements, such as the bases, undergo a certain migration; the soil-pH remains alkaline. These characteristics are related to the low annual rainfall and the marked water shortage recorded at least during dry seasons.

(3) Under subtropical and semi-humid tropical climates, the soils (fersiallitic soils, mediterranean soils and tropical ferruginous soils) generally show a strong development of 2/1 and 1/1 clay minerals, and a marked release of oxidized iron (Duchaufour, 1968; Paquet, 1969; Lamouroux, 1971; Bocquier, 1971), which gives the typical rubefacient aspect of these soils. The iron oxides are linked with clay minerals (fersiallitic soils) or accumulate in concrete masses and crust layer, forming the B horizon (ferruginous soils). Actually, the hydrolysis process is intensive, owing to the high average temperature, while the leaching process is less marked than in the equatorial soils. The removal of bases is not total, the soil-pH is about neutral and the removal of silica is only partial, especially in the fersiallitic soils, whereas in the ferruginous soils the leaching of silica is more intense and the migration of Fe often very spectacular: on low gradients and poorly drained tropical slopes (glacis landform), the hydromorphic conditions of argillaceous deep horizons favour the mobility of this element which accumulates in thick superficial horizons (ferruginous crust).

In brief, two major groups of pedogeneses may be distinguished: (1) the pedogeneses of rainy regions, giving acid soils where the bases, and to lesser extent silica, released are more or less leached and where the sesquioxides tend to accumulate with or without silica (subtractive pedogeneses, corresponding to the "pedalfers" of Marbut, 1928); (2) the pedogeneses of dry or arid regions, giving neutral alkaline soils, where the bases are retained and may accumulate near the surface and where the sesquioxides do not become individualized (redistributive pedogeneses, corresponding to the "pedocals" of Marbut). The main geochemical characteristics of the five major zonal pedogeneses are summarized in Table II: each pedogenesis is defined by the degree of release of elements from the primary minerals (which indicates the intensity of hydrolysis) and by the importance of their mobility during the pedogenetic process (which measures the leaching intensity). Elements with little mobility remain fixed or undergo limited redistribution, very mobile elements tend to be leached out of the profiles and moderately mobile elements behave intermediately.

From Table II and from data on the relative mobility of trace elements presented below (see p. 164), the behaviour of these elements in the major pedogeneses may be partially predicted, that is to say which elements are likely to accumulate with Fe and Al in acid soil profiles or with bases in alkaline ones. Therefore, the exact behaviour of trace elements during pedogenesis is not perfectly known. Some available data (Ermolenko, 1966; Nalovic, 1969; Nalovic and Pinta, 1971; Aubert and Pinta, 1971; Karpoff et al., 1973) give information about the contents of these elements in soils, in relation with the major pedogenetic factors; but the chemical weathering balances and studies dealing with the mechanisms which govern the mobility of these elements are very scarce.

Major factors of pedological differentiation

Let us now consider the pedological differentiation on the large scale of extensive areas and discuss the main environmental factors which govern the geographical distribution of the soils.

Climatic and biological factors: the zonal soils concept. The climatic and biological factors are the two prevailing variables which explain the disparities in the weathering types observed in the different climatic zones, as presented above.

Climate: the warmer and the wetter the climate, the more intense are the hydrolytic processes and the higher the degree of release of elements. The rainfall and the temperature also govern the subsequent behaviour of chemical elements in soil solutions: the mobility of the elements is conditioned by the pH of the solutions, which depends on the quantities of percolating rainwater; the solubilization rates and the solubility limits are influenced by the temperature which also controls the evapotranspiration and the concentration of solutions. Thus, rainfall and temperature are the two paramount parameters influencing the surface mobility of elements.

Vegetation and biological activity: these factors are closely related to the climate, and are fairly uniform over a given climatic zone, except locally where some very specific topographical or parent-rock conditions exist. They influence the pedogenesis either directly or indirectly (Lovering, 1959; Ponomareva, 1966; Duchaufour, 1968; Erhart, 1973a and b) by: (a) the more or less marked acidity of the humus they produce; (b) the role they play in the formation of the organo—mineral complexes, the nature and abundance of which condition the behaviour (solubilization and insolubilization) of numerous heavy cations and; (c) the specific ability of the vegetation to fix some soil elements and to release them later on (biogeochemical cycle).

In brief, the climate directly or indirectly (through the vegetation and the biological activity) controls the pedogenesis and this explains the *zonal distribution* of the major soil types.

TABLE II

Geochemical characteristics of the major zonal pedogeneses[1]

Climatic zones	Soil types	Behaviour of the elements during weathering processes (increasing stability ⟶)		
		alkaline and alkaline earth elements	combined silicium	iron and alumunium
			Increasing weathering rate ⟶	Increasing weathering rate ⟶
Neutral-alkaline pedogeneses, weak to inexistent drainage				
Dry continental zone, temperate–tropical	steppe isohumic soils	weak–moderate releasing, intense redistribution (caliche $CaCO_3$ crust)	very weak releasing, remaining fixed at the place of releasing	very weak releasing, remaining fixed
Subtropical–semi-humid tropical zone	fersiallitic soils and tropical ferruginous soils	moderate–intense releasing moderate leaching	moderate–intense releasing, weak leaching	moderate–intense releasing, remaining fixed or redistribution (ferruginous crust)
Acid pedogeneses, intensive drainage				
Cold humid zone	podzolic soils and podzols	moderate releasing intense leaching	moderate releasing moderate leaching	moderate releasing, intense redistribution or light leaching
Temperate humid zone	acid brown soils	moderate releasing moderate leaching	moderate releasing, moderate leaching	moderate releasing, remaining fixed, or light redistribution
Tropical humid zone	ferrallitic soils	very intense releasing, very intense leaching	intense releasing, intense leaching	intense releasing, intense redistribution and leaching

[1] The term *releasing* defines the proportion of the element extracted from the primary mineral network; the term *redistribution* defines the migration of the released element in the profile (generally from the eluvial A horizon to the illuvial B horizon); the term *leaching* defines the transportation of the element out of the soil profile

Parent rock and topography: the intrazonal and azonal soils concept. Parent rocks interfere essentially through the susceptibility to alteration of their minerals (see below p.
154: weathering sequence of common minerals). Under given chemical weathering conditions, the release and the removal of matter tend to increase together with this susceptibility. The geochemical balances established by Tardy (1969) in temperate and tropical
soil profiles derived from acid and basic parent rocks show sharp differences which are
related to the type of the parent rocks. The strong influence of the composition of the
parent rock in the weathering differentiation, up to the uppermost horizons of the soil
profiles, has been emphasized by Lelong (1969) for equatorial soils and by Souchier
(1971) for temperate soils. The nature of the secondary minerals developed during the
pedogenesis depends directly on the nature of the parent rock (Souchier, 1971; Dejou et
al., 1972).

Some rocks influence the pedogenesis processes so much that the soil types — even the
mature ones — derived from them, are radically distinct from the zonal soils occurring
under the corresponding climate. In temperate regions, for example, calco—magnesian
soils, which are formed on carbonate rocks, present characteristics which are more closely
related to base-rich soils of dryer regions, as long as the carbonate minerals are not
completely leached. Another example is given in wet tropical regions, where vertisols and
tropical brown soils, similar to those which develop on most of the rocks of the dry
tropical regions, occur on basic, highly vulnerable rocks. The modification of the normal
climate zonality of the weathering facies, in relation with changes in the petrographic
conditions, has been pointed out by Paquet (1969), in her concept of geochemical prevalence and deficiency.

Topography is also an important factor, inasmuch as the movement of ground-water
solutions depends on topography. On fairly steep slopes and relatively high land areas,
drainage is generally good, soils are well aerated and pedological evolution is normal. In
depressions and flat-lying areas, on the other hand, drainage is sluggish or inefficient, the
soil may be temporarily or permanently waterlogged and the pedological evolution becomes different; hydromorphic soil profiles with characteristic bleached gley horizons or
mottled pseudogley horizons are observed. The reducing conditions, which are more or
less marked in these horizons, may bring about an increase in the mobility of some
elements — such as Fe and Mn — and a decrease in the mobility of some others. Thus, the
reduction of sulfates by bacterial activity produces hydrogen sulphide which is liable to
precipitate, even in minute amounts, most of the metallic elements. Hydromorphic environments are also characterized by a strong development of clays and colloids, in which
many elements are likely to be entrapped (see p. 158).

The chemical composition of groundwaters is related with the topography. The salt
concentrations of these waters vary with the lateral flow and with the evapotranspiration
conditions from upstream to downstream. This explains a certain "ventilation" (frac-
tional separation) of ions along the slopes. From the combination of this phenomenon
with the climatic zonality of the weathering types, Tardy (1969) sketches an outline of

"chromatographic separation" of the elements in the landscapes. This outline sums up the great geochemical trends in relation with the climate and topographical position.

Thus, it is obvious that in some cases the influence of the parent rock and topographic factors is so determinative that the pedogeneses diverge to some extent from the normal bioclimatic evolution: the corresponding soils are termed "*intrazonal soils*".

Under extreme topographic conditions (mountainous reliefs and steep slopes) the erosion rate is higher than the chemical weathering rate and the pedological evolution becomes ineffective or even non-existent. The same phenomenon occurs in the accumulation zone of recent deposits (dunes, alluvium). Soils which remain immature because of mechanical erosion or continental sedimentation, whatever the climatic conditions may be, are termed "*azonal soils*".

Time of evolution. This factor also may explain very important geochemical differentiations in the surface of the earth. From simple calculations (Leneuf, 1959; Tardy, 1969), the unit of time required to form one meter of alterite leached of silicon and bases in equatorial zones, has been estimated to range from 50,000 to 200,000 years. In wet-temperate zones, the silicon content of subsurface waters and the intensity of drainage are not much lower (Davis, 1964; Tardy, 1969), which suggests that the unit of weathering time should not be much longer. However, the equatorial alterites are much thicker than those of temperate climates. This difference is probably due to the influence of the time factor: on old basements of equatorial regions such as West Africa, Guyana and Brazil, the chemical weathering process might have been going on over a very long period (hundred of thousands, or even millions of years) without being really impeded by any intense erosion, inasmuch as the alterites and the soils were preserved by a thick vegetal cover. In temperate regions, on the other hand, successive episodes of erosion (brought about by Quaternary glaciations) swept away the major part of the alterites formed during more ancient periods.

This example shows that some disparities between pedogeneses may result rather from differences in the degree of the chemical weathering process (duration of evolution) than from qualitative differences (nature of processes). The importance of the time factor has been recently emphasized by Icole (1973). However, as is shown below (p. 154), the chemical weathering and element-mobility mechanisms are so complex that differences in degree may finally bring about strongly marked geochemical contrasts.

Conclusion

(1) In well-drained areas, in which the conditions of weathering are oxidizing, at least two distinctly different major pedogenetic evolutions must be considered from a geochemical point of view: first, the pedogeneses giving rise to leached soils, which selectively concentrate Al- and Fe-oxides, sometimes silica, and all elements presenting a similar behaviour in superficial layers. They correspond to wet and to very wet climates, and the warmer the temperature is, the more rapid and intense is the differentiation they induce.

Secondly, one must consider the pedogeneses producing neutral or alkaline soils, which selectively concentrate the alkaline and alkaline-earth elements and which correspond to dry and to extremely arid climates.

These two major pedogenetic evolutions may be subdivided. Thus, the leaching processes are not exactly the same in the cold and warm wet regions: Pédro (1968) distinguishes the "acidolysis" corresponding to the podzolic pedogeneses of the high-latitude zone and the intense "hydrolysis" corresponding to the rainy tropical zones. Therefore, the acidity is also frequently very marked in the surface horizons of the tropical soils and both pedogeneses can be termed acid. The particularity of the wet tropical pedogenesis is chiefly the presence of a very thick, argillaceous weathered crust (lithomarge).

The geochemical evolution corresponding to the neutral or alkaline pedogeneses may also be subdivided into several types, according to the chemical composition of the environment and the degree of the ionic concentration of the soil solution. Referring to the general classification of the epigenetic processes (Perel'man, 1967, table 27, p. 157), one may distinguish: (a) the neutral carbonatic type, for which the main aqueous migrants liable to be redistributed and to concentrate are HCO_3^-, Ca^{2+} and Mg^{2+}; (b) the gypsiferous type and the chlor—sulfatic weakly alkaline type, for which they are respectively SO_4^{2-} and Ca^{2+} or Cl^-, SO_4^{2-} and Na^+; (c) the sodic, strongly alkaline type, for which they are CO_3^{2-}, OH^-, Na^+ and SiO_2. The type (a) occurs in steppe soils of contrasting climate conditions (chernozems, brown or reddish-brown soils, and vertisols). The types (b) and (c) occur in intrazonal saline soils developed under peculiar topographical and parent-rock conditions, mainly under arid—sub-arid climates where ionic concentration of the solutions may rise strongly through evapotranspiration processes.

(2) Apart from the oxidizing pedogeneses which are developed in well-drained eluvial soils, quite distinct evolutions are encountered in the hydromorphic environments of swamps, marshes and alluvial low-lands, where special geochemical differentiations are related to reducing conditions. These hydromorphic evolutions may be subdivided into several types, according to the dominant nature (organic or inorganic) of the soil constituent, the pH of the environment and the presence or absence of hydrogen sulphide (Perel'man, 1967; Duchaufour, 1970).

(3) The preceding distinction between the main geochemical trends corresponding to weathering and pedogenetic processes is not exactly similar to the distinction, often made, of different geochemical landscapes. A geochemical landscape is the result of the sum and of the interaction of soils, weathered crusts, surface and subsurface waters, erosional features and continental deposits (Millot, 1964; Perel'man, 1967). Typical geochemical landscapes are, for example, the wet-forest tropical peneplains or the dry steppic pediplains, which present specific geomorphological features, corresponding to more or less distinctive geochemical characteristics. But each geochemical landscape constitutes a rather complex system, in which the geochemical behaviour of the elements often varies from one locality to another, according to the parent rock and topographical conditions. Therefore, the concept of geochemical landscape, though very interesting in general,

seems too large in the perspective of the supergene metallogeny to determine the specific haloes of dispersion of the metallic elements.

ALUMINIUM ORE DEPOSITS: THE PROBLEM OF BAUXITES AND LATERITES

Generalities

Geochemical properties of aluminium. With a content of 8.1%, Al is the third most common element in the lithosphere, after oxygen (47.5%) and Si (28.8%). It is relatively concentrated in eruptive feldspathic rocks and argillaceous or schistose rocks, but relatively scarce in carbonate and sandstone rocks as well as in ultrabasics.

Al displays a marked affinity for oxygen and it is not found in the lithosphere in a native state. Its ionic radius (0.51 Å for sixfold coordination) is small compared to that of the other metallic elements, and is almost as small as the Si radius (0.42 Å). It is present in minerals, either in a tetrahedral form (prevailing in minerals originating from deep layers) or in an octahedral form (prevailing in minerals of surface origin).

Al is soluble in water at low pH (pH $<$ 4), where it occurs in the form of Al^{3+} ions, or at high pH (pH $>$ 10) where it assumes the form of AlO_2^- ions. Consequently, it proves to be only slightly mobile under common surface weathering conditions, and it tends to accumulate in secondary silico-aluminium minerals (clay minerals) or in purely aluminous ones (gibbsite, boehmite).

Aluminium-ore deposits: the bauxites. The major aluminium deposits are the bauxites, defined by Gordon et al. (1958) as "aggregates of aluminous minerals, more or less impure, in which Al is present as hydrated oxides". The most common impurities are kaolinite or similar clay minerals and the oxides and hydroxides of Fe and Ti. Different types of bauxite have been described, with regard to their depositional and assumed genetic conditions (Harder, 1952; Bracewell, 1962; Patterson, 1967; Valeton, 1972). Harder, for example, classified the bauxites according to the nature of the bed rocks on which they are found. Patterson distinguished "blanket deposits" (layers contained in residual weathered crusts), "interlayered deposits" (lens-like formations included in sedimentary series) and "pocket deposits" (accumulations at the surface of karstic landforms on carbonate rocks). Valeton subdivided the bauxites into autochthonous and allochthonous bauxites, depending on the localization of the source of Al. Here, a classification similar to that of Patterson will be adopted:

(1) Lateritic blanket bauxites. These are residual aluminous or aluminoferruginous layers resulting from the intense weathering of eruptive, metamorphic or sedimentary rocks. Such bauxites are said to be autochthonous, the Al present in the bed rock becoming preferentially concentrated in-situ owing to the removal of the other elements. The more common descriptions of bauxites, from various parts of the world, apply to this type of ore deposit (Fermor, 1911; Lacroix, 1913; Harrison, 1933; Harder, 1952; Wolfenden, 1961).

(2) Sedimentary bauxites. These are stratiform aluminous accumulations which are found intercalated in some sedimentary series. They are generally thought to be detrital deposits, colluvial or alluvial particularly and made up of elements or fragments coming from preexisting lateritic blankets, which have been eroded and then accumulated laterally by sedimentary processes (Gordon et al., 1958; Valeton, 1972). Some authors propose the possibility that aluminous sediment can be of chemical origin, formed by precipitation of Al from solutions formed during leaching under conditions of acid pedogenesis (Caillère and Pobeguin, 1964; Erhart, 1969). Both detrital and chemically precipitated bauxites are called allochthnous, the Al-concentration occurring at a more or less remote distance from its original source. However, the formation of sedimentary bauxites related to argillaceous deposits may involve two steps (De Vletter, 1963; Moses and Michell, 1963): (a) accumulation of clays; and (b) transformation in situ of these clays into bauxites. Consequently, these bauxites may be considered as autochthonous, the sedimentary clays playing the role of parent rock.

(3) Karst bauxites. These lie, directly or indirectly, on a carbonate and more or less karstified bed rock. Their origin is still not well understood. The two problems discussed are the source of the Al and the site of the bauxitization. Some authors think that the Al was derived from the carbonate rocks themselves because these rocks always contain some significant quantities of clays or aluminous colloids which are likely to accumulate in-situ during the karstification process, resulting under warm and humid climates in residual bauxites similar to the lateritic blankets (De Lapparent, 1924; Hose, 1963). Therefore, they should be considered as true autochthonous bauxites. Some other authors think that Al comes from silico—aluminous rocks (schists, marls, volcanic deposits, etc.) lying on carbonate rocks and which have been subsequently removed through alteration and erosion, leaving a bauxitic residue (Burns, 1961; Bonte, 1965; Rousset, 1968; Valeton, 1972). In this case, bauxites may still be considered as more or less autochthonous. Other researchers do not agree that the Al was derived from the carbonate host rocks nor from overlying rocks, but rather it was imported by water in a detrital or chemical form, or even by the wind from nearby lateritized areas (Roch, 1956; Zans, 1959; Watterman, 1962; Caillère and Pobeguin, 1964; Nicolas, 1968; Erhart, 1969). The bauxites would then be considered true allochthonous bauxites. The problem related to the site or localization of bauxitization concerns only the allochthonous bauxites. Roch, Zans, and Valeton assume that the bauxitization takes place in-situ, at the site where the ore is now located after the preliminary formation of silica—aluminous deposits over the carbonate rocks. Nicolas, on the other hand, thinks that this bauxite formation has taken place at a more or less remote distance from the site of deposition, on the silico—aluminous basement surrounding the karst; therefore, the bauxite was already formed when deposited on the karst. According to this interpretation, karst-bauxites must be considered as being strictly sedimentary. However, most authors accept that bauxites result from a complex evolution which takes place both before erosion and transportation (lateritization over crystalline ancient basements) and after the accumulation of the materials involved within

the karst depositional site (progressive enrichment of the Al contents).[1]

Distribution of bauxites over the world. The foregoing distinction between lateritic blanket bauxite, sedimentary bauxite and karst bauxite may appear somewhat arbitrary because it probably corresponds more to differences in age and successive steps of evolution than to genetic differences in the conditions of formation of Al minerals. As a matter of fact, when analysing the distribution of bauxites over the world (Fig. 3), one may observe that "lateritic" bauxites are mainly located in the present-day intertropical zone, whereas karst bauxites occur predominantly in present-day temperate or subtropical zones. The former, which are of the Quaternary or the Tertiary Period, seem generally more recent than the latter which are of Tertiary age in the West Indies, of Cretaceous age in the Mediterranean zone, and of Jurassic or even Paleozoic age in the Urals and the Far East (Hose, 1963). Sabot (1954) has suggested that the further the bauxites are from the equator, the more ancient they appear to be. Therefore, one may think that karst-bauxites are also related to vast ancient lateritic blankets whose position moved during geological times, following the polar migration. Owing to their very long history, these lateritic blankets may have been successively and/or cyclicly eroded and redeposited, which would explain their sedimentary characteristics.

Other sources of Al. The other potential sources of Al are either less abundant or if they predominate, they contain only small amounts of Al. The metallurgical treatment of such materials is often difficult.

(1) The alunite, $KAl_3 (OH)_6 (SO_4)_2$, is fairly rich in Al, but is not abundant: it develops through either superficial or deep chemical weathering of aluminous rocks in contact with sulphuric solution (weathering solutions from sulphide lodes and volcanic fumaroles, for example).

(2) The Al-phosphates contain up to 40% Al. Ores of this type have been found in Florida (U.S.A.) and in Senegal, where they formed through lateritic weathering of Ca-phosphate deposits.

(3) Aluminous clays. These are more commonly clays of the kaolinite group, either well or poorly crystallized (dickite, nacrite, kaolinite, fire-clay, halloysite) or they are more exceptionally aluminous chlorites. They often contain small quantities of crystallized or amorphous Al-hydroxides. Their alumina content can amount to 40%. They are of eluvial (weathering of rocks) or sedimentary origin, and in this last case, they seem to have been inherited from continental soils (Millot, 1964). Important aluminous clay layers are known particularly in Florida, Carolina, Pennsylvania (U.S.A.) and Hungary (Keller, 1964).

(4) The dawsonite (Na-Al-carbonate) which is contained in some bituminous schist

---

[1] See Chapter 4 by Zuffardi, for other ore deposits in karstic host rocks.

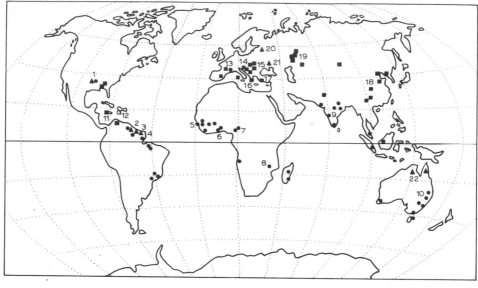

● Lateritic bauxites   ■ Karst bauxites   ▲ Sedimentary bauxites

Fig. 3. Principal deposits of bauxite and estimated total reserves in millions of tons (after Patterson, 1967, and Valeton, 1972). *Lateritic bauxites*: *1* = U.S.A. (Arkansas), *2* = British Guyana (80), *3* = Surinam (200), *4* = French Guyana (70), *5* = Guinea (1,200), *6* = Ghana (290), *7* = Cameroon (1,500), *8* = Malawi (60), *9* = India (64) and *10* = Australia (2,000), *Karst bauxites*: *11* = Jamaica (600), *12* = Haiti (85), *13* = France (70), *14* = Yugoslavia (200), *15* = Hungary (150), *16* = Greece (84), *17* = Turkey (30), *18* = China (150) and *19* = Russia (Urals) (300). *Sedimentary bauxites*: *1* = U.S.A. (Arkansas) (44), *2* = British Guyana, *3* = Surinam, *20* = Russia (Tikhvin), *21* = Russia (Ukrain) and *22* = Australia (Northern).

(e.g., in Colorado, U.S.A.) might be a by-product of petroleum. The potential resource of this low-grade ore would be considerable (Guillemin, 1974).

(5) Rocks exclusively or almost exclusively *feldspathic* (anorthosites, nephelinic syenites, for example) form a very abundant potential source, but with low content of Al (20—30%).

Exploitation of aluminium ores. Up to now, bauxites are nearly the only Al-ores being extracted. Their extraction is rapidly increasing: about 30 millions tons in 1961 and 65 millions in 1971. The respective parts of the three genetic types of bauxites in the production are not exactly known, since the genetic characteristics of the bauxites utilized are often not well established. Schellmann (1974) considers only two main types, the karst bauxites and the silicate bauxites (derived from the lateritization of silicate rocks). We may suppose that the latter include the "blanket" and most of the "interlayered" bauxites, even though many karst bauxites are now assumed having a sedimentary origin. The production of karst bauxites, which are chiefly located in or near the industrial countries, was dominant up till 1960–1965; but the minable reserves of silicate bauxites are more considerable and their part of the world production now reaches 55–60% (Schellmann, 1974). The principal producing countries are Jamaica, Australia, and Surinam.

Analytical study of bauxites

Nomenclature: laterites and bauxites. The term "bauxite", defined by Berthier (1821) after the locality of Les Baux (France), was initially applied to aluminous concentrations on limestones. It has then been extended to lateritic blanket bauxites. Indeed, bauxites and laterites are two closely related weathering formations in humid tropical and equatorial regions. But if the term "bauxite" is fairly precise, the term "laterite" covers many aluminous, ferruginous and argillaceous formations, and the concept of lateritization, considered by pedologists to be too indefinite, has been replaced by the concept of ferrallitization (Chatelin, 1972). Nevertheless, we will retain here the old term "laterite", with its wide meaning: weathering formation, indurated or loose, argillaceous or not, and containing Fe- and Al hydroxides. After Lacroix (1913), we will distinguish: (a) laterites sensu stricto with more than 90% hydrates (Fe–Al); (b) argillaceous laterites with 50–90% hydrates (Fe–Al); (c) lateritic clays with 10–50% hydrates (Fe–Al); and (d) clays with less than 10% hydrates (Fe–Al).

The laterites s.s. are subdivided into aluminous laterites or pure bauxites, alumino–ferruginous laterites or impure bauxites, and ferruginous laterites; the chemical composition of these formations is presented in Table III.

Types of bauxite ore deposits. These types are specified by distinctive morphological characteristics:

(1) Bauxites and laterites of lateritic blankets. Bauxites and laterites s.s. occur most commonly in the upper part of these blankets, as concretionary layers or crusts, more or less indurated, which cap hills and plateaus and which offer protection against erosion. These topographically high levels are fairly widespread in Africa, South America, the East Indies, Indonesia and Australia; they are often considered as remnants of ancient peneplains described in the East Indies as "high-level laterite" and in Africa as "African surface" (Fermor, 1911; King, 1948; Michel, 1970).

The profiles through these blankets present some differences, from one ore deposit to another, and only three fairly typical profiles will be described here, which correspond respectively to pure bauxites, alumino–ferruginous laterites and ferruginous laterites, without making reference to the superficial horizon of loose ground which covers the indurated level in some areas, particularly in forest regions.

(a) Pure bauxite on nepheline syenite. The vertical sequence is from top to bottom (Lacroix, 1913; Bonifas, 1959) in typical deposits of Iles de Los (Guinea): (i) bauxitic crust, massive, indurated, light coloured, with a compact or vesicular structure, and a breccia, pisolitic or homogeneous texture (up to 10 m thickness); (ii) granular bauxite, friable or little indurated, finely porous, with a structure inherited from that of the parent rock (very variable thickness); (iii) fresh bed rock (parent rock).

The vertical section of residual bauxites in Arkansas is about the same (Gordon et al., 1958). The contact between the granular bauxite and the parent rock is very sharp and irregular; in some places argillaceous saprolite formations are developed toward the base of the granular bauxite. All bauxites are generally very rich in Al (see samples 1, 2, 3, in Table III).

(b) Alumino—ferruginous laterites on greenschists, diorites, gabbros or similar rocks. The main levels from top to bottom are as follows: (i) alumino—ferruginous laterite crust, more or less dark-coloured, strongly indurated, massive or scoriaceous, with a homo-geneous, breccia or pisolitic texture; the thickness may be up to 10—15 m. The upper part of this crust often appears relatively enriched in Fe, whereas the lower part is relatively enriched in Al (occurrences of secondary crystallization of Al which cement the pore and line up vesicules and fissures); (ii) kaolinic clays (lithomarge), often mottled, with diffuse red spots of Fe-oxides, and in which the structure of the parent rock is sometimes recognizable; the thickness is quite variable and may sometimes reach 50 m or more; (iii) fresh parent rock.

These bauxites are generally impure, more or less rich in Fe (see Table III: sample 7, 8 and 9) and of heterogeneous composition; the Al-content varies from one point to an-other. Such bauxites are very common; many ore deposits are known in West Africa (Grandin, 1973; Boulangé, 1973), in Guiana (Harrison, 1933) and in Indonesia and Asia

TABLE III

Chemical composition of some bauxites and laterites

	(1)	(2)	(3)	(4)	(5)	(6)	(7)	(8)	(9)	(10)	(11)
SiO_2	4.6	2.2	4.1	1.8	1.4	5.4	3.3	0.6	0.8	1.5	0.3
Al_2O_3	62.0	55.8	61.5	60.0	60.5	44.3	46.8	46.2	37.7	9.0	5.3
Fe_2O_3	0.9	5.2	2.0	8.0	9.7	23.1	23.6	24.7	31.8	75.0	72.3
TiO_2	0.4	0.1	2.9	1.1	2.2	1.4	0.7	0.2	4.2	–	–
H_2O	31.6	30.4	30.7	29.4	25.5	24.5	22.9	26.3	24.5	12.0	14.5

(1) Typical bauxite on nepheline syenite (Gordon et al., 1958).
(2) Typical bauxite on nepheline syenite (Lacroix, 1913).
(3) Typical bauxite on schists (Boulangé, 1973).
(4) Typical bauxite on granite (Boulangé, 1973).
(5) Typical bauxite on argillite (Cooper, 1936).
(6) Alumino-ferruginous crust on syenite (Bonifas, 1959).
(7) Alumino-ferruginous primary laterite on dolerite (Harrison, 1933).
(8) Alumino-ferruginous crust on amphibolite (Boulangé, 1973).
(9) Ferrugino-aluminous crust on dolerite (Harrison, 1933).
(10) Ferruginous crust on dunite (Percival, 1965).
(11) Ferruginous crust on peridotite (Trescases, 1973b).

(Harder, 1952; Wolfenden, 1961). They occur on a large variety of rocks, especially basic or intermediary eruptive rocks, on lavas and tuffs, on schists and argillites, less commonly on acid, granitic or gneissic rocks. A thin and irregular level, where the parent rock is more or less totally bauxitized, is often differentiated under the lithomarge.

(c) Ferruginous laterite on ultra-basic rock (Percival, 1965; Trescases, 1973b). The main levels from top to bottom are: (i) ferruginous crust, dark-red—purplish, massive or vesicular but very dense and hard, with a brecciated or scoriated texture; under the crust or replacing it, there may be found a thick gravelly level, essentially formed of ferruginous concretions (from 5 to 10 m thick); (ii) soft layer, of lighter colour, brownish-yellow—red, sandy or silty but non-argillaceous, often preserving the parent-rock structure (variable thickness up to 50 m); (iii) more or less rubefacient, saprolized parent rock, with a coarse texture; and (iv) fresh parent rock.

From the study of these ore deposits, one may distinguish after Harrison (1933) and Erhart (1973b), two types of laterites: first, the "primary" or "typical" laterites in which the structures of the parent rock are preserved and where the Fe- and Al-concentration corresponds to a simple relative accumulation ensuing from the removal of the other constituents. Secondly, the "secondary" or "modified" laterites in which the initial structures vanish through phenomena of both relative and absolute accumulations (D'Hoore, 1954) and probably through some mechanical migratory processes. The absolute accumulation phenomena are confirmed by the presence in laterites of coatings and linings of Fe- or Al-hydrates on the fissure and vesicle surfaces.

Despite their chemical disparities, the first and third types of ore deposit prove to have a real morphological analogy: in both cases a thick level of primary, aluminous or ferruginous laterite lies immediately on the parent rock. On the other hand, the more frequent second type exhibits mostly secondary laterites, separated from the parent rock by a generally thick layer of argillaceous lithomarge.

(2) Karst bauxites. The bauxite ore deposits found on sedimentary limestone rocks in the south of France, generally show a vertical section which, if simplified, is as follows (Nicolas et al., 1967; Nicolas, 1968). (i) top: sandstone, lignitic, argillaceous or carbonate sediments (Upper Cretaceous or Tertiary); (ii) oolitic or pisolitic bauxite becoming richer in clays toward the top; (iii) brecciated bauxite; (iv) red kaolinic clays with fragments of bauxite; (v) bottom: karstified calcareous sediments (Upper Jurassic—Lower Cretaceous).

Nicolas points out the existence of breccias toward the base of the ore deposits, and sometimes the presence of fossils (*Rhynchonella, Terabratula*) in the bauxite itself.

The karst-bauxite deposits occurring along the northern border of the Mediterranean Sea are of the same type. Similar deposits are also found in the Urals and Vietnam, but the calcareous bed rocks are older; those occurring in the West Indies are of Eocene—Oliocene age (Hose, 1963).

(3) Sedimentary bauxites. The diagrammatic section giving the major types of bauxite in Arkansas (Gordon et al., 1958) shows, in addition to residual bauxites, bauxite deposits interbedded in transgressive Tertiary series. These deposits consist of detrital elements

of bauxite, about 1 mm in diameter, rounded—subangular shaped, included in an argil-laceous matrix, and occurring in ancient channels or small depressions within the basin which they fill. Intersecting stratifications and grain-sorting sequences may be observed.

Another example is afforded by stratified bauxites, in the recent sedimentary basin of the Guiana littoral zone: several levels of bauxite are reported (Boyé, 1963; De Vletter, 1963; Moses and Michell, 1963), lying either immediately on the weathered basement, or separated from it by sedimentary gravel, sand, clay or lignite layers of the Lower Quaternary or Tertiary Periods.

Mineralogy of bauxites

(1) Lateritic blanket bauxites. Al mostly assumes the form of very fine crystallizations of gibbsite. In primary laterite, this mineral appears as a porous mass, replacing the feldspar minerals (Lacroix, 1913; Harrison, 1933). In the crusts and secondary laterites, the micro-crystalline gibbsite is closely mixed with Fe-oxides (goethite, hematite) and with variable quantities of kaolinite. There, it forms compact masses in which more or less abundant, angular or rounded particles (fragments, ooliths, pisoliths) are differentiated inside a homogeneous matrix. Crystals of larger size develop as coatings and linings inside fissures and vesicles. In pisolitic crusts, gibbsite is often replaced by boehmite.

(2) Karst bauxites. Some of these bauxites, and especially the relatively recent ones occurring in Jamaica, are chiefly composed of gibbsite. But most of the bauxites developed on calcareous rocks in Southern Europe predominantly contain boehmite or diaspore (dense dimorphous variety of boehmite). Fe assumes the hematite and sometimes the magnetite form, but it may also be in some proportion included in boehmite and diaspore (Hose, 1963; Caillère and Pobeguin, 1964; Valeton, 1965). Diaspore is the major constituent of bauxites lying on ancient calcareous sediments, as, for example, in some Russian or Asian deposits. In regions where the bauxites have undergone general or contact metamorphism, the recrystallization of boehmite or diaspore in anhydrous minerals (corundum) may occur.

In brief, gibbsite is the most common mineral in recent or little altered bauxite facies (particularly primary laterites). Boehmite and diaspore are, on the other hand, the most common species in more ancient bauxites which have undergone a more or less marked diagenesis. Under metamorphic conditions, corundum develops. Kaolinite which accompanies Al-hydrates seems to be less susceptible than these latter to the transformations affecting the deposits.

Genesis of bauxites and laterites

Since all bauxites, sedimentary and karst bauxites included, seem to involve concentration processes related to weathering under a humid-tropical climate, we will consider here only the problem of the genesis of the lateritic blanket bauxites.

The first point to emphasize is the relative scarcity of bauxite ore deposits, although

Al is quite abundant. This might be explained by the very low mobility of this element under surface conditions. To arrive at significant accumulations, all the other elements must be removed, which implies a very unusual intervention of additional processes.

Bauxites are commonly defined as the ultimate residual product obtained through the weathering of silico—aluminous rocks exposed to extremely intense leaching conditions. These conditions are encountered in warm and humid climates, in relatively high and well-drained locations, and under a dense cover of vegetation allowing the chemical weathering effects to prevail over mechanical erosion. Under such conditions, the pedogenesis gives rise to ferrallitic soils with a profile showing some features resembling the lateritic differentiations:

Ferrallitic soil profile:

leached superficial horizon (A)

horizon of Fe- and Al-accumulation
   (2—3 m thick) (B)

horizon of mottled clays[1]

lower leached horizon[1] (C)

fresh parent-rock

Lateritic bauxite profile:

residual soil[1]

alumino—ferruginous crust (5—15 m)

argillaceous lithomarge[1]

primary laterite

fresh parent rock

The massive and thick accumulation of Fe and Al in bauxites and laterites can, therefore, be considered as the ultimate term of a very prolonged ferrallitic evolution, allowing the accumulations of the B horizons to aggregate progressively as the profile goes deeper and the horizons to migrate downward conjointly with the topographical surface. Induration of these accumulations into laterite crust is often explained by the existence of a climatic change with marked trend toward aridity, involving the destruction of the vegetal cover, the ablation of the superficial A horizon and the induration of the illuvial B horizon.

Primary bauxites and laterites. The concept of primary laterite explains fairly well the formation of very pure bauxites on silico—aluminous rocks free of quartz and poor in Fe, such as the nepheline syenites or even the aluminous clays. As reported below (p. 153), during the leaching process of silico—aluminous minerals and at a pH within the range of the more common pH (4—8), bases are removed first and then combined silica, whereas Fe and Al accumulate residually as crystallized products (gibbsite, boehmite, goethite). These residual minerals have been obtained in leaching experiments, at moderate temperature, from all kinds of crystallized or vitreous rocks and from silico—aluminous minerals, clay minerals included (Pédro, 1964; Trichet, 1969; Pédro et al., 1970). Nevertheless, in bauxites of this type, the bottom part of the profiles sometimes exhibits an argillaceous horizon, with neoformed kaolinite, a mineral which is hardly ever obtained experimental-

---

[1] The corresponding horizons do not occur systematically.

ly under normal pressure and temperature conditions (Wollast, 1961; Lerz and Borchert, 1962).

The same concept explains the formation of ferruginous laterites on ultra-basic parent rocks, poor in Al. The origin of these ferruginous laterites and bauxites is strictly residual: they are *"eluvial laterites and bauxites"*. Such laterites are essentially primary ones, even if redissolution and re-precipitation processes in the crustal layer progressively obliterate the parent-rock structures, giving facies of "secondary" laterites.

Secondary bauxites and laterites. The development of bauxites is not exclusively limited to some particular rocks: they are also found on acid rocks, containing quartz, such as granites and gneiss, or on intermediary or basic rocks which are more or less rich in Fe, of eruptive, metamorphic or sedimentary origin. The problem of the genesis of these bauxites raises the question of the elimination of the silica and/or of the separation of Fe and Al:

(1) Separation of aluminium and silica. Bauxites derived from feldspathic rocks, either quartzitic or not, generally lie on a kaolinic lithomarge in which the feldspars have been destroyed while the quartz persists. The bauxite formation process seems to proceed in two steps:

step 1 = feldspar → kaolinite;
step 2 = kaolinite → gibbsite.

The chemical reaction corresponding to step 2 is:

$$Al_2(OH)_4 \ Si_2O_5 + 5H_2O \rightarrow 2\ Al(OH)_3 + 2\ Si(OH)_4$$

The equilibrium is displaced towards the right side through the $Al(OH)_3$ precipitation (gibbsite), but this incongruent dissolution involves a very low concentration in $Si(OH)_4$ (less than 1 ppm SiO_2), and an intermediate range of pH (pH from 4 to 7) in which gibbsite is stable. In very acid or alkaline solutions, the reaction of dissolution of kaolinite becomes congruent and gives ionic Al^{3+} (acid solution) or $Al(OH)_4^-$ (alkaline solution) and the precipitation of gibbsite does not occur.

Thus, the proceeding of the second step of bauxitization normally requires, in the slightly acidic environment of tropical soils, that all the quartz and, a fortiori, the feldspars have been removed (see the diagram of the solubility of kaolinite, gibbsite and quartz, Fig. 4, and the thermodynamics data from Wollast, 1961, 1963; Tardy, 1969; Curtiss, 1970; Gardner, 1970 and Schellmann, 1974). Therefore, a complete bauxitization requires a particularly long evolution which would permit the very slow dissolution of the quartz and also a complete hydrolysis of the kaolinite (a mineral much more resistant than the primary silico—aluminous ones). This two-stage evolution results in bauxitic crusts of the secondary laterite-type. The structure of these bauxites does not generally bear any relationship to the rock structure, for this structure disappears progressively at the level of the lithomarge: they may be termed *"secondary eluvial bauxites"*.

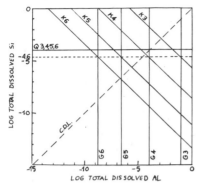

Fig. 4. Solubility of quartz, kaolinite and gibbsite contoured at pH values from 3 to 6. K = kaolinite; Q = quartz; G = gibbsite. Associated numbers indicate pH countours. $C.D.L.$ = congruent dissolution line of kaolinite. (After Gardner, 1970.)

However, some aluminous lateritic crusts developed on feldspathic rocks keep the parent-rock structure, although it has disappeared at the level of the subjacent lithomarge (Wolfenden, 1961). Bauxitic crusts on granite, with hardly any lithomarge, but with a preserved granitic structure, are also known (Boulangé, 1973). These ore deposits may be explained by assuming that on feldspathic rocks, either quartzitic or not, the bauxitiza- tion may be a direct process, without any argillaceous phase, like on nepheline syenite rocks. This is illustrated by many ferrallitic soil profiles exhibiting a gibbsite horizon lying immediately on a granitic parent rock (Lelong, 1969) and is also confirmed by the results of leaching experiments in the laboratory (Pédro, 1964). The direct process of bauxitization in presence of quartz and feldspar minerals may be explained by the consideration of reaction kinetics. Under intense leaching conditions, the solubility limits of the silicate minerals are not necessarily reached and the evolution may lead directly to the bauxite stage before the feldspars and the quartz are totally destroyed. The temperature may also interfere: the thermodynamic calculations made by Fritz and Tardy (1973) from Hel- geson (1969), show that at acid pH a rise in temperature tends to reduce the solubility of gibbsite and to increase that of kaolinite. This favours the direct bauxitization of rocks rich in silica.

Therefore, the aluminium–silicon separation proceeds either indirectly in two stages (rock → lithomarge → bauxite) or directly (rock → bauxite), providing that exceptional drainage and temperature conditions are fulfilled. In the second case, the occasional occurrence of an argillaceous lithomarge underlying the primary bauxite may be due either to subsequent processes of resilicification of the bauxite or the result of incomplete desili- cification of the parent rock, related to reduced drainage conditions (e.g., raising of the base level or deepening of the weathering front).

(2) Separation of Al and Fe. Except on parent-rocks very deficient in Fe, the forma- tion of bauxites requires an Al–Fe separation. In profiles of lateritic bauxites of West Africa, Grandin (1973) quoted a more marked enrichment for Fe than for Al in the

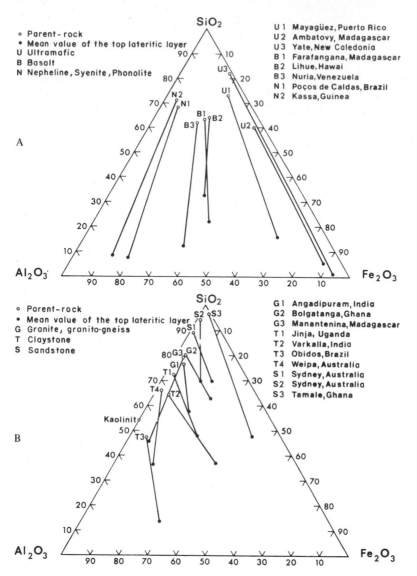

Fig. 5. Variations of Al_2O_3-, Fe_2O_3- and SiO_2-contents during lateritic weathering of: (A) quartz-free parent rocks; and (B) quartz-bearing parent rocks (after Schellmann, 1974).

surface lateritic crust. Schellmann (1974) compared the chemical trends during the lateritization of quartz-free rocks and of quartz-bearing rocks (Fig. 5). On the former rocks, Al_2O_3 and Fe_2O_3 seem to be equally stable; they are enriched in the same proportions in the lateritic crust, whereas on the latter, Fe_2O_3 is more enriched because some quantities of Al_2O_3 are dissolved during desilication processes.

As seen above, in the ferrallitic soil profiles on granitic rocks (see Fig. 1), Fe accumulates more easily than Al at the level of the B horizon, but it is relatively depleted at the level of the hydromorphic lithomarge, as is attested by the frequent bleaching of this formation.

Thus, the removal of Fe, which is required for the bauxite differentiation, may be related to the preferential leaching of this element in the more or less reducing conditions occurring at the lithomarge level. This process of removal, already well marked in the deep layers of relatively high areas, is probably more intense in soil profiles of low, flat or basinal areas, where hydromorphic conditions are more severe. But in the hydromorphic lithomarge, Si is partly retained with Al and the bauxite genesis implies a subsequent stage of desilication, with improved drainage conditions. Therefore, the resulting bauxites correspond to the secondary eluvial type, the formation of which requires two different steps of evolution with changes in the geomorphological conditions. Concerning the formation of the largest bauxite deposits of this type, which occur as perched crusts capping highlands, the supposition is sometimes made that the first stage of differentiation took place in low, flat or basinal areas, where the most propitious conditions for Fe removal exist, and that their preeminent position is due to relief inversion as a result of intensive differential erosion.

However, all the bauxite profiles which are developed on iron-rich parent rocks do not show a thick level of lithomarge, favouring the removal of Fe. Thus, we now have to consider the mechanisms of separation of Fe and Al which are able to operate in the superficial layer of the soil profiles, where these two elements have been accumulated together.

The solubility diagrams of aluminous and ferriferous oxides in relation to pH and Eh (see Norton, 1973) show that for rather low pH (pH $<$ 4), Al is more soluble than Fe, when the environment is oxidizing enough (Eh $>$ 0.4) (see Fig. 8). Therefore, some dissociation of both elements is plausible, Al being relatively leached in the superficial, acid and well-aerated layers of the profiles, before being redeposited in deeper layers or further away laterally, owing to small increases of pH. The complexing action of acid organic compounds may also lead to the same result. The sesquioxides are then likely to be "cheluviated" (see p. 157) and their mobility is increased in the range of pH existing in the soil profiles, but this action is selective; Al is generally recognized as moving further than Fe^{3+} (Duchaufour, 1968). In this way, one could explain the occurrences of bauxite layers, intercalated between a superficial crust relatively rich in Fe and deeper (argillaceous or not) horizons. This type of bauxite profile on silicate rocks with a high content of Fe, which has been described by Harder (1952), Boulangé (1973) and Schellmann (1974), resembles illuviated soil profiles in which selective leaching and reprecipitation processes may give secondary accumulations originated from the eluvial layer.

In the presence of carbonic acid, the dissociation of Fe and Al is also possible: experimental leaching on various rocks, in a bicarbonic, moderately acid environment, show that the Fe becomes relatively soluble while Al remains immobile (Pédro, 1964; Trichet, 1969). Then the differentiation is the reverse of the preceding one, the Fe being likely to accumulate downward or farther away.

Thus, the mechanisms governing the separation of Fe and Al possibly operate either in the lithomarge layer, where the mobility of Fe is increased, or in the superficial well-aerated layers, where Al or Fe may be selectively leached owing to several factors (Eh, pH, organic complexes, CO_2, etc.). In the first case, the bauxite genesis implies two stages of formation, and the resulting bauxites belong to the secondary eluvial type. In the second case, the bauxites must not be considered as strictly residual, since the Al-accumulations likely result from vertical and lateral redistribution of this element, either related or unrelated to the ferrallitic pedogenesis; for the purpose of simplification, these bauxites would be termed "bauxites of the *illuvial–eluvial* type".

Bauxitization and geomorphology. Except for the primary-type bauxite, developed on particularly propitious rocks, the bauxite genesis is not easily explained. It implies a succession of processes corresponding perhaps to different pedological conditions. For a better understanding of this genesis and of the relationships between the different types of deposits, we will attempt to situate them in the context of space and time, that is to say, to define the corresponding geomorphological evolution.

The forest equatorial pedogenesis normally corresponds to a peneplain landscape, with convex hills cut by a dense network of valleys, where the concretionary and, a fortiori, the superficial crust formations are scarce. Lateritic crusts and bauxites occur instead in the form of more or less indurated residual levels topping some reliefs or lying on vast plateaus. Such occurrences imply a complex history, during which weathering and erosion conditions could have changed several times. Different interpretations are plausible and Fig. 6 schematizes the possible history of some ore deposits in West Africa, by taking into account morphoclimatic changes that have occurred in that area since the Tertiary Period (Grandin and Delvigne, 1969; Boulangé, 1973; Grandin, 1973). The different phases are:

(1) Period of tectonic calm (Cretaceous, Tertiary?), equatorial climate with forest vegetation; a thick lateritic blanket developed, exhibiting a relatively superficial accumulation of Fe and Al (primary alumino–ferruginous laterite) in surface layers and a lithomarge in deeper poorly drained zones.

(2) Deeper penetration of weathering and slow erosion, under steady climatic conditions; Fe and Al may be progressively dissociated, either by a preferential leaching of superficial Al or by the mobilization of Fe in deeper hydromorphic layers. The mobilized elements redeposited more deeply and laterally in illuvial lateritic crusts developed on slopes or above the groundwater table.

(3) Occurrence of erosion due to a drier climatic phase: formation of a crust of primary laterite on the massifs, wearing away of the peneplain, and formation of pediplain with vast plane surface ("glacis") rich in lateritic debris (Pliocene, Lower Quarternary).

(4) Reappearance of a humid-forest pedogenesis, resulting in re-deepening of weather-

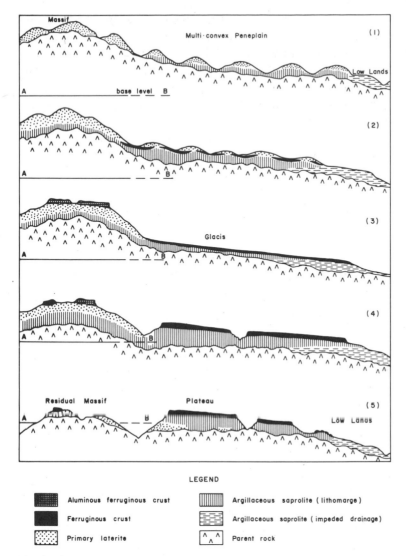

LEGEND

Aluminous ferruginous crust	Argillaceous saprolite (lithomarge)
Ferruginous crust	Argillaceous saprolite (impeded drainage)
Primary laterite	Parent rock

Fig. 6. The possible stages of the formation of West African bauxites and laterites. The stages (1)–(5) are described in the text.

ing profiles and lateritization of the glacis (forming secondary ferrugino—aluminous laterites).

(5) Erosion of tectonic or climatic origin leading to incision, dismantling of residual massifs and breaking up of the lateritized glacis, becoming lateritic crusts of plateaus. The present evolution in West Africa is approximately at this stage.

At the end of the above-mentioned evolution, a complete inversion of the relief may occur; the former massifs together with their primary laterites are eroded, whereas the former peneplain has become a highland covered with secondary lateritic crusts. These are, in turn, likely to be dismantled and destroyed. It is in this way perhaps that bauxite and laterite ore deposits, differently distributed in the landscape and in time, take form, become differentiated, and then are reduced and finally disappear. However long this history may be (in the range of pedological time), it is not endless if considered from a geological viewpoint and all these ore deposits are destined to disappear, unless their residues are recovered through erosion and re-introduced in sedimentary series where they could be entrapped. It is quite normal, therefore, that the stratified bauxites within sedimentary series, generally seem more ancient than lateritic blanket bauxites.

SUPERGENE CONCENTRATION OF MANGANESE[1]

Generalities

Manganese in rocks.

(1) In the decreasing order of element abundance (by wt.%), Mn stands in twelfth place, with a Clarke value of 0.1% or 1000 ppm. It is widely dispersed in the various types of rocks in the earth's crust. The contents in eruptive rocks range between 500 and 2500 ppm (Borchert, 1970). Endogenous concentrations are rare. In sedimentary rocks, where one finds the major accumulations, the contents are more diversified and Green (1953) gives average values of 620 ppm in shales and 300 ppm in limestones, the sandstones containing only some traces. Goldschmidt (1954) indicates contents of 2000–5000 ppm in soils. But the range of variations is greater: some concretionary tropical soils contain up to 5%, while in very acid soils, the Mn is almost totally leached, which may result in deficiencies leading to unfavourable effects on agriculture.

(2) The dispersal of Mn results from the ability of Mn^{2+} to replace other elements. The difference between the ionic radii of Mn^{2+} and Fe^{2+} is only 8%, so substitutions are easy in ferromagnesian minerals (pyroxenes, amphiboles, olivines). Substitutions of six-fold coordinated Al, $3\ Mn^{2+} \leftrightarrow 2\ Al^{3+}$, permit the penetration of manganese in phyllitic minerals (micas, chloritoid, montmorillonite). Substitutions for Mg^{2+} and Ca^{2+} (which have ionic radii differences exceeding 15%) and for Fe^{2+} exist in the garnet and manganesiferous carbonate groups. The trivalent Mn is less common and occurs mainly in the oxisilicates, oxides and hydroxides (braunite, haussmanite, manganite) present in some lodes and in their weathering products. Tetravalent Mn is more commonly found, and occurs mainly in superficial alterites and soils, as oxides and hydroxides (polianite, psilomelane, wad) which are often not well crystallized.

---

[1] For additional information on ancient and recent manganese ore deposits, see Chapter 9 by Roy, Vol. 7, and Chapter 8 by Callender and Bowser, Vol. 7, respectively.

Manganiferous ore deposits.

(1) The distribution (age of formations, geological location and size of ore deposits) of Mn-concentrations is very irregular (Varentsov, 1964; Bouladon, 1970). The Precambrian and Paleogene formations contain more than 85% of the reserves actually known, the remaining part being mainly distributed between the Lower Cambrian, the Carboniferous and the Cretaceous. Most of the ore deposits occur in areas of low tectonic activity and especially in littoral zones of continental shelves, or in subsiding troughs exhibiting submarine volcanism and in confined basins within these troughs. Five ore deposits (Nikopol and Tchiatoura, U.S.S.R.; Kuruman—Kalahari, South Africa; Moanda, Gabon; and Amapa, Brazil) provide nearly two thirds of the world production; Nikopol alone supplying about one third (Fig. 7).

(2) This irregular distribution and the large size of some ore deposits differentiate the Mn-concentrations from those of Fe or Al. The ratio of ore-content in mined deposits to the average rock contents is about 10 for Fe and almost 5 for Al, whereas it is 300—500 for Mn. The Mn ore deposits are generally the result of a series of geological histories. For example, the formation of the Moanda ore deposits is explained by "first a volcanic occurrence which raises the geochemical level in Mn, then a sedimentary concentration

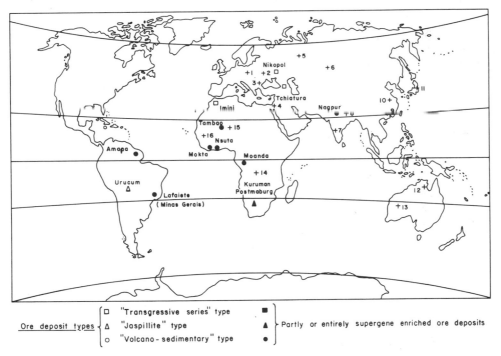

Fig. 7. Mn ore deposits in the world (after Varentsov, 1964, Putzer, 1968, and Bouladon, 1970). Main ore deposits: see the legend. Other ore deposits: *1* = Urkut; *2* = Jacobeni; *3* = Varna; *4* = Timna; *5* = Polunochnoe; *6* = Usink; *7* = Mysore; *8* = Orissa; *9* = Kwangtung; *10* = Kiangsi; *11* = Kotschki; *12* = Groote Eyland; *13* = Peak-Hill; *14* = Kisengue; *15* = Ansongo; *16* = Ziemougoula.

within a carbonate facies and at last, a superficial weathering which enriches and oxidizes the ore" (Weber, 1969). The concentration factors during these three stages are 3, 30–60, and 3–4, respectively.

(3) The main processes of concentration are exogenous, either sedimentary or volcano–sedimentary, and result in primary ores with carbonates and/or oxides. These primary ores may be more or less metamorphosed, with formation of silicates, such as spessartine, garnet and rhodonite. A disparity in the role played by supergene processes appears between the different types of ore deposits: (a) ore deposits of the "transgressive series"-type, are generally directly minable (Nikopol and Tchiatoura, U.S.S.R.; Imini, Morocco); (b) ore deposits of the "gondite"-type (considered in the broad sense) and of the "associated with black carbon schists" type become generally only minable after weathering concentration (Amapa, Brazil; Nsuta, Ghana; Nagpur, East-Indies; Moanda, Gabon); (c) ore deposits of the "jaspillite"-type are intermediate, some being directly minable (Corumba, Brazil) and some others having been superficially enriched (Kuruman–Kalahari, South Africa). With the exception of the large Paleogene ore deposits in the U.S.S.R., about 3/4 of the ore deposits actually mined proved to have required some supergene enrichment (Thienhaus, 1967).

Another type of concentration of minor importance, which will not be taken further into account, is the mixed iron–manganese-type. This ore deposit occurs when the contents of both elements are high in the bed rock undergoing meteoric action, which does not favour an efficient separation. This is the case for ferruginous and manganiferous itabirites in the Minas Gerais State (Van Door et al., 1956). The supergene leaching gives mixed ores, richer in Fe towards the surface than downwards. Another illustration is given by the ore deposits formed through the filling of karst such as the Dr. Geier Mine (Germany), which originated from the weathering of dolomites containing Fe and Mn in a 2/1 ratio (Bottke, 1969). On the other hand, some bog ores favour the concentration of Mn in addition to that of Fe (Ljunggren, 1953; Shterenberg et al., 1969). But this is the limit between the supergene concentration and the sedimentation phenomena.

Geochemical processes.

(1) Like Fe, Mn is a transition element. The properties of these two elements are very similar and particularly so when occurring in the form of bivalent ions, which is the most common one in the rocks. As a result of supergene processes (oxidation, hydratation, etc.) both elements will undergo chemical changes and will likely be redistributed in alterites and soils. Mn is generally present in much smaller amounts than is Fe and when submitted to the chemical reactions of supergene processes, it often plays a passive role only (Michard, 1969). Reactions, such as dissolution or precipitation of MnO_2 or $MnO(OH)$, do not extensively modify the pH and Eh of the solutions, compared with the effects induced by similar reactions when Fe is involved. Thus, the changes in Mn state, related to pH and Eh variations, seem subordinated to changes in the Fe state. But when Fe is lacking, Mn can play an active role. In supergene processes, therefore, the Mn

behaviour will be governed either by the influence of more abundant elements, particularly Fe, or by its own chemical properties.

(2) When in aqueous solution, however, Fe and Mn show notable differences in behaviour. If the leaching and transport of the two elements proceed similarly, the Eh at a given pH must be higher for the deposition of Mn^{4+} than for Fe^{3+} (see Fig. 8, after Norton, 1973). As soon as the pH goes below the neutrality point, a redox potential exceeding 0.6 V is necessary to permit the precipitation of MnO_2 from a solution containing 1 ppm of Mn (Fig. 9, after Hem, 1963). In fact, in the presence of oxygen, the redox potential always exceeds this value, and Fe and Mn are totally oxidized; but the Mn-oxidation is much slower than that of Fe and, therefore, it is mainly for kinetic reasons that the Mn can migrate further than the Fe in a reduced form (Michard, personal communication, 1974). Thus, in soils and alterites, the Mn may be mobile in the relatively acid upper parts of the profiles, where the environment is nevertheless oxidizing enough to favour the rapid precipitation of Fe. It may be carried downward and accumulated at the lower part of the profiles, or along the slopes, where the pH is higher.

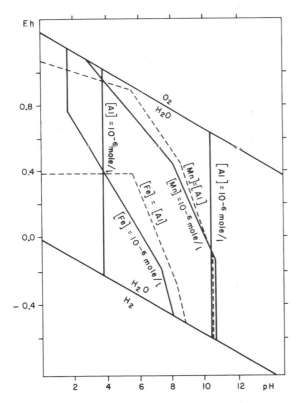

Fig. 8. Eh–pH diagram showing the solubility limits (10^{-6} mole/l) of Al, Fe and Mn, in the systems $Al_2O_3-H_2O$, $Fe_2O_3-H_2O$ and $MnO-H_2O$, at 25°C and 1 atm. The lines of equal solubility of Fe–Al and Mn–Al are dashed. (Modified after Norton, 1973.)

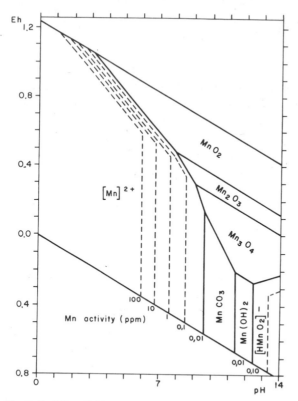

Fig. 9. Stability fields of the Mn-compounds in aqueous solutions with 100 ppm of bicarbonate (after Hem, 1963).

(3) It is also for kinetic reasons that the auto-catalysis phenomena are important in the precipitation of Mn. The oxidation reactions are rendered easier by the presence of propitious solid surfaces: already deposited Mn-oxides constitute such surfaces. When the Mn precipitates alone, it is generally in the form of compact oxide masses (concretionary, stalactitic and massive psilomelanes with a hardness exceeding 6), and redissolution is then very difficult.

Finally, in addition to the chemical processes, the biochemical processes related with the evolution of organic matter in the soil play an important role.[1] Soluble residues of fermentation favour the migration of Mn, in a negative complex-ion form, in environments unfavourable to cation migrations. The oxidation may develop inside the complex so formed; the quantity of protecting matter required being proportional to the valency, the precipitation takes place when complexing matter is present in insufficient quantities after the oxidation process or after its destruction by micro-organisms (Béthremieux, 1951).

---

[1] See Chapter 5 by Saxby in Vol. 2 and Chapter 6 by Trudinger in Vol. 2.

Accumulation of manganese in alterites and soils

Types of accumulations.

(1) Before describing the types of secondary Mn-accumulations, three points must be emphasized: first, important superficial Mn-concentrations are only possible in alterites and tropical soils where the hydrolytic conditions are aggressive enough to permit the release of Mn from primary minerals. Second, the Mn-distribution in the alterites of ultra-basic rocks is not the same as in the alterites of other types of rocks: the former become enriched (if compared with the rocks), while the latter are impoverished (Schellmann, 1969). This fact seems to be due to higher pH conditions toward the base of the profiles on ultra-basic rocks, and consequently the Mn rapidly becomes insoluble and then behaves as a residual element (Trescases, 1973b). In alterites derived from more acid rocks, the Mn is more mobile and, therefore, it tends to be carried away and it also may form secondary concentrations elsewhere. Third, there is no manganiferous concentration of economic interest in alterites and soils derived from rocks in which the Mn-content does not exceed the Clarke value. Assuredly, some secondary Mn-accumulations are observed in some types of tropical soils on common rocks, but then this element is mixed with clays and other elements and its content is much lower than 10% Mn. Thus, in vertisols, the Mn accumulates with Ca and Mg in carbonate nodules, and in ferruginous soils it is concentrated with Fe in the concretionary or crustal horizons.

(2) The most common secondary concentrations of Mn observed on rock rich in manganese are: (a) Mn-oxide crusts. These are deep-lying indurated horizons of ferrallitic soils, formed below the primary mineralization levels, on the lower slopes of ridges typical of the relief encountered in humid-tropical climates. The crusts result from an oblique migration of manganiferous and more or less ferruginous solutions through very thick soils, the top part of which is well-drained and forms an oxidizing and acid environment. The Fe and a part of the Mn are fixed in the upper horizons. The remaining Mn migrates downwards and is deposited at the base of the mottled clays in contact with the relict-structured weathered rock (see below Fig. 10C). The crusts are several metres thick and the wt.% of Mn-oxides is about 80–95. These oxides are chiefly psilomelanes and pyrolusite, with a massive, botryoidal or fibrous fabric. There is no transition between the mottled clays and the crust. From the very first stages of concentration, some cutanes formed of quasi-pure oxides appear. These cutanes become thicker and become anastomosed, embodying some clay pockets which persist in the very rich crust with initial white or red colour. (b) Manganiferous pisolites and nodules. These occur in the upper horizon of the ferrallitic soils. Their Mn-content is variable but very rarely reaches the limit of minability. The Mn is in not well-crystallized oxide and hydroxide forms and is associated with Fe-oxides and clays and sometimes with Al-oxi–hydroxide. The induration of pisolites is low–moderate and the horizons containing them remain loose. In this case, the Mn deposition corresponds to phenomena of coprecipitation with the Fe and of

adsorption on clays. Both pisolite horizon and Mn crust may be observed in the same vertical section.

(3) In some ore deposits formed from tilted mineralized strata, one may also observe accumulations of blocks of manganese ores topping these strata. Sometimes these accumulations form an important part of the minable ores. They become concentrated at the top of ridges during long periods of erosion and during the slow lowering of the topographic surface of the mineralized strata which were formerly oxidized and enriched within the weathering layer. Since the enclosing formations are fairly loose and fluidal argillaceous alterites, the strata could have undergone an intense creeping down and fracturation without having actually been outcropping at any time. This explains why they spread in the form of residual block accumulations lying parallel to the topographic surface, predominantly towards the bottom of the slopes (Fig. 10C). Ulterior recurrences of erosion provoked incisions of some of these accumulations and also mechanical and chemical degradation (partial remobilization of Mn).

One example: genesis of the Mokta ore deposit.

(1) The Mn ore deposit of Mokta in the forest zone of the south of the Ivory Coast (Africa) is a small ore deposit, enriched through supergene concentration. The primary ores, containing garnet, braunite and rhodochrosite, have an Mn-content of 10–20%. They form a series of lenticles interstratified within Precambrian tilted sericite schists. These lenticles dot the convex tops of a chain of hills, a few kilometres from which there occur some relicts of plateaus capped with ferruginous crusts, that are remnants of a crust-covered peneplain system formed at the end of the Tertiary Period. The plateau level is about 20–40 m higher than the hill tops, where thick ferrallitic soils contain reworked debris of ferruginous crusts similar to the plateau crusts. At the foot of the ore-hills and of the nearby plateaus, a system of glacis developed under dry climates during the Quaternary Period. Partly occupied by ferruginous and slightly manganiferous crusts, the glacis have been incised during episodes of contrasting tropical climates, prior to the humid-tropical period which favoured the present-day rain forest.

(2) If the protore is of the Precambrian Period, the formation of the ore deposit (considered in its economic meaning) results from much more recent evolution stages. The sequence of the concentration processes may be as follows (Grandin, 1973): oxidized and enriched ores formed first in the interstratified lenticles under the Tertiary plane surface. The migration of Mn was mainly vertical. The top layers became poorer in Mn while accumulations formed in the deeper layers. A ferruginous crust capped the ore deposit which cannot be distinguished from the surrounding material neither by its relief, nor by the superficial indurated matter (Fig. 10A).

A first period of erosion transformed the surface into an elongated plateau. Then, a long evolution (under a humid-tropical climate), dissected this plateau into a series of low convex ridges where the original ferruginous crust became broken up into blocks and gravels. During this evolution, the elevation was lowered by about 30 m (Fig. 10B). A

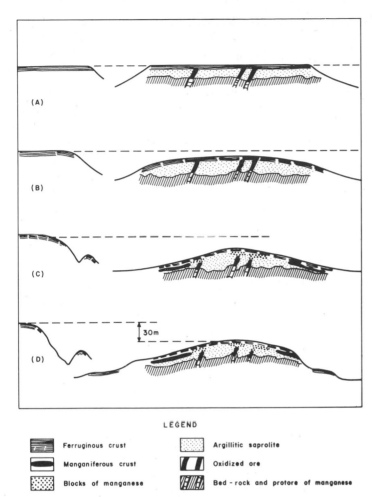

LEGEND

Ferruginous crust		Argillitic saprolite	
Manganiferous crust		Oxidized ore	
Blocks of manganese		Bed - rock and protore of manganese	

Fig. 10. The possible stages of the formation of the Mokta ore deposits (Ivory Coast). (A) Lateritized plateaus, resulting from Pre-Quaternary planation processes; oxidized protore in the weathering layers. (B) Destruction of ferruginous crust under humid climate; deepening of weathering and oxidation processes. (C) Formation of convex hills, under similar climate; formation of block accumulations and Mn-crusts (Lower and Middle Quaternary). (D) Erosion related with drier climatic phase; incision of some residual ore deposits and formation of piedmont glacis.

redistribution of the oxidized ores occurred concomittantly. The enriched lenticular ores, which existed between the ancient plateau and the newly-formed ridges, were dismantled giving: (a) some masses of blocks at the top of the ridges, which cap the oxidized lenticles and correspond to a relative accumulation of residual products; and (b) Mn-crusts under a barren cover, about 6–7 m thick, on the lower slopes of these ridges (Fig. 10C). These crusts result from the addition of Mn leached from the uppermost parts of the ore deposit, and particularly from the masses of blocks. The migration of Mn through solution was mainly oblique.

During the processes of redistribution, losses of Mn occur through chemical and mechanical erosion (losses restricted by the low relief). But these processes initiated a new type of ore, the crust-ore, which by its very high-grade (50–55%) contributes to increasing the average ore content of the oxidized deposits. At this stage, the balance proves positive. The supergene ores have reached their maximum development. Accumulations of blocks and Mn-crusts constitute about 90% of the minable ore and are located in a superficial stratum 15 m thick.

(3) The period of dissection, which was ended by the formation of the Upper Quaternary glacis, has not been long enough, nor of sufficient intensity, to destroy totally the low hills inherited from the previous evolution. The relative elevation of these hills has even been increased by the dissection, their slopes rendered steeper, and the rims of the Mn-crusts have been broken, creating locally steep cliffs on the slopes (Fig. 10D). In some places, the incisions may have attacked the interstratified mineralizations at the convex summits of the hills and may perhaps have removed some of them. At this stage of the geomorphological evolution, the impoverishment of the ore deposit started. But the resistance of the chain of hills and of the hard ores to the dissection, the low susceptibility of the compact blocks of Mn-oxides to leaching, the slow continuation of the protore oxidation and the dismantling in-situ of the lenticle tops prevented a too rapid degradation during the climatic fluctuations which were typical of the Quaternary.

Application to other ore deposits.

(1) In West Africa (Ivory Coast, Ghana, Haute-Volta), the end of the Tertiary Period appears as a period of particularly intense enrichment for most of the manganese deposits of any importance. The Pliocene peneplain-system and the Quaternary glacis landforms, which are encountered at Mokta, are known in all West Africa. It was during the time separating their respective formation that the major part of the supergene ores was formed. The ratio of superficial reworked or secondary ore to the enriched ore of the interstratified lenticles varies with the nature of the protore and the local morphoclimatic conditions. But most of the existing ore deposits are remnants of the reserves accumulated at the end of the Tertiary Period (Grandin, 1973).

For a given protore, the thickest supergene accumulations are formed where the weathering process has been fairly long and deep, and the subsequent erosion moderate. In the intensively eroded zones, the hills dating from the Pliocene peneplanation period are more deeply dissected, and most of the time contain only poor ore deposits or merely simple ore indications, which are residues of the ancient ore accumulations. In West Africa, the climatic zonality showed several fluctuations since the end of the Tertiary (extension of equatorial conditions toward the north or reversely progression of dry climates toward the south). Ore deposits presently under relatively dry climatic conditions, such as Tambao (Haute-Volta), resulted from the supergene concentration due to the humid phase of the uppermost Tertiary Period. But such ore deposits are rare, because the erosion conditions under subsequent dry climates do not favour their preservation.

There are still some other Mn ore deposits associated with peneplains either of more ancient origin, such as the remnants of the Eocene "African Surface" with its bauxite crust, or more recent in age, such as the Quaternary glacis. But if the ores are of good quality, the tonnage is very low, and, consequently, these occurrences fail to be of economic interest.

(2) Some ore deposits of other tropical regions of Africa, Brazil or the East Indies and especially those of Moanda (Gabon), Amapa and Lafaiete (Brazil), or Nagpur (Ghana) which are enriched by supergene processes, are also associated with ancient peneplains to which the dissection periods gave some relief. These ore deposits contain interstratified ore-levels, secondary formations related to Mn-crusts, pisolitic horizons and sometimes accumulations of blocks (Fermor, 1909; Van Door et al., 1949; Van Door et al., 1956; Putzer, 1968; Weber, 1973).

These ore deposits, like those of West Africa, benefitted from a major enrichment period, following the completion of vast peneplains. But for lack of convincing arguments, the problem of knowing if the enrichment of the large Mn-deposits over the whole tropical zones have been caused in the same uppermost Tertiary Period, has not been solved as yet.

Conclusion

(1) The relative mobility of Mn in a superficial environment places this element between the Fe and the metals of alkaline earths, and explains its possible accumulation: (a) with Fe in ferruginous crusts; and (b) with Ca and Mg in montmorillonite soils of fairly dry climate, where the bases may accumulate. These accumulations occur even on bed rock which is non-enriched in Mn, but they are not of economic interest.

(2) To get the high-grade ore concentrations required by industry, the Mn must accumulate alone, without being associated with Fe or Si. This involves three conditions: (a) a climate sufficiently hydrolysing to favour the release of Mn from primary minerals and particularly from silicates; (b) deep soils, both oxidizing and acid, able to fix the Fe and to permit a Mn-enrichment through vertical or lateral migration; and (c) a protore with a high Mn/Fe ratio and with a Mn content high enough to form oxidized ores resistant to leaching and to prevent the Mn from becoming entirely fixed by co-precipitation with Fe.

Thus, the formation of supergene ores is related to humid-tropical climates and ferrallitic weathering. However, a good enrichment does not imply only an intense weathering but also a relief low enough to reduce to the minimum the subsequent mechanical erosion and removal of the materials. On the other hand, supergene ores are obtained only from primary ores which have an average Mn-content about one hundred times higher than the Clark value. Before the carbonated protores of Nsuta (Ghana) and Moanda (Gabon) were discovered, some attempts were made to explain the formation of such ore deposits through a simple "lateritic" alteration of rocks with a 0.01—1% Mn-content (Varentsov, 1964; Thienhaus, 1967). In fact, the concentration factor of supergene processes due to tropical weathering rarely exceeds 5–6.

(3) Among these processes, three must be distinguished: (a) oxidation of the primary ore lenticles or strata, with leaching of the superficial layers and enrichment of the deep layers; (b) residual accumulations of oxidized ore blocks, during a progressive evolution of the topography, without abrupt mechanical erosion; and (c) absolute chemical accumulation, giving high ore contents in the deep horizons of soils, after oblique migration.

The enrichment processes, however, prove also to have a negative aspect. As a matter of fact, they are accompanied by: (a) a washing away of a part of the Mn in solution through the surface drainage network; (b) a reprecipitation of Mn with Fe, in the form of low-grade pisolites; (c) a dispersal of ore debris during dissection periods (these debris are particularly common among the spreading materials of glacis).

The supergene manganiferous ores, which are located near the surface, are easily destroyed by erosion. They are only transitory accumulations, and the continuation of the supergene processes leads inevitably to the dispersal of Mn and to its introduction into the cycle of sedimentation.

(4) Consequently, the exploration of such deposits must take into account the geomorphological criteria as much as the geological criteria. In the same manner that exploration is restricted to certain rocks, so must it also be to certain forms of relief, resulting from the evolution of ancient peneplains. In dry tropical or sahelian regions, only the biggest protores, forming hills with a solid structure of mineralized strata, resisted recent dissection. The main criteria of exploration are the nature and the size of the primary mineralizations. In humid-tropical regions, and especially in forest regions, the moderate dissection and the durable action of the supergene processes favour the enrichment of protores containing relatively small mineralized lenticles.

In extending exploration from one region to another, the main objective will vary; sometimes it will be the location of structures and rocks which seem the most propitious for a large primary accumulation, and sometimes the location of sites and landforms which are the most likely to have benefitted from an intense supergene enrichment.

When an ore deposit is discovered, a good knowledge of the regional geomorphology (evolution of the relief and genesis of the related weathering formations) is also of very great help to estimate the ore reserves before undertaking expensive systematic drillings. According to the geomorphologic evolution, one must look not only for the oxidized protores, but also for the accumulations of mineralized blocks which cap these protores and are concealed under superficial ferruginous laterites, as well as for the manganiferous crusts lying below thick layers of loose and barren soils.

SUPERGENE CONCENTRATION OF NICKEL

Generalities on the geochemistry of nickel

Abundance in the different rocks. The average concentration of Ni in igneous rocks is 80

TABLE IV

Nickel content in the lithosphere and in the hydrosphere

Endogene zone (a and b)	Ni (ppm)	Exogene zone (a and b)	Ni (ppm)	Supergene zone (a and b)	Ni (ppm)
Granite	0.5– 2	sandstones	2– 8	laterites (c)	180
Gabbro-basalt	50 –100	shales	24–95	laterites on peridotites (d)	10,000
Peridotite	3,000	limestones	0	river waters from peridotite massifs (d)	0.06
				sea waters (a)	0.001– 0.005

After: (a) Green, 1953; (b) Krauskopf, 1967; (c) Rambaud, 1969; (d) Trescases, 1973b.

ppm. This element is very irregularly distributed in the lithosphere and hydrosphere. As shown by Table IV, Ni is well represented only in ultrabasic rocks. The weathering processes concentrate this element, which becomes a major element in tropical soils derived from peridotites, for example.

Mineralogical forms. The Ni may occur in rocks as microcrystals of Ni-sulfides (pentlandite, millerite, etc.) which generally form inclusions in ferromagnesian silicates or ferrous sulfides (pyrrhotite). It can also be found in the form of alloys of native metals (awaruite; Fe, Ni_2). But the Ni is also camouflaged through substitution for ferrous Fe and Mg in silicates (particularly peridote) or oxide lattices (magnetite)

In the supergene residual formations (saprolites and soils), the Ni is more or less substituted for Mg occurring in nickeliferous silicates, namely the garnierites which are a mixture of serpentine (Ni, Mg) and hydrated talc (Ni, Mg), often interstratified with saponites or stevensites (Ni, Mg) (Brindley et al., 1972; Trescases, 1973b). Nickeliferous sepiolite and chlorite are also known. Ni may be fixed between the structural layers or even in the octahedral layer of ferriferous smectites (nontronite). Ni is also entrapped, probably as cryptocrystalline $Ni(OH)_2$ through adsorption, either in the cavities and cleavage planes of hypogene silicates before they become totally destroyed (orthopyroxenes, chlorites and particularly serpentine), or in secondary products (silico–ferric gels, opal and chalcedony, goethite, asbolane).

Chemical behaviour of Ni. Hypogene conditions favour the $Fe^{2+}-Mg^{2+}-Ni^{2+}-CO^{2+}$ association resulting from their similarity in ionic radius. The ionization potentials of these elements are quite different and their association is partially destroyed under supergene conditions. The solubility of the elements is governed, under given temperature and pressure conditions, by the pH and Eh. In the weathering zones with sufficient drainage, the oxidation of Fe^{2+} is very easy, but the oxidation of $Ni^{2+}-Ni^{3+}$ and $Co^{2+}-Co^{3+}$ is not common since it requires high Eh (Garrels and Christ, 1965; Norton, 1973). Mg is only bivalent. The sequence of the successive hydroxide precipitations is: Fe^{3+} (pH 2–3), Fe^{2+} (pH 7), Ni^{2+}–Co^{2+} (pH 8–9), Mg^{2+} (pH 10–11). Thus, under normal conditions (pH 4–9), Ni and Co are a little more mobile than Fe, but appreciably less so than Mg. The standard values of free energies of formation (ΔG_f°) for ferric and Mg-silicates show that under normal temperature and for normal concentrations in natural environment, the silication of Fe^{3+} is easier than that of Mg (Tardy et al., 1974). Similarly, the silication of Ni seems easier than that of Co and of Mg.

Nickel concentration – importance of secondary concentrations

Different types of ore deposits. Two types of minable accumulations may be distinguished (Lombard, 1956; Boldt, 1967), according to their genesis, namely the hypogene type and the supergene type.

(1) The *hypogene ore deposits* are essentially sulphureted masses and are associated with basic rocks (norites); they nearly always formed conjointly with the Precambrian orogens (Routhier, 1963). The ore deposit of Sudbury (Canada) is the biggest representative deposit of this type. Some others occur also in South Africa (Bushveld), Rhodesia (Great Dyke), Australia (Kambalta), the U.S.S.R. (Petsamo, Norilsk), Finland (Kotalahti), etc. The hypogene mechanisms (late crystallization of the sulphureted phase) locally concentrate the Ni disseminated in the noritic magma, by 300–600 times, the Ni-content increasing from 50 to 30,000 ppm. The mineralization often takes place at the contact of norites with a more acid enclosing formation. Ni occurs in association with Cu, Pt and Pd. Some ore deposits, containing sulphides, arsenides, antimonides and related to granitic pluton, are also known; they contain associations of Ni–Co and Ag, Bi, U (Erzgebirge group). The hypogene ores fall beyond the subject matter of the present chapter.

(2) The *supergene ore deposits* are associated with the thick lateritic blankets covering peridotites in intertropical areas. The most important ore deposit is found in New Caledonia. There are some others in South Africa, Australia (Queensland), Borneo, Brazil, Celebes, Guinea, the Philippines, Porto Rico and Venezuela. Ancient nickeliferous laterites also occur in Western U.S.A. (California, Oregon), in the U.S.S.R. (Urals, Ukraine), Greece and Yugoslavia. The weathering processes form economic concentrations of Ni in residual alterites, because, first, the Clarke value of the parent rock is high (2000–3000 ppm of Ni) and, second, the solubility of Si and Mg is high in tropical climates while Fe and Ni are only slightly mobile. Since the total Si and Mg represent 80% of peridotites, the concentration ratio of the Ni due to supergene processes ranges generally from 5 to 10.

Descriptions of alteration deposits.

(1) Deposits of New Caledonia: these deposits (Lacroix, 1942; De Chetelat, 1947; Ammou Chokroum, 1972; Trescases, 1973b) are representative of weathering blanket

deposits. The area covered by peridotites is about 5000 km^2, with almost 4000 km^2 extending continuously in the southern part of the island. These rocks which are supposed to be of the Oligocene Period (Routhier, 1953; Guillon, 1973) are mainly harzburgites (peridote, orthopyroxene and accessorily chromiferous spinel) partially serpentinized during a phase of retromorphism.

The weathering of the peridotites has been going on since at least the Pliocene Period, under a humid and warm climate. The low percentage of slightly mobile elements (Fe, Ni, Al, see Table V) in the parent rock explains the quasi-solubility of peridotites. This property favours the development of a karstic relief (Wirthmann, 1970; Trescases, 1973b) including dolines and poljes, which look like marshy depressions. A tectonic uplift has been affecting New Caledonia since the Pliocene (Dubois et al., 1973) and the remnants of several "erosion surfaces" which are levelled and covered by a thick alterite blanket lie at different levels on the landscape. These flat areas are separated by steep slopes bearing thin soils. It is generally the lowest surface that offers the largest number of marshy poljes.

The weathering profiles of low-gradient slope surfaces (Fig. 11) are often 20—30 m thick. Below the average level of the weathering front, and sometimes in more than 10 m thickness, diaclases, faults and crushed zones are occupied by some secondary quartz and chalcedony, with garnierities at some places. Pseudo-lodes and even real quartzo—nickeliferous box-works, which constitute the richest ores (up to 35% NiO), are formed. At the

TABLE V

Mean physical and chemical characteristics of weathering formations on peridotite in New Caledonia[1,2]

	Plateau soil profiles				Glacis soil profiles		
	rock	coarse saprolite	fine saprolite	gravelly horizon	coarse saprolite	fine saprolite	gravelly horizon
Thickness (m)	–	0.2–6	10–40	2–10	0.1–2	5–20	1–5
Mean density	2.8	1.6	0.9	–	1.6	0.9	–
Fire losses	10.5	13.0	13.7	13.5	13	13	13
SiO$_2$ (%)	38	33	1.2	0.5	33	3.5	2
FeO (%)	5	2	1	0.3	2	1	0.3
Fe$_2$O$_3$ (%)	3.5	17	72	74	18	71	73
Al$_2$O$_3$ (%)	0.4	2	4.5	5	1	4	6
CaO (%)	0.1	<0.1	<0.1	<0.1	<0.1	<0.1	<0.1
MgO (%)	41	29	0.9	0.5	29	1	0.7
Cr$_2$O$_3$ (%)	0.4	0.8	4.0	5.5	0.8	3.8	3.8
MnO$_2$ (%)	0.14	0.29	1.0	0.5	0.5	1.0	0.4
NiO (%)	0.40	2.5	1.0	0.4	2.5	1.0	0.3
CoO (%)	0.02	0.08	0.2	0.07	–	–	–

[1] After Trescases, 1973b.
[2] Localization of the weathering profiles: Kouaoua (plateau profiles) and Prony (glacis profiles).

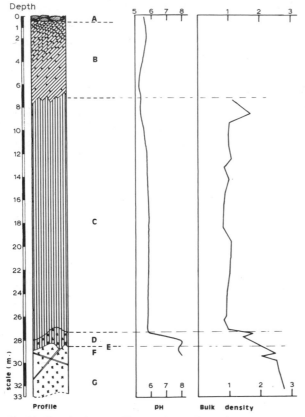

Fig. 11. Weathering profile on ultramafic rock (Kouaoua Plateau, New Caledonia), after Trescases, 1973b. A = ferruginous crust; B = gravelly ferruginous horizon; C = fine saprolite; D = coarse saprolite; E = weathering front; F = quartz and garnierite in diaclases; G = parent rock.

bottom of the weathered blanket the rock is transformed into a plastic mass where numerous rock fragments, ranging in size from some centimeters to some decimeters, still persist and the initial structure is preserved (horizon of coarse saprolite facies). The contact with the fresh rock is extremely irregular (lapiez aspect). In this horizon, some centimeters to some meters thick, the retromorphic serpentine is inherited, and entraps Ni through a physical adsorption; olivine alters into silico—ferric gels (zone of well-drained plateaus) or into nontronite (piedmont zone) and the orthopyroxene is replaced by talc and secondary quartz. The horizon of the coarse saprolite constitutes the *silicated Ni-ore* mined at present.

Over this horizon, a very thick formation occurs (from some meters to some tens of meters), in which the serpentinous framework of the bedrock is replaced by goethite and where the granulometric fractions $<50\ \mu$ represent 80—90% (fine saprolite facies). Si and Mg, which are only partially leached from the coarse saprolite horizon, have almost completely disappeared in the fine saprolite. This increase in intensity of leaching of soluble elements is accompanied by moderate settling. Some silicates neo-formed in the subjacent horizon (quartz, talc) persist in it, and accumulations of asbolane (crypto-crystalline hydroxides of Mn, Co, Ni) may be observed. The horizon of fine saprolite constitutes a lateritic *oxidized Ni-ore,* which is not exploited at present because of its too-low metal content. The profiles are capped by a reworked gravelly ferruginous horizon and sometimes by a ferruginous crust. These last two horizons are considered unproductive. The weathering of chromiferous spinel is very slow, and proceeds along the whole profiles. The average characteristics of these horizons are given in Table V, for a plateau profile and a piedmont profile. The quantitative mineralogical evolution of 100 cm^3 of harzburgite (Kouaoua profile) is given in Fig. 12.

On steep slopes, the profile is not thick and resembles the coarse saprolite horizon occurring at the bottom of the profiles developed in low-gradient zones.

The silicifications and the pseudo-veins of garnierite are much more frequent in sectors of plateaus than in low zones. Also, the Ni-content in the coarse saprolites varies considerably according to the location of the plateau profiles: it is high (up to 4 or 5% NiO) on

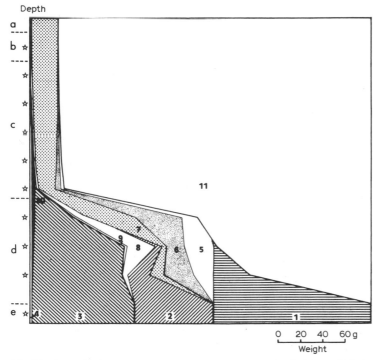

Fig. 12. Quantitative mineralogical evolution of 100 cm^3 (about 300 g) of harzburgite (Kouaoua Plateau, New Caledonia), after Trescases 1973b. *Horizons: a* = ferruginous crust; *b* = gravelly ferruginous horizon; *c* = fine saprolite; *d* = coarse saprolite; *e* = fresh rock. The stars show the emplacements of the samples. The scale is disturbed in depth. *Minerals: 1* = olivine; *2* = orthopyroxene (enstatite); *3* = serpentine; *4* = chromiferous spinel; *5* = silicic gel; *6* = ferric gel; *7* = goethite; *8* = talc; *9* = quartz; *10* = asbolane; *11* = soluble phase (Mg^{2+}, H$_4$SiO$_4$).

the plateau edges, in the breccia zones occurring along the large faults or in the cavities of the rock substratum and low (2—2.5% NiO) in the high parts of the plateaus. In low areas, in contrast, the Ni-content of the coarse saprolite horizons does not differ much from the mean value given in Table V.

(2) Comparison with other weathering ore deposits. The extent of the New Caledonian peridotites is quite exceptional and the other supergene ore deposits known in the world are much smaller. They are generally associated with uplands (remnants of "peneplains") and they can be compared to the New Caledonian ore deposits on plateaus.

The ore deposits of the Western U.S.A., especially that of Nickel Mountain, Oregon (Hotz, 1964), closely resemble those of New Caledonia. Like them, they contain a silicate horizon of coarse saprolite where the Ni is concentrated and quartzo—nickeliferous pseudo-veins embedded in the bedrock. In South Africa (De Waal, 1971), Australia (Zeissink, 1969), and Brazil (Farina, 1969), the parent rock is strongly serpentinized. The silicate horizon of coarse saprolite is thick, but the garnierites are not well represented. All these deposits occur in fairly dry climatic zones. They are called "silicate ore deposits".

In other deposits, in contrast, the silicate horizon of coarse saprolite at the bottom of the profiles is very thin or even absent. This is the case in Cuba (De Vletter, 1955), Guinea (Bonifas, 1959; Percival, 1965), the Philippines (Santos Ynigo, 1964), and Venezuela (Jurkovic, 1963). All these ore deposits occur in very humid regions. The weathering profiles show only ferruginous horizons, similar to those of fine saprolites and to superficial gravelly layers reported in New Caledonian deposits. Sometimes the Ni-contents of 1—2% Ni remain constant in almost the entire profile, but it is more common to find the Ni concentrated in the bottom part, where the content reaches or exceeds 2% Ni. All these ore deposits are named "oxidized ore deposits".

(3) Sedimentary ore deposits. In all supergene ore deposits, formed through weathering processes, the horizons with the highest Ni-content are those altered in-situ where the rock structure is preserved or little modified. In the superficial, pedologically more or less reworked levels, the Ni-content is not really higher than that of the parent rock. Thus, the reworking processes do not seem to favour the concentration of Ni.

The problem of sedimentary processes is still to be discussed. A deposit of oolitic Fe, containing 0.5—3% Ni, occurs at Lokris in Greece; this formation, which is probably of marine origin, lies on Triassic limestones and is covered by Cretaceous limestones (Petraschek, 1953). Nickeliferous laterites developed in neighbouring ultrabasic rocks are likely to have supplied the sedimentation. The Ni-concentration does not seem to have been favoured by this translocation, since the metal contents are generally lower than in weathering ore deposits.

In New Caledonia (Trescases, 1973b), mechanical erosion affects the weathering profiles of the different surfaces, and results in the transportation of these materials toward the low areas, e.g., "intermontane", marshy, depressions (poljes) or littoral mangroves. The Fe, which is transported in the goethite form, combines with the Si dissolved in aquifers to give nontronite. Lenticular levels, very rich in organic matter, are intercalated between the neoformed nontronite. The nontronite deposits contain 2—3% Ni and up to

5% in levels rich in organic matter. In these environments, Ni is essentially supplied in a detrital form, with the goethite. During the transformation process of hydroxide into nontronite, Ni becomes incorporated either into the lattice or between the sheets of the silicate. In addition, the nontronite can fix the small quantities of Ni mobilized by the complexing action of the organic matter. Thus, there is no possibility of concentration during the sedimentation processes except locally through organic matter and the Ni seems generally stabilized in the sedimentary environment. Actually, the silication occurring during the formation of these deposits induces an evolution which is opposed to that produced by weathering and the Ni-content turns out to be lowered.

Conditions of formation of weathering ore deposits

Mineralogical evolution. The secondary minerals derived from the three essential minerals contained in peridotites (namely, peridote, orthopyroxene and serpentine) during weathering in tropical zones, are schematized as follows:

Rock	Coarse saprolite	Fine saprolite
orthopyroxene →	talc + quartz + goethite →	goethite
peridote →	opal + ferric gel (humic climate, rapid drainage) nontronite (dry climate, slow drainage)	goethite
serpentine →	inherited serpentine →	goethite

Moreover, the drier the climate, the slower will be the weathering of the different minerals. Desilication even may fail to reach completion at the surface. On the other hand, on steep slopes, erosion inhibits the completion of the chemical weathering.

Geochemical separations. The weathering of peridotite is characterized by the high ratio of products removed in solution, particularly under humid climates. The Mg leached out is almost totally transported in solution, except for the small amounts entering into talc and garnierite neoformations. Si is also leached, but not so quickly and its removal is not total. The Fe, occurring in a ferrous state in the rock, becomes oxidized into insoluble Fe^{3+}, as soon as the chemical weathering starts. Therefore, the Fe-accumulations in profiles are relative ones. The Si and Fe separation is effective in well-drained zones and under very humid climate, but becomes very poor in low zones or when the climate turns drier.

In addition to Ni, several accessory elements are present in the parent rock: Al, possibly in clinopyroxenes and particularly in chromiferous spinels, which is associated with Mn and Va, and Co which is associated with Ni. All these accessory elements exhibit a residual behaviour during weathering.

The Al, and much more rarely the Cr, contribute to the formation of nontronite, when this process takes place within the profiles. On Al- and Cr-rich parent rocks, minerals of the kaolinite group generally form after the nontronite phase: such minerals may also be chromiferous (Maksimovič and Crnkovic, 1968). If the desilication goes on (humid climates, well-drained high lands) these elements become adsorbed to the goethite. As for Fe, the maximal concentration of Cr and Al takes place in the upper part of the profiles where the contents may exceed 5% for Cr_2O_3 and also 5% for Al_2O_3.

Mn, oxidized into Mn^{3+} and then into Mn^{4+}, and Co partly oxidized into Co^{3+}, are poorly soluble and never undergo silication. They segregate into asbolane concretions. However, these oxides are destroyed near the topographic surface: the most important accumulation (residual) of Mn and Co generally takes place in the middle part of profiles. The contents may be up to 1% CoO and 3% MnO_2.

The Ni is retained, either by the silicates (inherited or neoformed) in the horizon of coarse saprolite (silicate ore deposits), or by Fe- or Mn-hydroxides in the horizon of fine saprolite (oxidized ore deposits), or even by both silicates and hydroxides (New Caledonia, Western U.S.A.).

Thus, the distribution of chemical elements in profiles and landscapes is related to the stability of their host minerals and their own solubility. The Fe—Cr—Al separation is always negligible and that of Fe—Ni generally remains incomplete. Despite the high Fe-content (80% Fe_2O_3), the products resulting from peridotite weathering do not generally constitute a good Fe-ore. The ore deposit of Conakry (Guinea) is an exception: its parent rock is poor in Ni and Cr and the absence of a coarse saprolite horizon at the bottom has probably restricted the retention of Ni within the profile.

Mechanisms of nickeliferous accumulation. The presence of garnieritic pseudo-veins partly embedded in the bedrock once favoured some hydrothermal hypotheses (New Caledonia, Western U.S.A.). But for more than 30 years now, all authors have been in agreement on the point that the genesis of ore deposits associated with the lateritic blankets of peridotites is only governed by the weathering processes.

Most authors who have proposed an explanatory scheme for the concentration mechanism have emphasized the low solubility feature of Ni. Through elimination of Si and Mg, the Ni and Fe become relatively concentrated in the goethitic levels. After this first concentration phase, the Ni is thought to undergo a slow leaching downwards through the profiles, leaching being either vertical (De Vletter, 1955; Schellmann, 1971) or lateral (De Chetelat, 1947; Avias, 1969). This metal would then reprecipitate with Si and Mg, near the bedrock, where the pH is higher. The ore deposits which have a nickeliferous silicate layer at the bottom of the profile would, therefore, be the oldest, and the older they are, the richer they should be (Hotz, 1964).

Nevertheless, Trescases (1973a) notes that in New Caledonia, the Ni-content remains almost constant through the whole thickness of the fine saprolite ferruginous formations.

The leaching of this metal occurs only at the top of the profiles, together with the superficial pedological effects. In addition, the presence of a silicate horizon downward and its Ni-content depend mainly on the morphoclimatic environment. Consequently, Trescases (1973b) put forward another explanation, involving three mechanisms:

(1) If a silicate horizon exists at the bottom of the profile, the first type of trap for the Ni released from peridotes, is constituted by the neoformed or inherited and relatively little weathered silicates. In the first stage, the accumulation is relative. As the weathering front penetrates deeper, these minerals generally become weathered. The released Ni is then subjected to physico–chemical conditions quite different from those which provoked its immobilization; Mg has been leached, the pH becomes more acid and some organic acids may be present. Thus, the Ni tends to migrate down toward the bottom of the profile where it meets again the initial immobilization conditions (primary or secondary silicates still non-weathered). If the weathering goes on long enough, the quantity of Ni accumulated at the bottom of the profile becomes quite considerable. This accumulation follows the descending motion of the weathering front: it is a "compacted" relative accumulation.

(2) The total amount of Ni released by the weathering of the silicates does not migrate downward because one fraction (about 1/3 in New Caledonia) is entrapped in more or less crystallized asbolane and goethite hydroxides. This trap is very effective and immobilizes the Ni in the major part of the oxidized soil profile; only the near-surface horizon becomes impoverished in Ni. It is a true relative accumulation. When a silicate horizon is missing at the base, this mechanism occurs as soon as the weathering starts (oxidized ore deposits).

(3) The third type of accumulation is subsequent to the first two. The weathering balances established through isovolumetric or isochromium considerations with many samples of various soil profiles of New Caledonia (Trescases, 1973b) indicate that some zones of the lateritic blanket have lost Ni, while some others became enriched. The enriched zones are localized wherever groundwater circulations concentrate, e.g., in tectonic zones, at low sites, and along the edges of plateaus. The accumulation is absolute and implies lateral migrations; it can form in inherited silicates (serpentine, chlorite), in the products resulting from the weathering of peridotes (silico–ferric gels, nontronite), in the nontronite neoformed in swamps, and even in true nickeliferous silicates (garnierite) filling the diaclases of bedrock.

The result of these three processes of accumulation is a complex distribution of Ni between the minerals. Table VI and Table VII give the Ni-distribution of a plateau profile and a piedmont profile, in New Caledonia.

Factors controlling the mineralization.

(1) Parent rock. The weathering processes are likely to develop economic concentrations of Ni only on rocks with very high Ni-contents, i.e., on ultrabasic rocks. Among these, the Ni-richest rocks should normally be the most favourable for the formation of

TABLE VI

Nickel distribution in a plateau profile of New Caledonia (Kouaoua)[1]

Horizons	Olivine	Enstatite	Serpentine	Si–Fe gels	Asbolane	Goethite	Σ	(a)
Gravelly	0	0	0	0	0.9	95	–	A
ferruginous	–	–	–	–	3	0.5	–	B
horizon	–	–	–	–	0.03	0.47	0.50	C
	–	–	–	–	6	94	100	D
Fine	0	0	0	0	1.2	94	–	A
saprolite	–	–	–	–	6	0.8	–	B
(top)	–	–	–	–	0.07	0.73	0.80	C
	–	–	–	–	9	91	100	D
Fine	0	0	0	14.3	5.2	71	–	A
saprolite	–	–	–	0.7	9.2	0.8	–	B
(bottom)	–	–	–	0.09	0.48	0.57	1.14	C
	–	–	–	8	42	50	100	D
Coarse	0	0	24	36	1.6	27.	–	A
saprolite	–	–	2.1	0.6	8	0.7	–	B
(top)	–	–	0.51	0.22	0.13	0.19	1.05	C
	–	–	49	21	12	18	100	D
Coarse	0	0	49	26	0.6	2	–	A
saprolite	–	–	1.5	0.7	8	0.6	–	B
(middle)	–	–	0.76	0.18	0.05	0.01	1.0	C
	–	–	76	18	5	1	100	D
Coarse	16	10	40	24	0.4	6.5	–	A
saprolite	0.50	0.10	1.9	0.7	8	0.7	–	B
(bottom)	0.08	0.01	0.77	0.17	0.03	0.05	1.11	C
	7	1	69	15	3	5	100	D
Rock	47	22	29	0	0	0	–	A
	0.50	0.10	0.40	–	–	–	–	B
	0.25	0.02	0.13	–	–	–	0.04	C
	63	5	32	–	–	–	100	D

[1] After Trescases, 1973b.
(a) Results in: A = g mineral for 100 g of sample; B = NiO % in each mineral; C = NiO % of the sample; D = Ni % of the total Ni of the sample.

supergene ore deposits, but such a relationship is only valid for strictly residual and oxidized-type ore deposits.

The role of the parent-rock mineralogy is important, and more especially its degree of hypogene serpentinization. As a matter of fact, the complete weathering of the serpentinites and the total release of the Ni they contain are only possible under the most

TABLE VII

Nickel distribution in a piedmont profile of New Caledonia (Prony)[1]

Horizons	Olivine	Serpentine	Enstatite	Nontronite	Asbolane	Goethite	Σ	(a)
Fine	0	0	0	0	2	94	–	A
saprolite	–	–	–	–	5	1.1	–	B
	–	–	–	–	0.10	1.03	1.13	C
	–	–	–	–	9	91	100	D
Coarse	20	48	9	15	2.5	3	–	A
saprolite	0.40	4.0	0.10	6.8	12	0.7	–	B
(bottom)	0.08	1.92	0.01	1.02	0.27	0.02	3.32	C
	2.4	58	0.3	30.7	8	0.6	100	D
Rock	37	58	4.4	0	0	0	–	A
	0.40	0.35	0.10	–	–	–	–	B
	0.15	0.18	–	–	–	–	0.33	C
	45	55	–	–	–	–	100	D

[1] After Trescases (1973b).
(a) Results in: A = g mineral for 100 g of sample; B = NiO % in each mineral; C = NiO % of the sample; D = Ni % of the total Ni of the sample.

humid-tropical climates. The probability of nickeliferous supergene ore deposits occurring is, therefore, lower for serpentinites than for peridotites which are little or only moderately serpentinized. In contrast, a strongly serpentinized zone in a peridotite massif constitutes an impervious obstacle inhibiting water circulations and favouring a nickeliferous accumulation just above it. Finally, the most favourable conditions correspond to moderately serpentinized environments where the hydrolytic process is relatively slow, and where the presence of a silicate horizon at the bottom of the profiles constitute an effective Ni-trap.

The structural features of peridotites take a fundamental part in the genesis of nickeliferous ore deposits. Weathering proceeds more rapidly in fractured zones. In addition, the tectonic trends constitute preferential drainage-axes and are the seat of the most important absolute accumulations, such as garnierites. Finally, tectonism controls the formation of relief and so indirectly governs the distribution of the Ni (see below).

(2) Climate. Disparities in the behaviour of minerals become more strongly marked when the climate becomes drier and less hydrolyzing. They tend to disappear under equatorial climates. Thus, under the most aggressive climates, all the primary silicates are weathered almost at the same time, and the fringe of fresh serpentine occurring at the bottom of the profiles is very reduced, while the removal of Si is too rapid to permit the neoformation of silicates. The Ni-accumulation can only be residual and oxidized, and the ore deposits are only of low content, inferior to 2% (Guinea, Cuba, etc.).

In a subtropical climate (New Caledonia and California, for example), in contrast, the silicate fringe above the weathering front retains a large part of released — or in transit — Ni and favours its precipitation, and consequently the ore deposits prove to be richer. If the climate is temperate or too dry, the weathering becomes very slow and the amount of Ni leached is too small to permit the development of rich supergene concentrations.

(3) Time. The effect of the time factor is a function of the climate. Under very humid climates, the weathering is rapid and the nickeliferous laterites are thick, except when the landscape is dissected into steep slopes through intense tectonic uplift, as in New Guinea (Davies, 1969). Owing to the intensity of the leaching, however, the Ni-contents tend to become poorer with time, and so the most important supergene concentrations are not related to the oldest alterites.

Under drier climates, the mobility of Ni is lowered and the weathering is less rapid: the formation of large concentrations implies long durations of time. The lack of dense vegetation favours an intense erosion which is likely to destroy the potential ore deposits before they have the opportunity to become of economic value.

(4) Landform and tectonic activity. The accumulation of Ni occurs differently according to the morphological conditions; this is particularly the case in New Caledonia where a very differentiated landform developed owing to the large extent of the peridotite massifs and to the effects of the tectonic activity (Routhier, 1963; Avias, 1969; Trescases, 1973b). This landform is related to an ancient peneplain, with features of the karstic type, which has been uplifted and faulted during the Pliocene and Quaternary. It exhibits the following distinctive units, from the top which is formed by remnants of the peneplain, downwards: steep slopes, glacis and marshy depressions.

On the steep slopes, the mechanical erosion removes the alterites before the Ni is released. The thicknesses of profiles, as well as the Ni-contents, remain very small. However, Ni translocated from the plateaus may contribute to the enrichment of the upper parts of these slopes. In glacis zones the ore deposits are large and the Ni-content is medium (2—3%). In the marshy depressions the Ni-content falls to 1—2% except in peat horizons where it can reach 5%. In faulted zones and along the edges of plateaus (old glacis) incised by erosion, the groundwater circulations are more intense, the oblique leaching of nickeliferous alterites already formed is more effective, and high-grade ores may be formed.

The formation of important supergene Ni-accumulations seems favoured by a tectonic activity, which must be neither too low nor too intense; when the uplift is not intense enough, the leaching and succeeding remobilization of Ni are reduced and the content remains low; when the tectonic activity is too intense, the relief undergoes a strong dissection and the nickeliferous concentrations are rapidly eroded. The most favourable conditions correspond to a moderate tectonic activity, permitting the progressive formation of a peneplain, as in New Caledonia where the uplift seems to be due to the sliding of the Australo—Tasmanian plate under the oceanic plate (Dubois et al., 1973). Thus, the peridotite massifs related to an orogenic arc are normally more propitious to the formation of nickeliferous ore deposits than the peridotites embedded in old basements.

Conclusion

In spite of its rather low Clarke value, nickel may reach important economic concentrations:

(1) Through hypogene mechanisms. Due to its chalcophile property, this element concentrates in the sulphide phase of basic magmas. The local concentration ratios are extremely high, but the ores are restricted to the edges of massifs. These deposits have not been discussed in this contribution.

(2) Through supergene mechanisms. Ni then behaves as a lithophile element. The concentration ratio remains small, and the formation of ore deposits by this process implies a previous hypogene preconcentration; only the rocks derived from the upper mantle (peridotites), which are much richer in nickel than the other igneous rocks, are likely to form supergene ore deposits. On the other hand, such deposits may be very large, since they extend over the whole surface of massifs.

The world production of Ni was about 600,000 metric tons (m.t.) of metal in 1970. About 2/3 of this production came from hypogene ore deposits. The supergene ore deposits of New Caledonia supplied 100,000 m.t. and those of Cuba 35,000 m.t.

The supergene ore deposits seem to be promising reserves for the future. The ore resources in New Caledonia alone would ensure a supply for hundreds or even thousands of years, provided that a metallurgical process appropriate to ferruginous levels ("poor" lateritic ore) could sufficiently reduce the cost price of the metal.

BEHAVIOUR OF CHEMICAL ELEMENTS IN SUPERGENE CONDITIONS

In the previous sections, the geochemical differentiations corresponding to the present-day soil profiles (major elements) and to some residual ore deposits (Al, Fe, Mn and Ni) have been described. The formation of these deposits, which are considered strictly residual, actually implies a rather long and complex history, including pedogenesis, reworking, erosion and eventually continental sedimentation processes.

Consequently, the metallogenic role of the pedogenetic and chemical weathering processes should not be studied without referring to the other supergene (s.l.) processes. In this review, we attempt now to approach the general problem of the behaviour of chemical elements in supergene conditions, during the surface evolution of the landscapes, with special emphasis on the behaviour of trace elements. The knowledge of the supergene mobility of these elements is particularly useful from a theoretical and practical point of view: i.e., understanding of the genesis of supergene ore deposits and determination of the haloes of dispersion of the elements.

The plan will be as follows:

(1) Study of the distribution of chemical elements in minerals and rocks. This feature conditions their geological distribution and their susceptibility to be released into the weathering products and soils.

(2) Empirical determination of element behaviour by field investigations or laboratory tests.

(3) Theoretical study of the successive mechanisms controlling the element mobility: vulnerability of rock- and soil-minerals to chemical weathering, solubility of elements in aqueous solution, ability of the elements to enter into secondary-mineral composition, adsorption of elements on clay and humus surfaces, and fixation by plants and living organisms.

Distribution of elements in minerals

The localization of trace elements in rocks may vary greatly (Goldschmidt, 1954; Goni, 1966; Pédro and Delmas, 1970):

(1) They are often incorporated in the network of the essential minerals, especially silicates; they replace major elements presenting either a similar or different charge, provided that the radii of the ions are not appreciably different. Such elements are termed *camouflaged*. In feldspars, for example, Rb^+, Cs^+, Ba^{2+}, Sr^{2+} and Pb^{2+} may replace K^+, while in ferromagnesian silicates Mn^{2+}, Co^{2+}, Zn^{2+}, Li^+, Cr^{3+} and V^{3+} may replace Fe^{2+}, Fe^{3+}, Mg^{2+}, or Al^{3+}.

(2) They may also be located outside the structures of the essential minerals, and enter into the composition of accessory minerals. Such elements are termed *autonomous*. Thus, S, Cu and chalcophile elements are found in sulphide minerals, B is found in tourmalines, Zr in zircons, Cr in chromites, rare earths in apatites and monazites. These accessory minerals may be well established individually in rocks between the essential minerals, but they also may form inclusions inside them.

(3) The trace elements may be linked to the network surface and retained in an adsorbed form, on the intercrystalline surfaces, inside the physical discontinuities of the crystals or inside the fissures. Such elements are said to be *fissural:* Ni, Co and Mn, for example, may concentrate inside the microfissures in antigorite and Cu, Cr, W and V inside the micro-faults in biotite (Goni, 1966).

This qualitative distribution of elements conditions their behaviour during the chemical weathering processes; the release rate of camouflaged elements depends on the degree of destruction of the containing minerals as opposed to fissural elements which are generally very rapidly released.

From a quantitative viewpoint, some sources of data give the average contents of trace elements in the earth's crust (Ahrens, 1968), but the true contents depend greatly on the nature and the mineralogical composition of the rocks. For example, the sulphur and the sulphur-like elements are particularly concentrated in basic rocks, while Zr, B and rare earths are mainly found in acid crystalline rocks. Concerning the camouflaged elements,

TABLE VIII

Chemical composition of essential minerals[1]

Mineral	X (%)	0.X (%)	0.0X (%)	0.00X (%)	0.000X (%)
Plagioclase	K	Sr	Ba, Rb, Ti, Mn	P, Ga, V, Zn, Ni	Pb, Cu, Li, Cr, Co, B
K-feldspar	Na	Ca, Ba, Sr	Rb, Ti	P, Pb, Li, Ga, Mn	B, Zn, V, Cr, Ni, Co
Quartz				Al, Ti, Fe, Mg, Ca	Na, Ga, Li, Ni, B, Zn, Ge, Mn
Amphibole		Ti, F, K, Mn, Cl, Rb	Zn, Cr, V, Sr, Ni	Ba, Cu, P, Co, Ga, Pb	Li, B
Pyroxene	Al	Ti, Na, Mn, K	Cr, V, Ni, Cl, Sr	P, Cu, Co, Zn, Li, Rb	Ba, Pb, Ga, B
Biotite	Ti, F	Ca, Na, Ba, Mn, Rb	Cl, Zn, V, Cr, Li, Ni	Cu, Sr, Co, P, Pb, Ga	B
Olivine		Ni, Mn	Ca, Al, Cr, Ti, P, Co	Zn, V, Cu, Sc	Rb, B, Ge, Sr, As, Ga, Pb

[1] After Wedepohl (1967).

Table VIII shows that feldspars are enriched in Na, Ca, K, Sr, Ba, Rb, Ga, and Pb, while the amphiboles, pyroxenes and biotites are enriched in Ti, Mn, Fe, Mg, Cr, V, Ni, Cu, Co and Zn. Olivine prefers the concentration of Fe, Mg, Ni, Mn, Cr and Co.

Empirical determination of supergene behaviour of elements

Field investigations. The concept of element mobility during weathering processes was introduced by Smyth (1913) and then by Polynov (1937). These authors proposed a mobility classification, based upon a comparison between the element contents of river waters and of igneous rocks. The classification obtained by Polynov is shown in Table IX.

This concept has been the subject of many studies (Anderson and Hawkes, 1958; Miller, 1961; Udodov and Parilov, 1961; Feth et al., 1964; Tardy, 1969; Cernajev, A.M. and Cernajev, L.E., 1970; and Trescases, 1973b). Except for a few variants, the relative mobility sequences obtained by these authors generally agree with the Polynov sequence.

There is another method, based on the comparison between the element contents in weathering products and in fresh rocks. Results are also very numerous and deal with incipient weathering processes (Dennen and Anderson, 1962; Isnard and De la Roche, 1966; De la Roche et al., 1966) or with well-differentiated alterites (saprolites) and soils under temperate-climate conditions (Butler, 1953, 1954; Dejou, 1959, 1967; Dejou et al., 1972; Harris and Adams, 1966; Lelong and Souchier, 1970) and humid-tropical conditions (Harrison, 1933; Leneuf, 1959, Delvigne, 1965 and Lelong, 1969) or with alterites

TABLE IX

Relative mobility of common elements during weathering processes[1]

Group	Element	Relative mobility
Sesquioxides	Al_2O_3	0.02
	Fe_2O_3	0.04
Silica	SiO_2	0.20
Basic cations	K_2O	1.25
	MgO	1.30
	Na_2O	1.40
	CaO	3.00
Anions	CO_3	57
	Cl	100

[1] After Polynov (1937).

related to different climates (McLaughlin, 1957; Tardy, 1969). Again, the results are similar. The sequence commonly found for incipient weathering and temperate climate weathering is: Ca > Na > Mg > K > Si > Al and Fe; in tropical regions the release rates are appreciably increased, and it is not always possible to classify the most mobile elements such as Ca, Na and Mg.

Investigations giving a classification of the mobilities of trace elements are less numerous, however. Two examples only will be presented here. The first example groups data which have been recently obtained by Tardy (1969) and Trescases (1973b), with a hydrogeochemical method for sharply defined rocks and climates (Fig. 13). The reference elements used for granite and ultra-basic rocks are Na (base 1,000) and Mg (base 100) respectively. Ba and Sr, as well as Na and Ca, obviously appear as the most highly leached elements, whereas Ti, as well as Al and Fe, are the most retained. The other elements range generally in intermediate positions.

The second example is a more general scheme of the migrational ability of elements from siliceous rocks under temperate climate, given in Perel'man (1967), after Polynov (1937), (Table X). The coefficient Kx of migrational ability is:

$$Kx = \frac{mx\ 100}{a\ nx}$$

where mx is the x element's content in water (in mg/l), nx is its content in the rock (in %) and a is the mineral residue contained in the water (in %). The comparison of the oxidizing and reducing environments is very interesting and shows the migrational contrast of some elements, such as U and Zn. Other elements, such as Fe, Mn, Co and Ni also

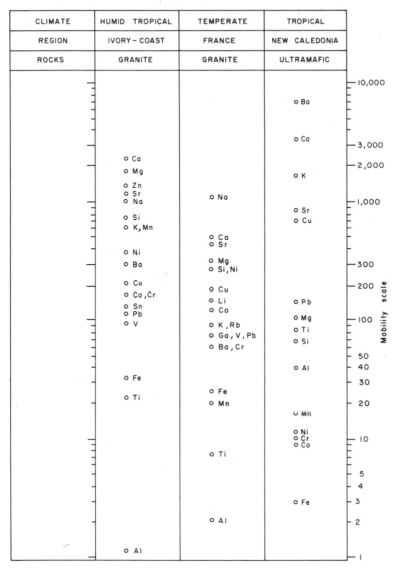

CLIMATE	HUMID TROPICAL	TEMPERATE	TROPICAL
REGION	IVORY – COAST	FRANCE	NEW CALEDONIA
ROCKS	GRANITE	GRANITE	ULTRAMAFIC

Fig. 13. Mobility scale of the elements on granitic and ultramafic rocks, under temperate and tropical climates (after Tardy, 1969; and Trescases, 1973b).

behave differently according to the values of the redox potential, but their strongest mobility corresponds generally to the more reducing environments.

Experimental studies. It appears from the above results that at least three important factors influence the mobility of elements, namely, the climate, the nature of the parent rock and the characteristics, either oxidizing or reducing, of the environment. But these variables are complex and involve additional parameters: temperature, drainage intensity,

TABLE X

Migrational series during the weathering of siliceous rocks in temperate climates[1]

Migrational intensity	Oxidizing environment K_x							Contrast of migration (weak ↔ strong)	Strongly reducing environments K_x						
	1000	100	10	1	0.1	0.01	0.001		1000	100	10	1	0.1	0.01	0.001
Very strong	Cl, I Br, S							←—— Cl, Br, I ——→		Cl, I Br					
Strong			Co, Mg Na, F Sr, Zn U					←— Ca, Mg, Na, F, Sr —→ ; ←— Zn, U			Ca, Mg Na, F Sr				
Medium				Ca, Si, P, Cu Ni, Mn K				←— Si, P, K ; ←— Cu, Ni, Co				Si, P, K			
Weak and very weak					Fe, Al, Ti, Y, Th, Zr, Hf, Nb, Ta, Ru, Rh, Pd, Os, Pt, Sn			←Al, Ti, Zr, Hf, Nb, Ta, Pt, Th, Sn→							Al, Ti, Sc, V, Cu, Ni, Co, Mo, Th, Zr, Hf, Nb, Ta, Ru, Rh, Pd, Os, Zn, U, Pt

[1] In Perel'man, 1967, following Polynov.

acidity, chemico–mineralogical composition and rock texture, for example.

Only experimental studies of chemical weathering performed in a laboratory allow the mechanism of each parameter to be analysed separately. Such studies are numerous, and only the most important ones, carried out under pressure and temperature conditions simulating natural surface conditions, will be mentioned here (Pickering, 1962; Correns, 1963; Wollast, 1961, 1963; Pédro, 1964; Trichet, 1969).

(1) Correns (1963) studied the role played by the pH in the chemical weathering of K-feldspars through successive water-filtration experiments. When the pH is lowered from 6 to 3, the Al becomes relatively more abundant than the Si in the solution, and its mobility is increased. This result is in agreement with the curves representing the water solubility of these two elements as a function of the pH. These curves are well known today and are available in many works (see, e.g., Loughnan, 1969, p. 32, fig. 15 and Perel'man, 1967, p. 40, fig. 11). Most of the metallic elements behave like Al in acid solution, and their mobility must increase rapidly when the pH decreases. However, Correns quotes earlier experiments dealing with some other minerals, which do not always provide similar results.

(2) Pickering (1962) presented leaching tests carried out on silicate rocks free of quartz (latite, andesite and peridotite) either by an immersion or by a filtration method at several pH values falling within the common range of pH and at different temperatures (from 0 to 35°C). The ratio of Si in solution is much higher for glassy rocks than for crystalline rocks. This ratio increases considerably when the pH passes from 7 to 4.8 (hence, the acidity is less marked than in the experiment carried out by Correns) and also when the temperature rises. On the other hand, no significant variation was observed in the ratio of Al and Fe in solution, under these conditions. From this, the author concludes that laterite formation can develop under fairly acid pH conditions.

(3) Pédro (1964) studied some rock fragments with the help of a Soxhlet apparatus designed for continuous leaching, and proceeded to separate investigations of the roles played by the parent rock (granite, basalt and acid lava), by the drainage and temperature conditions, as well as by the pH in relation to the nature of the major weathering agent (pure water pH = 6; sulphuric water pH = 4.5; carbonic water pH = 4; acetic acid pH = 2.5). These investigations produced the following results: (a) Bedrock. The weathering and leaching rates increase in the following order: granite, acid lava, basalt. Hence, they increase together with the basicity and the glassy characteristics of the tested rock. But the composition of the leachate does not change appreciably; in the presence of pure water, bases and silica are preferentially leached, the Fe after being oxidized remains stable, and the Al behaves in an intermediate way. (b) Drainage and temperature. The reduction of drainage limits the leaching intensity of all elements, but the effect is more marked on Al than on Si and bases. Any drop in temperature has the same effect, and below a certain temperature limit, the Fe does not form any more distinctive oxide or hydroxide products, but instead tends to become incorporated into residual argillaceous minerals of the smectite type. (c) pH. A decrease in pH and an increase of the partial

pressure of CO_2 favour the dissolution rate of minerals and increase the solubility limits of most metallic elements (bases, Fe, Al, etc.). Consequently, the relative content of Si in the leachates decreases (the pH is lower than in the experiments carried out by Pickering, which may explain the discrepancy). (d) Chemical weathering agent. The role played by CO_2 has just been noted. In a hydrogen-sulphide solution, the Eh (oxidation–reduction potential) decreases together with the pH and this favours the relative leaching of the Fe and elements such as Mn, Ni and Co, if they are not precipitated in the form of sulphides.

Mechanisms controlling the mobility

The successive stages which influence the element mobility are: weathering of primary minerals, the rate of which is a function of their alterability; dissolution of the elements; formation of secondary-mineral products; adsorption on colloids of soils; and fixation by the vegetation and living organisms.

Susceptibility of the minerals to chemical weathering. Goldich (1938) was the first author to have defined a full stability sequence of the silicate minerals formed in crystalline rocks, in accordance to their susceptibility to superficial weathering. This sequence,

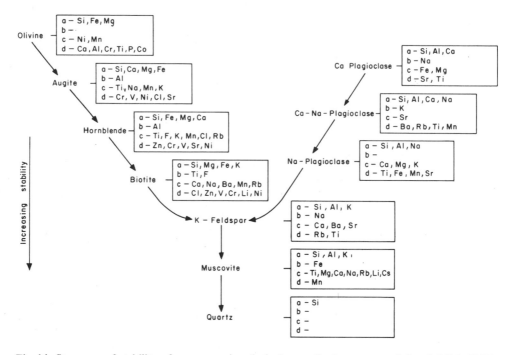

Fig. 14. Sequence of stability of common minerals during weathering processes (after Goldich, 1938). The element content in each mineral is given in order of relative abundance (a = 10x%; b = x%; c = 0.x%; d = 0.0x%). (After data from Wedepohl, 1967 and Deer et al., 1963.)

which has often been confirmed by more recent studies, is presented in Fig. 14, with complementary data about the element contents camouflaged in the minerals.

This sequence may be completed with the results obtained by Pettijohn (1941) by comparing the degree of persistence of minerals in sediments. The anatase, rutile, zircon and tourmaline are placed close to the muscovite. Consequently these minerals are preserved in soils and weathering crusts and they may be accumulated, through erosional processes, in alluvial fans, to form placer deposits. The monazite, garnet, apatite and ilmenite are close to the biotite; the epidote and the sphene are close to the hornblende. Jackson (1968) has also classified the fine minerals found in soils, in relation to their susceptibility to the leaching of bases and silica. The weathering stability is higher for most clay minerals — especially for aluminous chlorites and kaolinite — than for quartz. On the other hand, calcite and, particularly, gypsum are much less resistant than the least-resistant silicates.

In very wet-tropical climates, most of the minerals are rapidly hydrolyzed; only those most resistant persist (quartz, kaolinite, gibbsite, goethite and anatase). In dry climates, almost all minerals persist and in intermediate climates, temperate climates, for example, the weathering stability differences of moderately resistant minerals appear. Several attempts have been made to interpret this differential stability; the causes commonly put forward (Loughnan, 1969) are: (1) disparities in compactness of structure and packing of crystalline networks; the most stable structures are those of tectosilicates, such as quartz, and the least stable are those of neosilicates, such as peridotes; (2) discrepancies in the formation energies of the tetrahedral bondings $Si-O$ and $Al-O$ or of the octahedral bondings $M-O$ and $M-OH$, which are more or less hydrolysis resistant, depending mainly on the electro-negativity of the M elements and on the more or less ionic characteristics of the bondings; highly ionic bondings seem to be less resistant to hydrolysis than the covalent ones.

To these factors, the differences in stabilization degree of ions in the electronic field of crystalline polyhedra (Burns, 1970) may also be added. All these factors explain, at least partly, the variations observed in the dissolving rates of elements according to the structures in which they are present. Hence, Si is less rapidly dissolved from quartz than from feldspar and a fortiori from peridote; the rate reaches the maximum with amorphous Si-gel. Also, in a particular crystalline polyhedron, an element such as Mg with a fairly strong electro-positivity and a predominantly ionic bonding, will be of comparatively lower stability than Fe or Al.

Solubility of elements in aqueous solution. The element mobility depends, in the second place, on their solubility limits which are a function of the pH and, for some elements, of the Eh. The solubility also depends on the presence of other ions or organic complexes (Charlot, 1963; Krauskopf, 1967; Pédro and Delmas, 1970):

(1) In pure aqueous solutions, the solubility limits are determined by the value of solubility products (*SP*) of the hydroxide compounds of the $M(OH)n$ type:

$SP = -\log Kc$

where $Kc = [M^{n+}] [OH^-]^n$

The elements may be classified in several categories (Table XI) according to their ionic potentials ϕ (ratio of the charge to the ionic radius) as defined by Goldschmidt (1954). The following divisions may be proposed: (a) elements forming soluble hydrated or simple cations in the full range of pH; these are alkaline and alkaline-earth elements (ionic potential below 2); (b) elements forming insoluble hydroxides in a certain range of pH; these are the metallic cations Fe^{2+}, Mn^{2+}, Ni^{2+}, Co^{2+}, Pb^{2+}, Zn^{2+}, Cu^{2+}, Cr^{2+}, Cr^{3+}, Al^{3+},

TABLE XI

Solubility products (S.P.) of the hydroxides and of some compounds[1]

Category	Elements	S.P. hydroxides	S.P. carbonates	S.P. sulphurs	S.P. varied compounds	Category	Elements	S.P. hydroxides
A ($\phi < 2$)	Na^+	− 2.9				C ($3 < \phi < 7$)	Be^{2+}	20.8
	Cs^+	− 2.8					Ce^{3+}	22.3
	K^+	− 2.6					Hg^{2+}	25.5 (HgO)
	Li^+	− 1.4					Sc^{3+}	26.3
	Tl^+	0.2		20.3			Cr^{3+}	30
	Ba^{2+}	2.3	8.8	−	10.0		Bi^{3+}	30.4
	Sr^{2+}	3.5	9.6	−	(BaSO$_4$)		Al^{3+}	32.5
	Ca^{2+}	5.3	8.4	−	4.5		In^{3+}	33.2
					(CaSO$_4$)		V^{3+}	34.4
							Ga^{3+}	35
							Fe^{3+}	38
							Ti^{4+}	40
							Th^{4+}	44.7
							Tl^{3+}	45
B ($2 < \phi < 3$)	Ag^+	7.6 (Ag$_2$O)	−	49.2	9.8 (AgCl)		U^{4+}	45
	Mg^{2+}	11.0	5.1	−			Mn^{4+}	56
	Mn^{2+}	12.7	10.2	12.6			Sn^{4+}	56
	Cd^{2+}	13.7	11.3	27.8				
	Ni^{2+}	14.7	6.9	25.4				
	Co^{2+}	14.8	12.8	20.4				
	Fe^{2+}	15.1	10.5	17.2		D ($7 < \phi < 11$)	V^{5+}	14.7 (V$_2$O$_5$)
	Pb^{2+}	15.3	13.1	27.5	8.0		Si^{4+}	8.0 (SiO$_2$)
	Zn^{2+}	17.0	10.8	22	(PbSO$_4$)		Ge^{4+}	2.3 (GeO$_2$)
	La^+	19.0		12.7				
	Cu^{2+}	19.7	9.3	36.1		E ($\phi > 11$)	B^{3+} As^{5+} P^{5+}	< 0

[1] In Pédro and Delmas, 1970 (after data from Charlot, 1963; Krauskopf, 1967: and Perel'man. 1967).

V^{3+}, Fe^{3+}, Ti^{4+}, Th^{4+}, etc., which have an ionic potential between 2 and 7. Most of them are only soluble at acid or very acid pH; some, like Al, are emphoteric and are soluble at both very acid and alkaline pH; (c) elements forming simple or complex anions (hydroxi- or oxi-anions) more or less soluble within a large range of pH; this are Cl^-, Br^-, $(CO_3H)^-$, $(SO_4)^{2-}$, $(NO_3)^-$, $(PO_4)^{3-}$, $(SiO_3)^{2-}$, etc., which have an ionic potential above 7.

(2) In the presence of anions, such as HCO_3^-, HSO_4^-, SO_4^{2-}, HS^-, etc., the preceding values of solubilities are no longer valid. Table XI, established by Pédro and Delmas (1970), shows in each category of elements, different ranges of solubility product, which correspond to different compounds. The solubility products of the sulphides, for example, are very high and this fact explains that, as soon as the solutions contain some traces of HS^-, the immobilization of most metallic elements is almost total. Similarly, in the presence of important quantities of SO_4^{2-} and CO_3^{2-}, such as in strongly mineralized waters of arid areas, the precipitation of the very mobile elements may occur in the form of carbonates and sulphates.

(3) In the presence of complexing agents, such as organic acids, an increase of mobility is noted for slightly soluble or insoluble elements, such as metallic elements belonging to the Fe- or Al-group. This phenomenon of mobilisation by organic compounds, termed chelation or cheluviation (Swindale and Jackson, 1956; Lehman, 1963), favoured the congruent dissolution of primary and secondary silicate minerals[1] (Huang and Keller, 1970, 1971).

From these data, it is clear that the pH has a determinant influence on the limits of solubility of the metallic elements and, consequently, on their mobility, once they have been released by hydrolysis. For elements with variable valencies, the limits of solubility also depend on the oxidation reduction potential· for Fe and Mn, and also, to a lesser degree, for Ni and Co, the solubility increases when Eh decreases. On the other hand, some other elements, such as V, Cr and U, are more soluble in an oxidizing environment than in a reducing one (Garrels and Christ, 1965; Perel'man, 1967; Norton, 1973).

Under natural conditions, the solutions contain the elements as different soluble species. In addition to the free ions (Na^+, K^+, Mg^{2+}, Ca^{2+}, Fe^{2+}, Fe^{3+}, Al^{3+}, etc.), major elements are found in complex forms ($Mg(OH)^+$, $Ca(OH)^+$, $Al(OH)^{2+}$, $Al(OH)_2^+$, $Al(OH)_4^-$ etc.) and that is also frequently the case for the trace elements (e.g., uranyl cations $(UO_2)^{2+}$, $(UO_2)OH^+$, etc.). As the anions are very numerous (OH^-, SO_4^{2-}, SH^-, CO_3^{2-}, CO_3H^-, Cl^- and many complex anions of Si, P, V, As, etc.), the order of precipitation is not easily determined since the exact nature of all possible compounds is not always known, particularly for trace elements. The less mobile elements are those for which the solubility products are reached first, the most mobile elements are those for which the solubility products are reached last.

Mechanical migrations. The problem of the mechanical transport and redeposition of

---

[1] See also Chapter 5 by Saxby, Vol. 2.

heavy or coarse particles by erosional forces is beyond the scope of this study, but the possibility of migration of some substances in a colloidal state must be considered. Perel'man (1967) reports that elements, such as V, Cr, Ni, Be, Ga and Zn, are transported mainly in suspensions or adsorbed on colloidal matter. The common elements of the soils, Al, Fe, Mn and P, which are concentrated in the fine, organic and inorganic fractions (humus, colloids, and clay minerals), are selectively displaced by erosion in the form of suspended matter in the runoff waters. This "selective erosion" is very efficient, even in the soils protected by a dense forest cover, and this explains the presence of "impoverished horizons" (horizon depleted in colloids, Fe, Al, etc.) at the top of the soil profiles (Roose, 1973). Thus, the possibility of distant migrations also exists for the less chemically mobile elements, in landscapes well protected against the erosional forces and a fortiori in arid countries, without plant-cover.

Entrapping within secondary mineral products. The mobility of the elements released and being made soluble by the hydrolysis of primary minerals, also depends on the formation of secondary-crystalline or amorphous mineral products, into which the elements can enter.

The secondary minerals likely to be formed are quite numerous: sulphates, carbonates, phosphates, silicates (clays), oxides and hydroxides. A very large mineralogical variety of salts of metallic — even rare — elements occurs in the superficial chemical weathering zones, where the subsurface solutions are exceptionally rich in various kinds of ions, as in the gossan capping ore bodies. But, apart from this particular and very localized case, the only common minerals produced by chemical weathering, which are characteristic of alterites and soils, are clay minerals, oxides, hydroxides, carbonates and, locally, sulphates.

These common minerals may be formed in two ways (Millot, 1964; Millot et al., 1965):

(1) By "transformation" of pre-existing minerals: clay minerals are frequently formed in this way. They may be produced from primary phyllites which evolved during the hydrolysis process through leaching and exchange of some elements with soil solutions, without destruction of the mineral structure. It is in this way that vermiculite, Al-chlorite, or montmorillonite of certain acid soils, are formed from micas or chlorites derived from crystalline rocks; the process is termed degradation or aggradation, according to whether the subtraction or the addition of elements prevails in the transformed minerals. This evolution is typical of cold—temperate climate soils, where the chemical weathering processes, which are slow and moderate, are nevertheless accompanied by a fairly marked mobility of Fe- and Al-oxides favoured by organic compounds.

(2) By "neoformation" of new mineral species, which crystallize directly from elements in solution or after aging of amorphous precipitates. To each chemical weathering-type there corresponds a certain number of more or less typical neoformed minerals which have, as previously mentioned, a development process predominantly conditioned by climatic conditions:

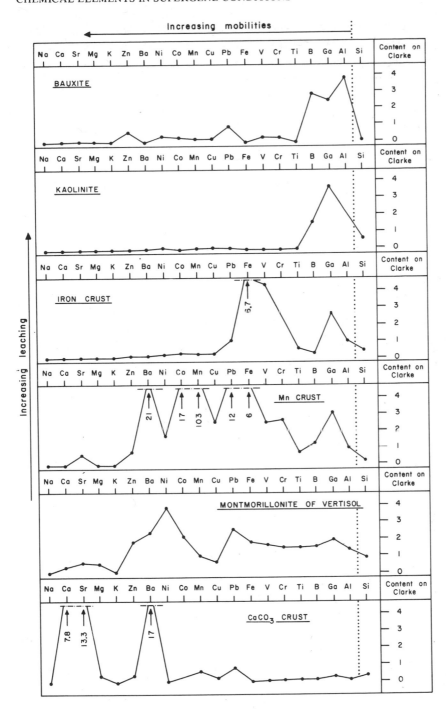

Fig. 15. Trace-element contents in the main weathering facies (after Tardy, 1969).

(a) Under tropical, very humid leaching climates, solutions are always highly diluted. Si is leached as easily as bases, Fe and Al which are not very mobile accumulate relatively, and neoformed minerals, such as gibbsite, $Al(OH)_3$, or boehmite, $AlO(OH)$, and goethite, $FeO(OH)$, are produced.

(b) Under tropical but less intensely leaching conditions, solutions are not so highly diluted. Si is not released so easily, a part of it combines with Al to give 1/1 clays (kaolinite), and the Fe continues to assume the hydroxide or oxide forms.

(c) Tropical but drier conditions favour 2/1 clays, such as montmorillonites of steppe soils or vertisols in which Fe and Mg can enter.

(d) In true arid regions, where the solutions can reach high concentrations through evaporation, siliceous neoformations, calcareous crusts and fibrous clays are mainly found (Millot, 1964; Millot et al., 1969; Dutil, 1971), with salt deposits, such as Ca-sulphates, Na-carbonates, Na-chlorides and silicates (Hardie, 1968; Eugster, 1970; Maglione, 1974). These deposits are observed on the fringes of salt lakes, playas and sebkhas, where the most mobile elements leached from the surrounding landscapes may accumulate.

In actual fact, the neoformed mineral zonality is not sharply defined: gibbsite and kaolinite, for example, co-exist in most ferrallitic soils, kaolinite and 2/1 clays occur in most fersialitic soils, 2/1 clays and carbonate minerals in most steppe soils. Indeed, the climatic zonality is greatly complicated by the action of the other factors of soil formation, such as parent rock and topography.

Development of amorphous-mineral products often bears a relationship with a particular parent rock. So, for example, many amorphous silico–aluminous products (allophanes) are observed in soils derived from volcanic rocks having an unstable glassy structure (Bates, 1962; Sieffermann, 1969). Also, amorphous products rich in Fe, Mn, Co and Ni (limonites, wads, psilomelanes) are found in alterites from ultra-basic rocks. In soils with very acid humus (podzols), favouring an intense Fe and Al cheluviation, amorphous oxides are also more or less linked with organic matter.

Trace elements become selectively concentrated in these different minerals and weathering products, as shown by Fig. 15, established by Tardy (1969). In bauxites and kaolinites, Ga and B are the most retained elements; in ferruginous crusts, they are V and Cr; in manganiferous crusts, Ba, Co, Pb; in vertisol montmorillonites, Ni, Ba, Pb and Zn; and in calcareous crusts, Ba and Sr.

These results confirm those published by some other authors. The elements concentrated in bauxites are often: Ti, Fe, V, Ga, Cu, Zr, B, Cr, Co and Ni (De Chetelat, 1947; Bardossy, 1958; Gordon et al., 1958; Maksimovič, 1968). In kaolinites there are very few trace elements enclosed in the network, except perhaps Ti (McLaughlin, 1957). Sherman, (1952), Lajoinie (1964) and Worthington (1964) report enrichments of Ti, Cr, V, Ni, Co and Mn in ferruginous concretions. Taylor et al., (1964) and Taylor (1968) report concentrations of Ni, Co, Cu, Ba and Li in manganiferous concretions. In vertisol montmorillonites and generally in ferriferous or Mg-phyllosilicates, Ti, V, Ni and Cr are relatively concentrated (Paquet, 1969; Trescases, 1973b).

Comparing the behaviour of U, Cu, Zn and Pb in different weathering processes, Samama (1972) concludes that U and Cu are concentrated under allitization processes (giving bauxite facies), Cu and Zn under monoziallitization (giving 1/1 clay minerals facies) and Zn and Pb under bisiallitization (giving 2/1 clay minerals facies). As and Sb are but slightly mobile and seem relatively enriched in oxidizing environments of lateritic soils and derived siderolitic sediments (Chakrabarti and Solomon, 1971; Guillou, 1971 and personal communication, 1974). In reducing environments, rich in organic matter, high concentrations of Pb, Zn, Cu, Ni, V, Co, Mo and Sn are also to be noted (Perel'man, 1967) and the fixation of U by organic matter is also well known[1] (Barbier, 1974).

The amorphous weathering products may also play an important role in the selective entrapping of some elements: Perruchot (1973) illustrates this by showing that a Si—Mg-gel obtained by an experimental weathering of a peridote is able intensively to concentrate transition elements, such as Mn, Fe, Co, Ni, Zn and particularly Cu.

Adsorption on the colloidal fractions of soils. Ion exchanges between the solutions and the adsorbing organo—mineral complexes of the soils (clay, humus, and amorphous products) interfere in a very complex way with the element mobility (Ermolenko, 1966; Duchaufour, 1970). Most of the adsorbing constituents of the soils are electro-negatively charged. The exchangeable elements are then cations, mainly Ca, Mg, Na, K, Al, and among the trace elements, B, Mn, Zn, Cu and Mo.

In neutral or alkaline soils, the adsorbed elements are mainly basic cations, while in acid soils, Al and H prevail in relation with the respective ionic activity in solution. The selective exchange reactions proceed through dynamic equilibria, following some general, mostly empirical, rules which apply only to very dilute solutions·

(1) All these ions are not adsorbed with the same intensity. The adsorption is less intense for hydrated ions (Na, for example) than for but little hydrated ones.

(2) The adsorption intensity increases generally together with the valency. The order of increasing adsorption is as follows:

$$K^+, Na^+ < Ca^{2+}, Mg^{2+} < Al^{3+}, Th^{4+}$$

(3) For a particular valency, the adsorption increases with the atomic weight of the element and the heavy cations Mn, Cu and Zn tend to be adsorbed more tightly than light cations. Thus, uranyl cation UO^{2+} and perhaps other complex ions of this element can be strongly adsorbed by clays or humic products.

Reversely, positively charged colloids, chiefly hydroxides and oxi-hydroxides of Fe, Al and Mn, are liable to adsorb anions, such as PO_4^{3-}, VO_4^{3-}, ASO_4^{3-}, SO_4^{2-}, (Ermolenko, 1966; Perel'man, 1967). In certain circumstances, the adsorbed elements may form stable compounds with the colloids and even progressively crystallize. Thus, the adsorption processes are an important mechanism of concentration of some elements. Therefore, the

---

[1] See also Chapter 5, by Saxby in Vol. 2.

adsorbed cations and anions may be transported in the surface and subsurface waters, in the form of colloidal suspensions of clays and humus, as seen above (p. 158).

Fixation by plants and living organisms. The role played by living organisms and particularly by plants in the supergene cycle of elements, must also be mentioned.[1] The total mass of organic matter over the globe is considerable and its relative importance compared with the mass of mineral matter in soils is not to be neglected in regions of dense vegetation. Moreover, the living organisms are perpetually renewed. The living matter concentrates mainly C, P, S, N and Ca and plants which accumulate Si, Fe or Al or even trace elements are also known (Lovering, 1959; Ermolenko, 1966). The ability of the plants to adsorb and/or absorb certain elements selectively is well defined by the coefficient of biologic accumulation (Perel'man, 1967). This coefficient is the ratio of the element content in the plant-ash to its content in the soil on which the plant grew. A classification of chemical elements on the basis of the value of this coefficient is given in Table XII (after Perel'man, 1967, p. 9).

The elements concentrated in living matter return to the soil during the decay of this matter, in a form permitting easy mobilization. Thus, there is a perpetual cycling of the elements from the soil to living matter and from living matter to soils (biogeochemical cycle), but the system is not closed; some losses of elements occur by erosion (wind, runoff) or by leaching, and the plants compensate for these losses by taking up the mineral reserve of the soil.

The biological mobilization of elements is one of the basic features of the pedogeneses (Duchaufour, 1968) and also of the supergene geological phenomena (Perel'man, 1967; Erhart, 1973a). It governs the distribution of many elements in the soil profiles, C, N and Ca, especially. The selective action of each plant or vegetal association must be taken into account in the establishment of rules of geochemical exploration. Sometimes, the biological mobilization favours a behaviour of elements entirely independent of the physicochemical laws. Therefore, the genesis of some either residual or sedimentary mineral deposits cannot be explained without referring to it. More essentially, by the continual consumption of the atmospheric CO_2 and by the release of O_2, plant activity is directly responsible for the major changes in the weathering conditions and during sedimentary processes from Precambrian times to the present.

Conclusion

At the end of this discussion, an order of the element mobility during chemical weathering proves difficult to establish because too many and, often contradictory, factors interfere during the successive processes of hydrolysis, solubilization, secondary-mineral formation and exchanges in the soil. Not the least of these is the foremost factor

---

[1] See Chapter 5 by Saxby, in Vol. 2.

TABLE XII

Biologic accumulation series of chemical elements[1]

		Coefficients of biologic accumulation					
		$100n$	$10n$	n	$0.n$	$0.0n$	$0.00n$
Elements of biologic capture	very strong	P, S, Cl					
	strong			Ca, K, Mg, Na, Sr, B, Zn, As, Mo, F			
Elements of biologic accumulation	moderate				Si, Fe, Ba Rb, Cu, Ge, Ni, Co, Li, Y, Cs, Ra, Se, Hg		
	weak					Al, Ti, V, Cr, Pb, Sn, U	
	very weak						Sc, Zr, Nb, Ta, Ru, Rh, Pd, Os, Ir, Pt, Hf, W

[1] After Perel'man (1967).

involved, that is the stability of the primary minerals in which elements are enclosed.

The problem of the relative mobility is complicated by the existence of some threshold effects and the effect of many interactions. For example, below a certain drainage rate, the hydrolysis mobilizes the bases rather than the Si, whereas above this rate, the Si may be preferentially mobilized, so that a slight quantitative variation is sufficient to modify the nature of the processes. The decrease or the increase in solubility of some elements when additional ions are present illustrates well the interaction phenomenon. Another example is afforded by the influence of the development of secondary minerals on the solubility equilibria between the primary minerals and the solution.

Owing to this complexity, only some general laws applicable to the common weathering environments can be offered. The most-mobile elements are those which are rapidly released from unstable primary-minerals and which are later entrapped in secondary minerals. The least-mobile elements are those which are retained a long time in resistant primary-minerals, and which, being less soluble, are consequently rapidly entrapped in secondary minerals. Therefore, however unsatisfacory it may be, only a simple classification of the elements into large mobility categories is possible. The more recognized categories are the four groups proposed by Polynov (1937) or the classes defined by Goldschmidt (1954), mentioned above (p. 150 and p. 156), or still from a naturalist view point the two Erhart (1956) "phases": the migrating soluble phase (alkaline cation and cations of the alkaline earths, combined silica) which is removed during the pedogenesis process and the residual phase (Si from quartz, Al, Fe, Ti) which becomes accumulated in soils. The partial accumulation of combined Si in the clay minerals of most soils corresponds well with the intermediate position of this element between the classes 2 and 3 of Goldschmidt. But the Erhart classification applies principally to the pedogenesis of humid-tropical type.

The most general classification of the migrational ability of elements in the supergene processes has been established by Perel'man (1967), on the basis of the mode of migration (aerial or aqueous), of the intensity of migration and of contrast of the element behaviour. This classification scheme, given in Table XIII, is applicable to the supergene

TABLE XIII

Geochemical classification of elements on the basis of their supergene migration[1]

A. Aerial migrants

A_1. Active (forming chemical compounds)
  O, H, C, N, and I
A_2. Inactive (not forming chemical compounds)
  A, Ne, He, Kr, Xe, and Rn

B. Aqueous migrants

B_1. Very mobile anions, with K_x between $10\,n$ and $100\,n$
  S, Cl, B, and Br
B_2. Mobile, with $K_x = n$
  B_{2a}. Cations: Ca, Na, Mg, Sr, and Ra
  B_{2b}. Anions: F
B_3. Weakly mobile, with $K_x = 0.1n$
  B_{3a}. Cations: K, Ba, Rb, Li, Be, Cs, and Tl
  B_{3b}. Mostly anions: Si, P, Sn, As, Ge, and Sb
B_4. Mobile and weakly mobile in oxidizing media ($K_x = n$ to $0.1n$), and inert in strongly reducing media (K_x less than $0.1n$)
  B_{4a}. High mobility in acid and weakly acid oxidizing waters and low mobility in neutral and alkaline waters (predominantly cations): Zn, Ni, Cu, Pb, Cd, Hg, and Ag
  B_{4b}. Energetic migration in both alkaline and acid waters — more energetic in alkaline than in acid (predominantly anionic): V, U, Mo, Se, and Re
B_5. Mobile and weakly mobile reducing, colloidal media ($K_x = n-0.1n$) and inert in oxidizing media ($K_x = 0.01n$): Fe, Mn, and Co.
B_6. Poorly mobile in most environments ($K_x = 0.1n$ to $0.01n$)
  B_{6a}. Weak migration resulting in formation of chemical compounds: Al, Ti, Zr, Cr, rare earths, Y, Ga, Cb, Th, Sc, Ta, W, In, Bi, and Te
  B_{6b}. Not forming or rarely forming chemical compounds: Os, Pd, Ru, Pt, Au, Rh, and Ir

---

[1] After Perel'man, 1967. The coefficient of aqueous migration K_x is defined on p. 150.

continental zones. In each category, the elements are classed in decreasing order of abundance. In particular cases, important variations in the behaviour of some elements may be observed. Perel'man (p. 153) indicates: "Co is placed in the B5 group together with Fe and Mn, but because of certain peculiarities of its behaviour in the supergene zone, it could be placed together with Cu, Zn and others in the group B4a. Cr, placed in the B6 group of poorly mobile elements, acquires considerable mobility in deserts, and is then analogous to the elements of the B4b group. Some poorly mobile elements like Zr or Y may move in the form of organic complexes. The intensities of their migration increase in tundra, swamps and similar environments. The migration of metals of the B4, B5 and B6 groups intensifies in the sulphatic waters of the oxidation zone of sulphide ore deposits".

The mobility of elements depends also on the scale considered: it might vary according to whether a local evolution restricted to a particular soil profile is considered, or a general evolution over a large area. The mobile elements leached out of a profile may be laterally fixed after a more or less prolonged migration together with drainage waters (lateral geochemical redistribution); these elements, which are mobile on the scale of the soil profile, must be considered rather stable on the scale of the landscapes.

In practice (e.g., during metallogenic exploration projects), the mobility of the chemical elements must be first considered at the scale of large areas. A significant geographical zonation of the weathering facies exists along the earth's surface, in relation to the bioclimatic zonation. Inside each zone, petrogenetic and topographical variations induce more or less marked changes of the weathering type. The main weathering facies are regularly ordered in a sequence between two extreme environments:

(1) The most intensely leached and chemically very acid environments (evolving through "subtractions" of mineral matter, Millot, 1964), which are located in the top layers of soil profiles in equatorial regions, where the least mobile elements, Al and Ga, concentrate.

(2) The most confined and chemically very basic environments (evolving through "addition"), which are located in the depressions and are temporarily water-logged, in the arid regions, where the more soluble elements Mg, Na, Ca, Rb and Li, etc., accumulate and precipitate with Cl^-, SO_4^{2-}, NO_3^-, etc.

Between the above two extremes, one finds a wide range of weathering facies which correspond to less and less leached and more and more confined environments, where the concentrated elements are of increasing mobility. The different facies are defined in Table XIV from data given by Tardy (1969). This weathering sequence may be observed entirely on the large scale of continents from the equatorial regions to the deserts. It may be partially observed on the limited scale of toposequences (succession of soil types along a slope) where the upper, relatively well-drained-part, shows the most intensively leached facies and where the lower and relatively poorly-drained-part shows the most confined facies.

This general scheme permits of predicting the haloes of dispersion of chemical ele-

TABLE XIV

Partition of chemical elements between the main facies of weathering: oxi—hydroxides, clay minerals, salts and alkaline silicates

Bauxite Al—Ga—B	Iron crust Fe—Ti—V— Cr—Mo	Manganese crust Mn—Fe—Co— Pb—Ba—Sn— Ni—Cu	Calcite crust Ca—Sr—Ba— Zn	Sulfates Ca—Sr—Ba	Sodium carbonates Sodium chlorides Na—Rb—Li
	Kaolinite Al—SiO$_2$—Ga—B		Montmorillonites SiO$_2$—Ti—Al—Fe—Mg—Ni— Zn—Pb—Cu		Zeolites Magadiite Mg-montmorillonite SiO$_2$—Mg—Na—Ca

Humid climate Upstream Top of the profiles	increasing mobilities —————————————————→	Arid climate Downstream Bottom of the profiles

ments according to the characteristics of the landscape and, inside each landscape, according to the topographical conditions. Nevertheless, the lateral geochemical differentiation is often unapparent, because of the superposition of the effects of mechanical erosion, the processes of which are entirely distinct from the chemical migration processes.

REFERENCES

Ahrens, L.H., 1968. *Origin and Distribution of the Elements.* Pergamon, London, 1178 pp.

Amou Chokroum, M., 1972. *Contribution à la Valorisation des Ferralites Nickelifères de Nouvelle-Calédonie. Distribution Minéralogique des Eléments et Etude de leur Comportement au Cours de la Réduction Solide—Gaz des Matériaux.* Thèse Univ. Nancy, I. 170 pp. (multigr.)

Anderson, D.H. and Hawkes, H.E., 1958. Relative mobility of the common elements in weathering of some schist and granite areas. *Geochim. Cosmochim. Acta,* 14: 204—210.

Aubert, H. and Pinta, M., 1971. Les éléments traces dans les sols. *Trav. Doc., O.R.S.T.O.M.,* 11: 103 pp.

Avias, J., 1969. Note sur les facteurs contrôlant la genèse et la destruction des gîtes de nickel de la Nouvelle Calédonie. Importance des facteurs hydrologiques et hydrogéologiques. *C.R. Acad. Sci. Paris,* 268-D: 244—246.

Barbier, M.J., 1974. Continental weathering as a possible origin of vein-type uranium deposits. *Miner. Deposita,* 9: 271—288.

Bardossy, G., 1958. The geochemistry of Hungarian bauxites. *Acta Geol. Acad. Sci., Hung.,* 2: 104—155.

Bates, T.F., 1962. Halloysite and gibbsite formation in Hawaii. *Proc. Nat. Conf. Clays Clay Miner.,* 9: 315—328.

Berthier, P., 1821. Analysé de l'alumine hydratée des Baux, département des Bouches du Rhône. *Ann. Miner.,* 6: 531—534.

Béthremieux, R., 1951. Etude expérimentale de l'évolution du fer et du manganèse dans les sols. *Ann. Agron., Paris,* 3: 193—295.

Bocquier, G., 1971. *Genèse et Evolution de Deux Toposéquences de Sols Tropicaux du Tchad. Interprétation Biogéodynamique.* Thèse Doc. Fac. Sci., Strasb., 364 pp; *Mém. O.R.S.T.O.M.*, 52: 350 pp.

Boldt, J.R., 1967. *The Winning of Nickel.* Longmans Canada Ltd., Toronto, 487 pp.

Bonifas, M., 1959. Contribution à l'étude géochimique de l'altération latérique. *Mém. Serv. Carte Géol. Als. Lorr., Strasb.*, 17: 159 pp.

Bonte, A., 1965. Sur la formation en deux temps des bauxites sur mur calcaire. *C.R. Acad. Sci., Paris,* 260-D: 5076–5078.

Borchert, H., 1970. On the ore-deposition and geochemistry of manganese. *Miner. Deposita,* 5: 300–314.

Bottke, H., 1969. Die Eisen-manganerze der Grube Dr. Geier bei Bingen (Rhein) als Verwitterungs-bildungen des Mangans von Typ Lindener Mark. *Miner. Deposita,* 4: 355–367.

Bouladon, J., 1970. Les principaux types de gisements de manganèse et leur importance économique. *Rev. Ind. Min.,* 1: 1–8.

Boulangé, B., 1973. Influence de la géomorphologie sur la genèse des bauxites latéritiques. *Congr. Int. ICSOBA, 3e, Nice,* pp. 215–221.

Boulet, R., 1974. *Toposéquences de Sols Tropicaux en Haute-Volta: Equilibres Dynamiques et Bio-climats.* Thèse Doc. Univ. L. Pasteur, Strasb., 329 pp.

Boyé, M., 1963. La géologie des plaines basses entre Organabo et le Maroni (Guyane française). *Mém. Carte Geol. Fr., Paris, Impr. Natl.,* 148 pp.

Bracewell, S., 1962. Bauxite, alumina, aluminium. *Overseas Geol. Surv. Miner. Resour. Div., London,* 235 pp.

Brindley, G.W. and Pham Thi Hang, 1972. The hydrous magnesium—nickel silicate minerals (so-called garnierites). *Int. Clay Conf.,* 1: 41–50.

Burns, D.J., 1961. Some chemical aspect of bauxite genesis in Jamaïca. *Econ. Geol.,* 56: 1297–1303.

Burns, R.G., 1970. *Mineralogical Applications of Crystal Field Theory.* Univ. Press, Cambridge, 224 pp.

Butler, J.R., 1953. The geochemistry and mineralogy of rock weathering (1): the Lizard area, Cornwall. *Geochim Cosmochim. Acta,* 4: 157–178.

Butler, J.R., 1954. The geochemistry and mineralogy of rock weathering (2): the Nord marka area, Oslo. *Geochim. Cosmochim. Acta,* 6: 268–281.

Caillère, S. and Pobeguin, T., 1964. Considération sur la génèse des bauxites de la France méridionale. *C.R. Acad. Sci., Paris,* 259-D: 3033–3035.

Cernajev, A.M. and Cernajev, L.E., 1970. Classification hydrochimique des éléments-traces rares et dispersés d'après les particularités de leur migration supergène. *Dokl. Akad. Nauk. SSR,* 195: 460–463 (Transl. SCD, O.R.S.T.O.M., Bondy, France).

Chakrabarti, A.K. and Solomon, P.J., 1971. A geochemical case history of the Rajburi antimony prospect, Thailand. *Cah. Inst. Min. Metall., Spec. Vol.,* 11: 121.

Charlot, G., 1963. *L'Analyse Qualitative et les Réactions en Solution.,* Masson, Paris, 442 pp.

Chatelin, Y., 1972. *Les Sols Ferrallitiques, t. 1: Historique, Developpement des Connaissances et Formation de Concepts Actuels.* Initiation, Doc. Techn., 20, O.R.S.T.O.M., Paris 98 pp.

Cooper, W.G., 1936. The bauxite deposits of the Gold Coast. *Gold Coast Geol. Surv. Bull.,* 7: 33 pp.

Correns, C.W., 1963. Experiments on the decomposition of silicates and discussion of chemical weathering. *Proc. Natl. Conf. Clays Clay Miner, 10th, 1961,* Pergamon, Paris, pp. 443–459.

Curtiss, C.D., 1970. Differences between laterite and podzolic weathering *Geochim. Cosmochim. Acta.,* 34: 1351–1352.

Davies, H.L., 1969. Note on papuan ultra mafic belt mineral prospects, Territory of Papua and New-Guinea. *Commonw. Aust., Bur. Miner. Resour., Geol. Geophys., Canberra-Rec.,* 67: 19 pp.

Davis, S.N., 1964. Silica in streams and ground water. *Am. J. Sci.,* 262: 870–891.

De Chetelat, E., 1947. Genèse et évolution des gîtes de nickel de Nouvelle Calédonie. *Bull. Soc. Géol. Fr., Paris,* 17: 105–160.

Deer, W.A., Howie, R.A. and Zussman, J., 1963. *Rock Forming Minerals.* Longmans, Green and Co., London.

Dejou, J., 1959. *Etude Comparative des Phénomènes d'Altération sur Granite Porphyroïde de Lormes et sur Anatexites à Cordiérite du Morvan du Nord et des Sols qui en Dérivent.* Thèse Doct. Fac. Sci. Clermont-Ferrand.

Dejou, J., 1967. L'altération des granites à deux micas du Massif de la Pierre-qui-vire (Morvan). *Ann. Agron.,* 18: 145–201.

Dejou, J., Guyot, J. and Chaumont, C., 1972. Altération superficielle des diorites dans les régions tempérées humides. Exemples choisis dans le Limousin. *Sci. Geol. Bull., Strasb.,* 25: 259–286.

De Lapparent, J., 1924. Sur la constitution minéralogique des bauxites et des calcaires au contact desquels on les trouve. *C.R. Acad. Sci., Paris,* 178: 181–185.

De la Roche, H., Lelong, F. and Francois, J., 1966. Données géochimiques sur les premiers stades de l'altération dans le massif granitique de Saint-Renan (Finistère). *C.R. Acad. Sci., Paris,* 262-D: 2409–2412.

Delvigne, J., 1965. *Pédogenèse en Zone Tropicale. La Formation des Minéraux Secondaires en Milieu Ferrallitique.* Dunod, Paris, *Mem. O.R.S.T.O.M.,* 13: 177 pp.

Dennen, W.H. and Anderson, H.J., 1962. Chemical changes in incipient rock weathering. *Geol. Soc. Am. Bull.,* 73: 375–383.

De Vletter, D.R., 1955. How Cuban ore was formed. A lesson in laterite genesis. *Eng. Min. J., New York,* 156 (10): 84–87.

De Vletter, D.R., 1963. Genesis of bauxite deposits in Surinam and British Guiana. *Econ. Geol.,* 58: 1002–1007.

De Waal, S.A., 1971. South Africa nickeliferous serpentinites. *Min. Sci. Eng., Pretoria,* 3: 32–45.

D'Hoore, J., 1954. *L'Accumulation des Sesquioxydes libres dans les Sols Tropicaux.* I.N.E.A.C., Sér. Sci., 62: 131 pp.

Dubois, J., Launay, J. and Recy, J., 1973. Les mouvements verticaux en Nouvelle-Calédonie et aux Iles Loyautés et leur interprétation dans l'optique de la tectonique des plaques. *Cah. O.R.S.T.O.M., Ser. Géol., Paris,* 5: 3–24.

Duchaufour, P., 1968. *L'Evolution des Sols. Essai sur la Dynamique des Profils.* Masson, Paris, 94 pp.

Duchaufour, P., 1970. *Précis de Pédologie.* Masson, Paris, 481 pp., 3rd ed.

Dutil, P., 1971. *Contribution à l'Etude des Sols et Paléosols du Sahara.* Thèse Doc. Fac. Sci. Strasb., 346 pp. (multigr.).

Erhart, H., 1935. *Traité de Pédologie, t. 1, Pédologie Générale.* Strasbourg, 260 pp.

Erhart, H., 1956. *La Genèse des Sols en tant que Phénomène Géologique.* Masson, Paris, 83 pp.

Erhart, H., 1969. Sur la genèse des sédiments bauxitiques et ferrifères engendrés par l'altération podzolique, au cours de périodes géologiques successives. *C.R. Acad. Sci., Paris,* 268-D: 2653–2656.

Erhart, H., 1973a. *Itinéraires Géochimiques et Cycle Géologique du Silicium.* Doin, Paris, 224 pp.

Erhart, H., 1973b. *Itinéraires Géochimiques et Cycle de l'Aluminium.* Doin, Paris, 256 pp.

Ermolenko, N.F., 1966. *Trace Elements and Colloids in Soils.* Akad. Nauk, Belorusskoi, SSR. Isr. Program Sci. Transl., Jerus., 1972, 259 pp.

Eugster, A.P., 1970. Chemistry and origin of the brines of lake Magadi, Kenya. *Mineral. Soc., Spec. Pap.,* 3: 213–235.

Farina, M., 1969. Ultrabasitos niqueliferos de catingueira, Paraiba. *Serv. Geol. Econ., Brazil,* 7: 53 pp.

Fermor, L.L., 1909. The manganese ore deposits of India. *Mem. Geol. Surv. India,* 37: 1294 pp.

Fermor, L.L., 1911. What is a laterite? *Geol. Mag.,* 5: 454–462.

Feth, J.H., Roberson, C.E. and Polzer, W.L., 1964. Sources of mineral constituents in water from granite rocks Sierra Nevada, California and Nevada. *U.S. Geol. Surv., Water-Supply Pap.,* 1535-L: 70 pp.

Fritz, B. and Tardy, Y., 1973. Etude du système gibbsite–quartz–kaolinite–gaz carbonique. Application à la genèse des podzols et des bauxites. *Sci. Geol. Bull.,* 6: 339–367.

Ganssen, R., 1965. *Grundsätze der Bodenbildung.* Bibliogr. Inst. Mannh., 135 pp.

Ganssen, R., 1968. Schema der Bödern und Bodennutzwung in typischen Bildungs Raumen der Erde. *Mitt. Dtsch. Bodenk. Ges.,* 8: 293–298.

Gardner, L.L., 1970. A chemical model for the origin of gibbsite from kaolinite. *Am. Mineral.*, 55: 1380–1389.

Garrels, R.M. and Christ, C.L., 1965. *Solutions, Minerals and Equilibria.* Harper and Row, New York, N.Y., 450 pp.

Goldich, S.S., 1938. A study in rock weathering. *J. Geol.*, 46: 17–23.

Goldschmidt, V.M., 1954. *Geochemistry.* Clarendon, Oxford, 730 pp.

Goni, J., 1966. Contribution à l'étude de la localisation et de la distribution des éléments en trace dans les minéraux et les roches granitiques. *Mém. B.R.G.M.*, 45: 68 pp.

Gordon, M., Tracey, J.I. and Ellis, M.W., 1958. Geology of the Arkansas bauxite region. *U.S. Geol. Surv., Prof. Pap.*, 299: 268 pp.

Grandin, G., 1973. *Aplanissements Cuirassés et Enrichissement des Gisements de Manganèse dans quelques Régions d'Afrique de l'Ouest.* Thèse Doc. Univ. L. Pasteur. Strasb., 410 pp. (multigr.). Mem. O.R.S.T.O.M., Paris, in press.

Grandin, G. and Delvigne, J., 1969. Les cuirasses de la région birrimienne volcano–sédimentaire de Toumodi: jalons de l'histoire morphologique de la Côte d'Ivoire. *C.R. Acad. Sci., Paris,* 269-D: 1474–1477.

Green, J., 1953. Geochemical table of the elements for 1953. *Bull. Geol. Soc. Am.*, 64: 1001–1012.

Guillemin, C., 1974. Les recources minerales et energetiques vont-elles manquer? *Rev. Palais Decouverte, Num. Spec.,* 3: 100 pp.

Guillon, J.H., 1973. *Les Massifs Péridotitiques de Nouvelle Calédonie. Modèle d'un Appareil Ultrabasique Stratiforme de Chaîne Récente.* Thèse Univ. Paris VI, 130 pp. (multigr.), Mém. O.R.S.T.O.M., Paris, in press.

Guillou, J.J., 1971. Quelques regularités dans la distribution de minéralisations sulfurées (en particulier en antimoine) dans les niveaux carbonatés du Paléozoïque inférieur du Géosynclinal asturien. *Ann. Soc. Geol. Belg.,* 94: 21–37.

Harder, E.C., 1952. Examples of bauxite deposits illustrating variation in origin. *Symp. Problem of Clay and Laterite Genesis.* Am. Inst. Min. Metall. Eng., New York.

Hardie, L.A., 1968. The origin of the recent non-marine evaporit deposits of saline valley, Inyo County, California. *Geochim. Cosmochim. Acta,* 32: 1279–1301.

Harris, R.C. and Adams, J.A.S., 1966. Geochemical and mineralogical studies on the weathering of granitic rocks. *Am. J. Sci.,* 264: 146–173.

Harrison, J.B., 1933. *The Katamorphism of Igneous Rocks under Humid Tropical Conditions.* Imp. Bur. Soil Sci. Rothamted Exp. Stn. Harpenden, 79 pp.

Helgeson, H.C., 1969. Thermodynamics of hydrothermal systems at elevated temperatures and pressures. *Am. J. Sci.,* 267: 729–804.

Hem, J.D., 1963. Chemical equilibria and rate of manganese oxidation. *U.S. Geol. Surv., Water Supply Pap.,* 1667-A: 64 pp.

Hose, H.R., 1963. Jamaïca-type bauxites developed on limestones. *Econ. Geol.,* 58: 62–69.

Hotz, P.E., 1964. Nickeliferous laterites in southwestern Oregon and north-western California. *Econ. Geol.,* 59: 355–396.

Huang, W.H. and Keller, W.D., 1970. Dissolution of rock forming minerals in organic acids : simulated first stage weathering of fresh mineral surfaces. *Am. Mineral.,* 55: 2076–2094.

Huang, W.H. and Keller, W.D., 1971. Dissolution of clay minerals in dilute organic acids at room temperature. *Am. Mineral.,* 56: 1082–1095.

Icole, M., 1973. *Géochimie des Altérations dans les Nappes d'Alluvions du Piémont Occidental Nord-Pyrénéen. Essai de Paléopédologie Quaternaire.* Thèse Doc. Univ. Paris VI, 328 pp. (multigr.).

Isnard, P. and De la Roche, H., 1966. Evaluation statistique du bilan chimique de l'altération naissante dans le granite du Sidobre. *C.R. Acad. Sci., Paris,* 262-D: 2573–2576.

Jackson, M.L., 1968. Weathering pf primary and secondary minerals. *Trans. Int. Congr. Soil Sci., 9th, Adelaïde,* 4: 281–292.

Jenny, H., 1941. *Factors of Soil Formation.* McGraw-Hill, New York, N.Y., 270 pp.

Jurkovic, I., 1963. Some geochemical aspects about the genesis of the nickel deposit Loma de Hierro (Venezuela). *Geol. Vjesn., Zagreb, Yugosl.,* 17: 103–112.

Karpoff, A.M., Bocquier, G., Isnard, P. and Tardy, Y., 1973. Géochimie d'une toposéquence au Tchad. Utilisation des méthodes statistiques. *Sci. Geol. Bull.*, 26 (4): 315–339.

Keller, W.D., 1964. The origin of high alumina clay minerals; a review. *Proc. Natl. Conf. Clays Clay Miner., 12th, 1963,* Pergamon, pp. 129–151.

King, L.C., 1948. On the ages of African land surfaces. *Q. J. Geol. Soc.,* 104: 439–459.

Kovda, V.A., Rozanov, B.G. and Samoylova, Y.E., 1969. Soil map of the world. *Sov. Soil Sci.,* 1: 1–10.

Krauskopf, K.B., 1967. *Introduction to Geochemistry.* McGraw-Hill, New York, N.Y., 721 pp.

Lacroix, A., 1913. Les latérites de la Guinée et les produits d'altération qui leur sont associés. *Nouv. Arch. Mus.,* 5: 255–356.

Lacroix, A., 1934. Les phénomènes d'altération superficielle des roches silicatées alumineuses des pays tropicaux; leurs conséquences au point de vue minier. *Publ. Bur. Etud. Geol. Min. Colon., Paris,* pp. 19–47.

Lacroix, A., 1942. Les péridotites de la Nouvelle Calédonie, leurs serpentines et leurs gîtes de nickel et de cobalt. Les gabbros qui les accompagnent. *Mém. Acad. Sci., Paris,* 66: 1–143.

Lajoinie, J.P., 1964. Etude des latérites du secteur de Tinkoto (Sud-Est Sénégal). Premier bilan des travaux de laboratoire. *Rapp. B.R.G.M., DS A11* (inédit).

Lajoinie, J.P. and Bonifas, M., 1961. Les dolérites du Konkouré et leur altération latéritique. *Bull. B.R.G.M., France,* 2: 1–34.

Lamouroux, M., 1971. *Etude de Sols Formés sur Roches Carbonatées. Pédogenèse Fersiallitiques au Liban.* Thèse Doc. Fac. Sci., Strasb., *Mém. O.R.S.T.O.M.,* 56, Paris, 266 pp.

Lehman, D.S., 1963. Some principles of chelation chemistry. *Proc. Soil Soc. Am.,* 27: 167–170.

Lelong, F., 1969. *Nature et Genèse des Produits d'Altération de Roches Cristallines sous Climat Tropical Humide (Guyane Française).* Thèse Doc. Fac. Sci., Nancy, *Mém. Sci. Terre,* 14, Nancy, 188 pp.

Lelong, F. and Souchier, B., 1970. Bilan d'altération dans la séquence de sols vosgiens, sols bruns acides à podzols, sur granite. *Bull. Serv. Carte Géol. Als.-Lorr.,* 23: 113–143.

Lelong, F. and Souchier, B., 1972. Comparaison de bilans d'altération sur roches granitiques en zone tempérée et en zone équatoriale. *C.R. Acad. Sci., Paris,* 274-D: 1896–1899.

Leneuf, N., 1959. *L'Altération des Granites Calco-Alcalins et des Granodiorites en Côte d'Ivoire Forestière et les Sols qui en sont Dérivés.* Thèse Doc. Fac. Sci. Paris, 210 pp., Publ. O.R.S.T.O.M., 1959, Paris, 210 pp.

Lerz, H. and Borchert, W., 1962. Verwitterung von Mikroklin unter atmosphärischen Temperature–Druck Bedingungen. Ein experimenteller Beitrag Zum Problem der Kaolinisierung. *Chem. Erde,* 22: 386–429.

Ljunggren, P., 1953. Some data concerning the formation of manganiferous and ferriferous bog ores. *Geol. Fören Förhandl.,* 75: 277–297.

Lombard, J., 1956. Sur la géochimie et les gisements du nickel. *Chron. Min. Outre-Mer, Paris,* 244: 20 pp.

Loughnan, F.C., 1969. *Chemical Weathering of the Silicate Minerals.* Elsevier, Amsterdam, 154 pp.

Lovering, T.S., 1959. Significance of accumulator plants in rock weathering. *Geol. Soc. Am. Bull.,* 70: 781–800.

Maglione, G.F., 1974. Un modèle de sédimentation évaporitique continental actuel: le lac Tchad et ses dépendances hydrologiques littorales. *Rev. Geogr. Phys. Geol. Dyn.,* 16: 171–176.

Maksimovič, Z., 1968. Distribution of trace elements in bauxite deposits of Herzegovina, Yugoslavia. *I.C.S.O.B.A., Zaghreb,* 5: 63–70.

Maksimovič, Z. and Crnkovic, B., 1968. Halloysite and kaolinite formed through alteration of ultramafic rocks. *Int. Geol. Congr. Prague, 23rd, Symp. Genesis of the Kaolin Deposits,* 14: 95–105.

Marbut, C.F., 1928. A schema for soil classification. *Proc. 1st Int. Congr. Soil Sci.,* 4: 1–31.

McLaughlin, R.J.W., 1957. Element partition in a kaolinic clay. *Clay Miner. Bull.,* 18: 184–188.

Michard, G., 1969. *Contribution à l'Etude du Comportement du Manganèse dans la Sédimentation Chimique.* Thèse Doc., Univ. Paris, 7: 195 pp.

Michel, P., 1970. *Les Bassins des Fleuves Sénogal et Gambie. Etude Géomorphologique.* Thèse Doc., Univ. Strasb., 3 tomes, 1169 pp. (multigr.).

Millar, C.E., Turk, L.M. and Foth, H.D., 1958. *Fundamentals of Soil Science.* Wiley, New York, N.Y. 3rd ed.

Miller, J.P., 1961. Solutes in small streams draining single rock types, Sangre de Cristo Range, New Mexico. *U.S. Geol. Surv., Water-Supply Pap.,* 1535-F: 23 pp.

Millot, G., 1964. *Géologie des Argiles.* Masson, Paris, 499 pp.

Millot, G., Lucas, J. and Paquet, H., 1965. Evolution géochimique par dégradation et agradation des minéraux argileux dans l'hydrosphère. *Geol. Rundsch.,* 55: 1–20.

Millot, G., Paquet, H. and Ruellan, G., 1969. Néoformation de l'attapulgite dans les sols à carapace calcaire de la Basse-Moulouya (Maroc Oriental). *C.R. Acad. Sci., Paris,* 268-D: 2771–2774.

Moses, J.H. and Michell, W.D., 1963. Bauxite deposits of British Guiana and Surinam in relation to underlying unconsolidated sediments suggesting two-steps origin. *Econ. Geol.* 58: 250–262.

Nalovic, L., 1969. Etude spectrographique des éléments en trace et de leur répartition dans quelques types de sols de Madagascar. *Cah. O.R.S.T.O.M., Pédol., Paris,* 7: 133–181.

Nalovic, L. and Pinta, M., 1971. Recherche sur les éléments en trace dans les sols tropicaux : étude de quelques sols du Cameroun. *Geoderma,* 7: 249–267.

Nicolas, J., 1968. Nouvelles données sur la genèse des bauxites à mur karstique du Sud-Est de la France. *Miner. Deposita,* 3: 18–33.

Nicolas, J., Lecolle, M. and Hieronymus, B., 1967. Précisions sur les modes de passage de la bauxite karstique du Var à ses différents toits et sur les variations de faciès. Interprétations sédimentologiques. *C.R. Acad. Sci., Paris,* 264-D: 240–243.

Norton, S.A., 1973. Laterite and bauxite formation. *Econ. Geol.,* 68: 353–361.

Novikoff, A., 1974. *L'Altération des Roches dans le Massif du Chaillu (République Populaire du Congo). Formation et Evolution des Argiles en Zone Ferrallitique.* Thèse Doc. Univ. L. Pasteur, Strasb., 298 pp.

Okamoto, G., Okura, T. and Goto, K., 1957. Properties of silica in water. *Geochim. Cosmochim. Acta,* 12: 123–132.

Ordway, R.J., 1972. *Earth Sciences.* Van Nostrand/Reinhold, 2nd ed., New York, 788 pp.

Paquet, H., 1969. *Evolution Géochimique des Minéraux Argileux dans les Altérations et les Sols des Climats Mediterranéens et Tropicaux à Saisons Contrastées.* Thèse Doc. Fac. Sci. Strasb., Mém. Serv. Carte Géol. Als. Lorr., 30: 212 pp.

Patterson, S.H., 1967. Bauxite reserves and potential aluminium resources of the world. *Geol. Surv.,* 1228: 176 pp.

Pédro, G., 1964. *Contribution à l'Etude Expérimentale de l'Altération Chimique des Roches Cristallines.* Thèse Doc. Fac. Sci., Paris, 344 pp.

Pédro G., 1968. Distribution des principaux types d'altération chimique à la surface du globe. Présentation d'une esquisse géographique. *Rev. Géogr. Phys. Géol. Dyn.,* 10: 457–470.

Pédro, G. and Delmas, A.B., 1970. Principes géochimiques de la distribution des éléments-traces dans les sols. *Ann. Agron.,* Paris, 21: 483–518.

Pédro, G., Berrier, J. and Tessier, D., 1970. Recherches expérimentales sur l'altération "allitique" des argiles dioctaèdriques de type kaolinite et illite. *Bull. Groupe Fr. Argiles,* 22: 29–50.

Percival, F.G., 1965. The laterite iron deposit of Conakry. *Trans. Inst. Min. Metall.,* 74: 429–462.

Perel'man, A.I., 1967. *Geochemistry of Epigenesis.* Plenum Press, New York, 266 pp.

Perruchot, A., 1973. Sur les propriétés d'échangeurs d'ions de gels p SiO_2, q Mo, r H_2O, où M est un élément alcalino-terreux de transition. *C.R. Acad. Sci., Paris,* 276-D: 2927–2930.

Petraschek, W.E., 1953. Die Eisenerz- und Nickelerz-Lagerstätten von Lokris in Ostgriechenland. *Miner. Wealth Greece, Athènes,* 3: 83–113.

Pettijohn, F.J., 1941. Persistance of heavy minerals and geologic age. *J. Geol.,* 49: 610–625.

Pickering, R.J., 1962. Some leaching experiments on three quartz-free silicate rocks and their contribution to an understanding of lateritization. *Econ. Geol.,* 57: 1185–1206.

Polynov, B.B., 1937. *Cycle of Weathering.* Murby, London, 220 pp.

Ponomareva, V.V., 1966. *Theory of Podzolisation.* Israël Program Sci. Transl., 1969, 309 pp. (Translation from Russian).

Putzer, H., 1968. Die Mangan-Erzlagerstätte Mokta, Elfen beinküste, als Beispiel für lateritische Verwitterungslagerstättern in den Tropen. *Z. Erzbergbau Metallhüttenwes.,* 21: 467–471.

Rambaud, D., 1969. *Etude sur la Répartition des Eléments Traces dans quelques Sols Ferrallitiques.* Thèse Doc. Univ. Paris, O.R.S.T.O.M., Paris, 161 pp. (multigr.).

Roch, E., 1956. Les bauxites de Provence : des poussières fossiles? *C.R. Acad. Sci., Paris,* 242-D: 2847–2849.

Roose, E., 1973. *Dix sept Années de Mesures Expérimentales de l'Erosion et du Ruissellement sur un Sol Ferrallitique Sableux de Basse Côte d'Ivoire. Contribution à l'Etude de l'Erosion Hydrique en Milieu Intertropical.* Thèse Doc. Ing. Univ. Abidjan., 148 pp. (multigr.).

Rousset, C., 1968. *Contribution à l'Etude des Karsts du Sud-Est de la France: Altérations Morphologiques et Minérales.* Thèse Doc., Fac. Sci., Aix-Marseille, 533 pp. (multigr.).

Routhier, P., 1953. *Etude Géologique du Versant Occidental de la Nouvelle Calédonie entre le Col de Boghen et la Pointe d'Arama.* Thèse Doc., *Mém. Soc. Géol. Fr.,* 32 (67): 271 pp.

Routhier, P., 1963. *Les Gisements Métallifères, 1 et 2.* Masson, Paris, 1275 pp.

Sabot, J., 1954. Les latérites. *C.R. Congr. Geol. Int. 19th, Alger, 1952,* 21: 181–192.

Samama, J.C., 1972. Ore deposits and continental weathering : a contribution to the problem of geochemical inheritance of heavy metal contents of basement areas and of sedimentary basins. In: G.L. Amstutz and A.J. Bernard (Editors), *Ores in Sediments. Int. Union Geol. Sci., Ser. A,* Springer, Berlin, pp. 297–265.

Santos Ynigo, L., 1964. Distribution of iron, alumina and silica in the Pujada laterite of Mati, Davao Province, Mindanao Island (Philippines), *Int. Geol. Congr. 22d, New Delhi,* 14: 126–141.

Schellmann, W., 1969. Lateritische Verwitterung und die Anreicherung von Mangan. *Schr. Ges. Dtsch. Metallh. Bergleute,* 22: 137–143.

Schellmann, W., 1971. Uber Beziehungen lateritischer Eisen, Nickel Aluminium und Mangan-Erze zu ihren Ausgangsgesteinen. *Miner. Deposita,* 6: 275–291.

Schellmann, W., 1974. Kriterien für die Bildung, Prospektion und Bewertung lateritischer Silikatbauxite. *Geol. Jahrb.,* D, 7: 3–17.

Sherman, G.D., 1952. The titanium content of Hawaiian soils and its significance. *Soil Sci. Soc. Am. Proc.,* 16: 15–18.

Shterenberg, L.E., Dimitrik, A.L. and Nesterenko, S.P., 1969. Rôle des microorganismes dans la formation des concrétions de fer-manganèse d'après des exemples du Lac de Punnus-Jarvi. *Trad. Isvest. Akad. Nauk., S.S.R., Ser. Geol.,* 1: 97–111.

Sieffermann, G., 1969. *Les Sols de quelques Régions Volcaniques du Cameroun. Variations Pédologiques et Minéralogiques du Milieu Equatorial au Milieu Tropical.* Thèse Doc. Fac. Sci., Strasb., 290 pp.

Smyth, C.H., 1913. The relative solubilities of the chemical constituents of rocks. *J. Geol.,* 21: 105–120.

Souchier, B., 1971. *Evolution des Sols sur Roches Cristallines à l'Etage Montagnard (Vosges).* Thèse Doc. Univ. Nancy, *Mém. Serv. Carte Géol. Als. Lorr.,* 33: 134 pp.

Swindale, L.D. and Jackson, M.L., 1956. Genetic processes in some residual pedzolized soils of New Zealand. *6th Congr. Intern. Sci. Sol, Paris,* E: 233–239.

Tardy, Y., 1969. *Géochimie des Altérations. Etude des Arènes et des Eaux de quelques Massifs Cristallins d'Europe et d'Afrique.* Thèse Doc. Fac. Sci., Strasb., *Mém. Serv. Carte Géol. Als. Lorr.,* 31: 199 pp.

Tardy, Y., Cheverry, Y. and Fritz, B., 1974a. Néoformation d'une argile magnésienne dans les ouadis du lac Tchad. Application aux domaines de stabilité des phyllosilicates alumineux, ferrifères et magnésiens. *C.R. Acad. Sci., Paris,* 278 D: 1999–2002.

Tardy, Y., Trescases, J.J. and Fritz, B., 1974b. Evaluation de l'enthalpie libre de formation de montmorillonites ferrifères. *C.R. Acad. Sci. Paris,* 278 D: 1665–1668.

Targulian, V.O., 1971. *Soil Formation and Weathering in Cold Humid Regions.* Nauka, Moscow, 266 pp.

Taylor, R.M., 1968. The association of trace-elements with manganese and cobalt in soils. Further observation. *J. Soil. Sci.,* 19: 77–80.

Taylor, R.M., McKenzie, R.M. and Norrish, K., 1964. The mineralogy and chemistry of manganese in some Australian soils. *Aust. J. Soil Res.,* 2: 235–248.

Thienhaus, R., 1967. Montangeologische Probleme lateritischer Manganerz-lagerstätten. *Miner. Deposita,* 2: 253–270.

Trescases, J.J., 1973a. Weathering and geochemical behaviour of the elements of ultramafic rocks in New-Caledonia. *Bur. Miner. Resour. Geol. Geophys., Canberra, Bull.,* 141: 149–161.

Trescases, J.J., 1973b. *L'Evolution Géochimique Supergène des Roches Ultra-basiques en Zone Tropicale et la Formation des Gisements Nickelifères de Nouvelle Calédonie.* Thèse Doc., Univ. Strasb., 347 pp. (multigr.), *Mém. O.R.S.T.O.M., Paris,* in press.

Trichet, J., 1969. *Contribution à l'Etude de l'Altération Expérimentale des Verres Volcaniques.* Thèse Doc. Fac. Sci., Paris, 232 pp.

Udodov, P.A. and Parilov, Y.S., 1961. Certain regularities of migration of metals in natural waters. *Geochem. Int.,* 8: 763–776.

Valeton, I., 1965. Fazies-Problem in Südfanzösischen Bauxitlagerstätten. *Beitr. Mineral. Petrogr.,* 11: 217–246.

Valeton, I., 1972. *Bauxites.* Elsevier, Amsterdam, 226 pp.

Van Door, J., Park, C.F. and Glycon de Paiva, 1949. Manganese deposits of the Serra do Novio district, Territory de Amapa, Brazil. *U.S. Geol. Surv. Bull.,* 964-A: 1–51.

Van Door, J., Coelho, I.S. and Horen, A., 1956. The manganese deposits of Minas Gerais, Brazil. *Int. Geol. Congr., 20th, Mex., Symp. Yacim. Mangan.,* 3: 279–346.

Varentsov, I.M., 1964. *Sedimentary Manganese Ores.* Elsevier, Amsterdam, 119 pp.

Watterman, G.C., 1962. Some chemical aspect of bauxite genesis in Jamaïca. *Econ. Geol.,* 57: 829–830.

Weber, F., 1969. *Une Série Précambrienne du Gabon : le Francevillien. Sédimentologie, Géochimie, Relations avec les Gîtes Minéraux Associés.* Thèse Univ. Strasb., *Mém. Carte Géol. Als.-Lorr,* 28: 238 pp.

Weber, F., 1973. Genesis and evolution of the Precambrian sedimentary manganese deposits at Moanda (Gabon). *Kiev Symp. Earth Sci., 1970,* 9: 307–322.

Wedepohl, K.H., 1967. *Géochimie.* Walter de Gruyter, Berlin, 221 pp.

Wirthmann, A., 1970. Zur geomorphologie der Peridotite auf Neukaledonian. *Tübinger. Geogr. Stud., Tübingen.,* 34: 191–201.

Wolfenden, E.B., 1961. Bauxite in Sarawak. *Econ. Geol.,* 56: 972–981.

Wollast, R., 1961. Aspect chimique du mode de formation des bauxites dans le Bas-Congo. *Acad. R. Sci. Outre-Mer, Nouv. Sér.,* 7: 468–489.

Wollast, R., 1963. Aspect chimique du mode de formation des bauxites dans le Bas-Congo. Confrontation des données thermodynamiques et expérimentales. *Bull. Acad. R. Sci. Outre-Mer,* 2: 392–412.

Worthington, I.E., 1964. An exploration program for nickel in the south eastern United States. *Econ. Geol.,* 59: 97–109.

Zans, V.A., 1959. Recent views of the origin of bauxites. *Geonotes,* 1: 123–132.

Zans, V.A., Lemoine, R.C. and Roch, E., 1961. Genèse des bauxites caraïbes. *C.R. Acad. Sci., Paris,* 252-D: 302–304.

Zeissink, H.E., 1969. The mineralogy and geochemistry of a nickeliferous laterite profile (Greenvale, Queensland, Australia). *Miner. Deposita* 4: 132–152.

Chapter 4

KARSTS AND ECONOMIC MINERAL DEPOSITS

PIERO ZUFFARDI

INTRODUCTION

The importance of karsts as loci in which certain natural resources (such as: bauxite, clay, kaolinite, laterite, ochers, sands, placer minerals, phosphate, oxidized ores of Pb, Zn, U, V, Mn, Cu, i.e., all those ores and minerals which are traditionally considered as "exogenic" or "supergene") can accumulate has long been known. It only recently has been recognized that many more natural resources (both metalliferous and non-metalliferous, and even those that were usually classified as "endogenic"/"primary" ores and minerals, such as sulfides, barite, fluorite) can form in karsts and by the same process of karstification. These new concepts derive both from better knowledge of supergene/sedimentary geochemical cycles and from more appropriate investigations of metallogeny bearing on karstic evolution.

It is noteworthy that the karstic cycle was already defined in detail in 1918 by Cvijic, but the metallogenic processes involved were delineated only recently, e.g., by Leleu (1966, 1969), Bernard (1972), Bernard et al. (1972). Increasing attention is now being attached to (paleo)-karsts as favourable environments for ore-mineral deposition by many modern authors (see below: Summary of current literature). A number of deposits, formerly considered as telethermal by metasomatism in carbonate country rocks, have been reinterpreted and definitely proved to be related to karstic formations. It seems probable that further investigations may demonstrate that more deposits, the genesis of which has been described in different ways, should be interpreted as being karstic accumulations.

The modern conceptual model will be dealt with in this chapter and some supporting examples will be discussed. The readers should thereby be aware that this contribution has been presented in the form of a concise review, supported by numerous references, so that it was not possible to go into details of the particular aspects of genesis.

SUMMARY OF CURRENT LITERATURE

Three hypotheses have been expressed about the geneses of many deposits dealt with in this chapter (particularly of sulfide, barite and fluorite), namely:

(1) Igneous-hydrothermal solutions were responsible both for the formation of the caves and for their filling.

This genetic model was supported by almost all early authors so far as sulfide, barite, and fluorite are concerned. The deposits were often classified as "telethermal" or "cryptothermal", when reasonably evident correlations with deep-seated igneous sources were lacking. Selective metasomatic replacement has been assumed to be responsible for their constant connection with soluble country rocks.

(2) The cavities had a karstic origin, but their filling was attributed to igneous-hydrothermal solutions. Again, these deposits were often classified as "telethermal" or "cryptothermal", but the hypothesis of metasomatic replacement was disregarded. This genetic model was stressed in more recent times; e.g., Walker, 1928; Mather, 1947; Hoagland et al., 1965; Callahan, 1965; Lebedev, 1967; Brown, 1967; Crawford and Hoagland, 1968; Morrissey and Whitehead, 1970; Soc. Economic Geologists, 1971.

(3) Karstification (by supergene and/or by artesian waters) is responsible both for the opening and widening of cavities and for the deposition of material in them. This genetic model is accepted in present-day literature; among others by Searls, 1952; Benz, 1964; Leleu, 1966, 1969; Zuffardi, 1968, 1969, 1970; Brusca and Dessau, 1968; Ford, 1969a; Lagny, 1969, 1971; Marcello, 1969; Padalino et al., 1972; Bernard, 1972.

The reasons why the first two hypotheses are less (or not at all) acceptable nowadays are the following: (a) the inconsistency of evidence supporting correlations between these deposits and deep-seated sources; (b) the consistency of evidence supporting correlations between these deposits and particular paleogeographic and paleoclimatic conditions.

As a matter of fact, the possibility of the formation and filling of solution cavities being related to hot ascending waters, more or less connected with a deep-seated source, cannot be denied (at least from a purely theoretical standpoint); but factual observations lead to the conclusion that very many (if not almost all) the major (filled and unfilled) solution cavities can be easily and with reasonable certainty, be explained as having been formed by supergene water (paleo)-circulations, whilst it is rather (sometimes very) difficult to establish convincingly connections with hydrothermalism of any type.

Materials of economic importance occurring in karsts, according to the current literature, are listed in Table I.

GENERAL CHARACTERISTICS OF KARST DEPOSITS

The deposits, which will be dealt with herein, show the following characteristics:

(1) They occur mostly in carbonate country rocks; other (soluble) country rocks (such as gypsum, anhydrite, salts) are far less frequent.

(2) They occur beneath (but not necessarily close to) an unconformity/disconformity surface (U.D.S.). The carbonate country rock may be either cut by the above-mentioned surface or separated from it by a permeable complex.

TABLE I

Economic karst deposits (extended from Quinlan, 1972b)

---

(1) Barite:
   Brusca et al., 1967; Dean and Brobst, 1955; Ford and Sarjeant, 1964; Kesler, 1950; Mather, 1947; Tamburrini and Zuffardi, 1969; Zuffardi, 1970.
(2) Bauxite:
   Anelli, 1958; Bardossy, 1970; Ginzburg, 1952b; ICSOBA, 1970, 1971; Nicolas, 1968; Roch, 1966; Routhier, 1963; Valeton, 1966; Zans, 1959.
(3) Ca-feldspar:
   Lukashev, 1970.
(4) Chalcedony, opal, agate:
   Lukashev, 1970.
(5) Clay, kaolin, and sand:
   Bretz, 1950; Calembert, 1945; Cavaille, 1960; Ford and King, 1969; Ginzburg, 1952b; Keller et al., 1954; Wash and Brown, 1971.
(6) Coal:
   Bretz, 1950; Perna, 1972.
(7) Diamonds and gold:
   du Toit, 1951; Lukashev, 1970.
(8) Fluorite:
   Ford, 1969b; Ginzburg, 1952b.
(9) Gold and antimony:
   Wolfenden, 1965.
(10) Halite:
   Lukashev, 1970.
(11) Iron:
   Bleifuss, 1966; Bretz, 1950; Burchard, 1960; Ginzburg, 1952b; Padalino et al., 1972; Perna, 1972.
(12) Iron, nickel and manganese:
   Ginzburg, 1952b.
(13) Lead, zinc, silver, copper, pyrite (barite, celestite, fluorite):
   Beales and Jackson, 1968; Benz, 1964; Bernard, 1972; Bernard et al., 1972; Brown, 1967; Brusca and Dessau, 1968; Callahan, 1965; Crawford and Hoagland, 1968; Cumming, 1968; Derry, 1969; Ford, 1969a; Ginzburg, 1952b; Gruszcyk, 1967; Hoagland et al., 1965; Lagny, 1969, 1971; Leleu, 1966, 1969; Lukashev, 1970; Marcello, 1969; Mills and Eyrich, 1966; Monaco, 1964; Morrissey and Whitehead, 1970; Padalino et al., 1971, 1972; Poljak, 1952; Searls, 1952; Soc. of Econ. Geologists, 1971; Walker, 1928; Zuffardi, 1968, 1969, 1970.
(14) Manganese:
   Borchert, 1970; Bottke, 1969; Larson, 1970; Perna, 1972; de Villiers, 1960.
(15) Mercury:
   Yates and Thompon, 1959.
(16) Ochers:
   Comel, 1943; Bosellini et al., 1967.
(17) Phosphates:
   Altschuler et al., 1964; Altschuler and Young, 1960; Gèze, 1949; Ginzburg, 1952b; Obelliane, 1963.
(18) Rhodochrosite:
(19) Saltpeter:
   Ginzburg, 1952b.
(20) Siderite:
   Ginzburg, 1952b.

TABLE 1 (continued)

---

(21) Sodium sulphate:
    Grossman, 1968; Hitchon et al., 1969.
(22) Sulphur:
    Davis and Kirkland, 1970; Moore, 1971; Sokolova and Karavaiko, 1968.
(23) Tin:
    Ingham and Bradford, 1960; Materikov, 1961.
(24) Uranium:
    Bowles, 1968; Hart, 1958; Routhier, 1963.
(25) Vanadium:
    Routhier, 1963; Verwoerd, 1957.

---

(3) Their shapes are:

(a) Branching pipes having their long axes more or less orthogonal to the U.D.S., and being controlled by the country-rock fracture pattern (see, for example, Fig. 13).

(b) Lenses and/or groups of pockets, scattered either along the U.D.S. or along and upon a surface of reduced permeability, or a (fossil) water-table (see, for example, Fig. 9).

Hanging- and foot-walls can have very different forms, in this case; namely (see Table II): a *hanging-wall* can be: (i) a regular and/or collapsed bed, coinciding with the lower bed of the complex overlying the U.D.S., or of a (permeable) non-soluble complex underlying it; (ii) a cupoliform irregular surface carved in the same soluble country rock; a *foot-wall* can be: (i) an erosional surface in the permeable country rock; (ii) a regular bedding-surface, coinciding with the hanging-wall of a complex with reduced permeability.

(c) More or less equidimensional bodies, localized along intensely sheared or fractured zones, and/or controlled by a permeable or less permeable strata distribution.

TABLE II

Some shapes of lenticular karstic accumulations (illustrated on p. 179)

---

(1) Regular flat hanging-wall, coinciding with an U.D.S., the foot-wall being an erosional surface carved in a soluble complex. This shape is typical of denuded karsts which have been completely filled before fossilization.

(2) Collapsed hanging-wall (coinciding with and U.D.S. or with a permeable non-soluble complex underlying it); foot-wall is an erosional surface carved in a soluble complex. This shape is typical of mantled karsts.

(3) Both hanging-wall and foot-wall are erosional surfaces, in a permeable complex, with (partial) collapse of hanging-wall. This shape is typical of deep karstification along the permanent circulation zone, open along a water table.

(4) Hanging-wall as in 2; foot-wall is a regular bedding plane. This shape is typical of deep karstification along the permanent circulation zone open upon an impermeable horizon.

(5) Hanging-wall is a cupoliform surface (maybe with irregular lateral extent) carved in a soluble complex; foot-wall as in 2 or 3. This shape represents the end product of hanging-wall collapsing in the permanent circulation zone.

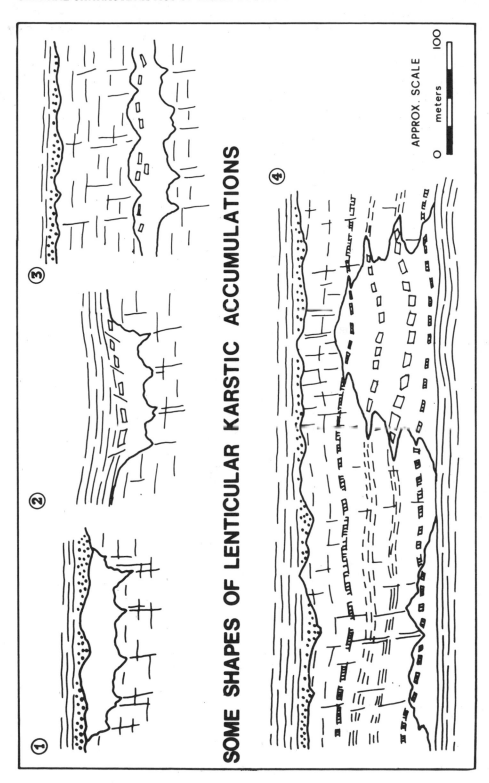

SOME SHAPES OF LENTICULAR KARSTIC ACCUMULATIONS

APPROX. SCALE

meters

0   100

(d) Vein-like bodies or groups of lenses running along fault breccias, tension-gashes or branching veins.

(e) Fine net-works of veinlets, controlled by the fracture pattern of the country rock.

(4) Combinations of the above-mentioned forms are not unusual; an aureole of fine veinlets being quite common around the other types of deposit.

(5) Ore/mineral accumulation textures are:

(a) collapse breccia type (see Fig. 1);

(b) crustified, each crust being radial fibrous, lining the wall sides and growing towards the center of the body (see Fig. 2);

(c) bedded (often: fine bedded): beds lie roughly according to horizontal (paleo)-surfaces; oolitic or pisolitic structures are frequent in this case (see Figs. 3 and 4).

(6) Compositions of the ore mineral accumulations are:

(a) "Autochthonous", that is qualitatively made up of the same components as the country rocks. From the qualitative standpoint, accumulations are richer in insoluble or slightly soluble components and are poorer in soluble ones than is the country rock. Concentration rates are sometimes as high as 100 : 1, and even higher. This process of accumulation is normally described as "by residual concentration".

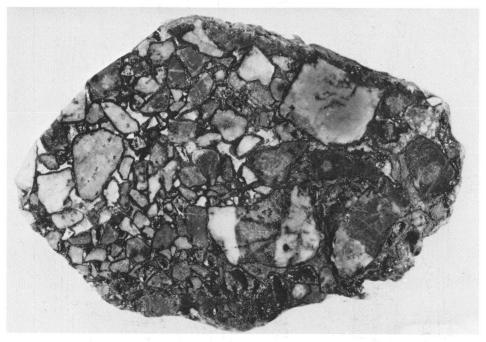

Fig. 1. (Small-size) collapse-breccia structure: polished section, about natural size, of a sample from the "silver-rich ore deposits" of S. Giovanni (Sardinia). Slightly rounded fragments of dolomite and calcite (whitish to greyish) are covered with a thin crust of highly argentiferous galena and silver minerals (black) and cemented with fresh calcite-aragonite (snow white). (Reproduced from Tamburrini and Zuffardi, 1967.)

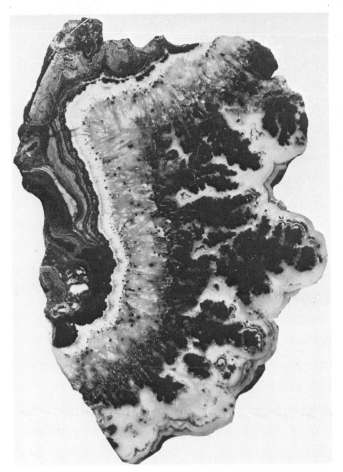

Fig. 2. (Small-size) crustified structure: polished section, about natural size, of a sample from Sidi Bou Auan (Tunisia). From left to right: grey (with various shades) to black = galena and sphalerite crusts. The whitish area included in them is calcite admixing fine-grained sulfides; milky white = crust of calcite-aragonite, with small black spots of galena; whitish, needle-shaped = transparant barite; black, dendritic = galena; milky white, enveloping galena = aragonite.

(b) "Allochthonous", that is (mainly) made up of constituents, which are "foreign" both to the host rocks and to their immediate surroundings.

(c) "Parautochthonous", that is (mainly) made up of constituents which are "foreign" to the country rocks, but the same as their close surroundings. This process of accumulation may also be considered as "by residual concentration".

(7) *On the hand-specimen scale,* the following textures are recognizable:

(a) idiomorphic crystalline; elongated forms with their long axes oriented towards the center of the body and/or hopper structures are common (see Fig. 6);

Fig. 3. Hand specimen of rhythmic barite (grey)–silica (black)–galena (white). The sample is in its top-bottom position; actual vertical dimension: about 7 cm. Barega barite deposit (Sardinia). (From: Padalino et al., 1972; courtesy Springer Verlag, Berlin.)

(b) colloform, both banded or globular, oolitic, pisolitic, reniform or stalactiform or stalagmiform (see Fig. 5).

The different textures mentioned above are controlled by depositional conditions and by mineralogical composition; namely: (a) textures are frequent in galena, and in depositions from quiet, tranquil waters (e.g., inside geodes, or isolated cavities, below the water table); (b) textures are general in sphalerite, marcasite, chalcedony, opal and in depositions from moving waters (e.g. above the water table).

(8) *On the microscale* the following characteristics should be mentioned:

(a) constituents of accumulations normally have absolute model-ages penecontemporaneous with, or older than, the country rock;

(b) sulfur-isotope composition is frequently (but not always) of biogenic type;

(c) framboidal pyrite is present;

(d) chalcedony crystals show lenght-slow orientation;

(e) fluid inclusions in paleo-karst often show temperatures of homogeneization ranging from 70 to 140°C. This same temperature in recent karst and in the outcrop zones of paleo-karst is near to room temperature (30–50°C).

KARST TYPOLOGY AND ENVIRONMENTAL CHARACTERISTICS

Three different sections have to be distinguished in a karst system when it reaches its complete development (the so-called "mature stage"), namely:

(1) The section close to the inlets; this is the *percolation zone,* which is nearly vertical, made up of fissures and caves and flooded with fresh acidic aggressive waters, which mainly run along the walls or drop from them, like rain. The abundance of fresh (partially dissolved) air and the strong turbulence of the water (that allows its release), help to maintain oxidizing conditions. Carbonate waters can also decompose alkali and Ca-silicate, through complex conditions of equilibrium/disequilibrium, that lead to the final deposition of $CaCO_3$ and of various forms of silica (chalcedony, opal, agate or microgranular black quartz). The (sometimes widespread) silicification of some karst is explained in this way. Mechanical erosion largely prevails over chemical leaching. Coarse detrital sedimentation, crusts along the walls and stalactite or stalagmite depositions are frequent. The pH of the flowing water gradually increases, because of the continuous dissolution of carbonates, thus favouring chemical deposition of dissolved salts. Strong oxidizing conditions enhance the dissolution of many elements (e.g., V, Cr, Mo, As, Se, Te), which readily precipitate, in various compounds, in less-oxidizing environments. Oxidation of soluble bisulfide, which may be partially biogenic, leads to deposition of sulfate: $BaSO_4$ crystals and crusts are formed in this way.

(2) The part of the nearly horizontal section, *overlying* the outlets, and connecting the above-mentioned percolation zone with them. This is the so-called *permanent circulation zone,* made up of fissures and galleries; collapse caves are also frequent. Water flow is either free or (mostly) forced, the mean velocity ranging between 3 and 100 m per hour. The waters are neutral, or, at least, less oxidizing than in the percolation zone. Intense chemical dissolution (and hence collapse of hanging-walls) and fine detrital deposition are dominant. Basification of circulating water becomes complete.

(3) The part of the section mentioned in 2 above, *lying below* the level of the outlets, is the *general imbition zone,* occurring below the water table. It has to be pointed out that it may or may not exist at this stage of karstification (that is: when the karst reaches its maturity), according to local geologic conditions. As a matter of fact, if waters flow in the permanent circulation zone directly upon an impermeable complex, no stagnant waters can exist below the permanent circulation zone. Should the general imbition zone occur (and we shall see afterwards that it does occur, at least in the *senile stage* of karstification), its environmental characteristics are those of neutral or reducing conditions, with ultra-detritic and (bio)-chemical sedimentation. Biogenic sulfuration plays an important role, in this zone.

Additional evolutions occur in the karst system, after it reaches its mature stage; these further evolutions are controlled by the erosion of the block holding the inlets and the percolation zone; as a consequence, the neighbouring basins will be filled up, and the outlets will tend to rise, and so will the permanent circulation zone. The imbition zone

Fig. 4. Different appearances of a same karst deposit. Fig. 4A shows a front in a barite quarry (Barega, Sardinia), open in the percolation zone of a karst system: a stock-work of barite veins (white) is held

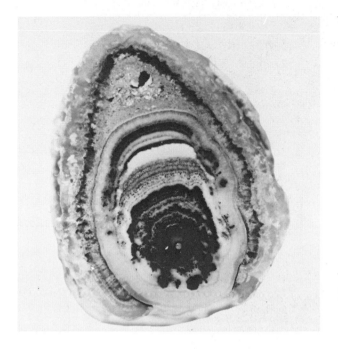

Fig. 5. Polished surface of a cross-section from a composite stalactite, about 2.5 × natural size (Djebel Hallouf, Tunisia). Whitsh = calcite—aragonite; dark grey to black = galena; white = empty spaces.

Two formational stages are recognizable; (1) a first, almost round portion, with a perfectly round central opening; (2) a second, almost elliptical portion enveloping the first one; its central hole is moulded along a sector of the first portion.

will thus develop and grow to a vertical extent equal to the rising of the outlets during the aging of the system.

This stage of karst evolution is currently called *the senile stage*. This time-space evolution is schematized in Fig. 7.

Karstification can go on as long as water flow along the karst system takes place. Those systems, in which karstification was interrupted are called *fossil karsts*. They may be fully or partially filled; Llopis Llado (1953) suggested the terms *olofossil* and *mero-*

---

in shattered and corroded limestone (gray). (From Tamburrini and Zuffardi, 1967.)

Another front of the same barite quarry is shown in Fig. 4B, open along the permanent circulation zone of a karst system. Rhythmic deposition of barite (white) and silica with galena (narrow dark seams interbedded with barite) held in limestone (dark gray to blackish).

The deposition is parallel to the foot-wall, well visble in the lower right and left corners of the photo. The hammer in the center helps in evaluating the sizes. The sample of Fig. 3 was picked from this front. (From Tamburrini and Zuffardi, 1967.)

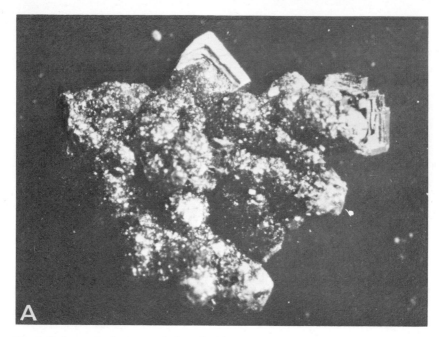

Fig. 6. Embryonal galena crystals deposited by descending waters.
A. Hopper structure of galena (two crystals are visible on the upper side of the sample) deposited on botryoidal sphalerite in a crevisse in the gossan of the lead–zinc mine of Montevecchio (Sardinia); × 2. (From Salvadori and Zuffardi, 1964.)

fossil to describe these two cases. Normally, fossilization is the natural conclusion of the senile stage, but it may intervene at any earlier stage; as a matter of fact, rapid (allochthonous, parautochthonous) infilling, or deposition of an impermeable cover over the inlets, or alteration of the surface/underground hydrology can prevent water from flowing in the karst system. Nothing hinders a fossil karst system from being reactivated, maybe after a long time-interval: this phenomenon is called *karst rejuvenation.*

Another possibility is that a buried fossil karst system underwent thermal metamorphism. The mineralogical complexity of karstic accumulations leads then to the formation of complex new secondary mineral associations, with possible migration of some components and drastic changes of structures and textures, so that recognizing the original karst pattern may be a rather difficult problem. Examples of both rejuvenation and metamorphism are given by the lead–barite deposit of Arenas (see Fig. 12).

The sizes of karst systems are extremely variable, according to: (1) the relative positions of the controlling topographic factors (vertical and horizontal distances between inlets and outlets); (2) the intensity of karstifying agents (water flow rate, its aggressivity,

B. Embryonal crystals of galena (black with geometrical outline) immersed in "terra rossa" (gray, with various shades) of a small karst from the Iglesiente district (Sardinia).

Metasomatic transformation to cerussite (whitish) is almost complete in the central-lower crystal; it is incipient in the other. Note zonations galena–cerussite in the lowest samples; × 7. (From Tamburrini and Zuffardi, 1967.)

etc.); (3) the favourable conditions of the rocks undergoing karstification (solubility, fracture systems, etc); and (4) the time span during which karstification could take place.

The classic Italian–Yugoslavian Karst region covers an area of more than 2000 km^2, is very rich in every type of cave and there is an underground section of a river (Timavo River) 34 km long. Another example of a very well developed karst system is the Mammoth Cave (in central Kentucky), which has numerous chambers and galleries situated on five levels, extending over a total length of 240 km.

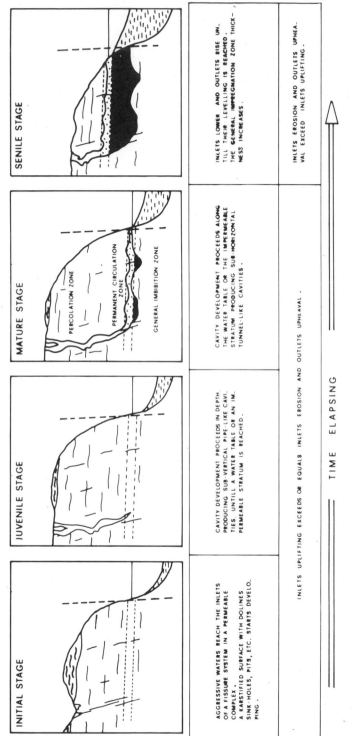

Fig. 7. Time sequence in karstification by supergene waters. Four stages are defined (initial, juvenile, mature, senile); changing of shapes and controlling events are described in each of them.

FORMATION OF KARSTIC CAVITIES

Two stages will be distinguished in discussing the genesis of karstic ore-mineral accumulations; namely: (1) formation of karst cavities; and (2) deposition within them.

It has to be pointed out that this subdivision is made only for convenience in exposition and is without any chronological implication. As a matter of fact, the two stages overlap partially, each one being present even in a same section of a karst system.

A karst is (on an ethymological basis) a cavity created by dissolution, corrasion and collapse controlled by meteoric waters moving down and along soluble rocks, the Italian-Yugoslavian Karst being the type area. Of course, nature has other possibilities to form solution cavities and to fill them, namely: (a) by warm or hot artesian waters (e.g., compaction fluids from sedimentary basins); and (b) by hot fluids more or less related to deep-seated sources. As was already pointed out above, the latter model is less (or not at all) acceptable to recent authors.

Suitable rocks for karst formation are: (a) carbonates (especially limestones and dolomites but also sandstones and conglomerates cemented with carbonate or gypsum/anhydrite; (b) gypsum and anhydrite; and (c) salts.

Limestones and dolomites are by far the most interesting, according to factual data, because of the enormous extent of such complexes (certainly larger by far than those of the other rock types quoted) that allows the development of conspicuous karstic systems. According to Maksimovich's calculations (Maksimovich, 1947, in Lukashev 1970), the global *area* occupied by soluble rocks is as follows: about 40,000,000 km^2 of carbonate rocks, 7,000,000 km^2 of gypsum and anhydrite, 4,000,000 km^2 of halites. Taking into account that the average thicknesses of (some) carbonate rocks are far higher than those of the other two types, one can realize the validity of the above-made statements.

Carbonate solubility in pure water is very low; the following figures can be quoted for calcite: at room temperature: 0.014 g/l; in boiling water: 0.019 g/l. Aragonite has almost the same (or even very slightly higher) solubility. Natural waters, however, are not pure waters, and they have remarkably higher solvent effects on carbonate rocks. This fact is related to their contents of: (1) dissolved CO_2 and/or humic acids, as may be the case in regions with thick vegetation; (2) H_2SO_4 and of $Fe_2(SO_4)_3$ as may be the case for waters trickling across sulphide (especially FeS_2)-bearing complexes; and (3) alkali chlorides as may be the case for sea water and/or brackish waters, which facilitate the formation of highly soluble Na–Ca complex ions.

(Paleo)-climatic conditions are very important in karst development. First of all it is obvious that the more abundant the rainfall is, the higher is the rate of dissolution of rock. Weather temperature and its cycles play a role as controlling factors of:

(1) Vegetation (and in general, biological) growth and decay, humic and organic acid

development. The present-day average annual temperature in tropical regions is 24–26°C, and the average annual rainfall 1200–3000 mm. According to data reported by Strakhov (1967), the increment of organic material in tropical forests amounts up to 100–200 tons a year per hectare. Correspondingly huge quantities of humic materials and of CO_2 are fed to the weathering crust.

(2) Solubility of natural reagents in water, particularly of CO_2, through its partial pressure in the atmosphere which is linked to temperature. The decomposing effect of water is appreciably enhanced by increasing temperatures. Quoting Raman (1911, from Lukashev, 1970), the indexes of the degree of water dissociation are the following:

temperature (°C)	0	10	18	34	50
degree of dissociation	1	1.7	2.4	4.5	8.0

(3) Fracturing of rock, both mechanical (insolation, freezing and thaw) and connected to vegetation (tree roots, organic acids).

(4) Oxidation capability (especially on sulfides).

Water temperature is also important; as a matter of fact, the higher it is, the stronger is the solvent capability of the solution, particularly in acidic waters. However, in the case of natural meteoric waters, the influence of temperature is rather restricted, first of all because of its narrow range (more or less from 0°C to 45°C) and, secondly, because after a short penetration in depth, all waters reach the same temperature (more or less 20°C). Therefore, supergene water temperature plays a role only in the processes that take place on, or close to, the earth's surface (dissolution of natural acids and of CO_2, reactions with sulfides).

One thus may conclude from the foregoing that *tropical (paleo)-climates, especially in coastal regions (where brackish waters are frequent), are the most suitable for (paleo)-karst development.*

Favourable climatic conditions are necessary, but are not alone sufficient for the formation of deep and large karst systems. Indeed three additional conditions have to be fulfilled, namely: (1) possibility of penetration of water in depth; (2) concentration of large quantities of water in confined rock volumes; and (3) permanence of such conditions over a sufficient (geologic) time interval.

Well-developed shear, fracture or fault zones, and particularly their intersections, with long distances between water inlets and outlets are favourable to the development of large karst systems; the vertical components of the systems are then limited by the altitude difference between inlets and outlets and will be particularly high in (continuously) uplifted blocks, such as horsts.

Should the penetration depth be small, *karstified surfaces,* that is areas in which sinkholes, such as dolines, poljes, lapies, pits, basins, irregular flutings, karren and "furrowed fields" (intricate erosion forms), would develop rather than true karst systems.

The permanence of conditions over a sufficient geologic time interval is also necessary; extensive developments of karst systems were thus possible during each of the peneplana-

FORMATION OF KARSTIC CAVITIES

tion periods which separated the orogenic periods. As a matter of fact, should the hydrological conditions change continuously (as may happen during orogenesis), the surface and underground water flow patterns would also change continuously, thus preventing the development of a small number of large karst systems and leading instead to the formation of many scattered and small karsts.

One may conclude that *singular portions of tectonized complexes* (e.g., intersections of two or more shear, fracture or fault systems, boundaries between two different permeability complexes, and so on), *in blocks continuously uplifted by epeirogenesis, without tilting and bending, are mostly favourable for deep and large karst development.*

Water inlets in the soluble complex undergoing karstification may be either at the surface or underground. Karst is called *denuded* in the first case, and *mantled* in the second.

Underground karstification may be quicker than surface karstification; indeed, the presence of a permeable complex lying upon, and in hydrological connection with, the underground inlets may act as a pressure water reservoir: this is a condition that enhances the solubility of gases in waters, and, consequently, their powers of corrosion of soluble rocks. An ice cover, that (partially) melts in connection with cyclic atmospheric temperature changes, also promotes karstification.

It is possible to estimate roughly the time span required for karstification. As reported above (p. 189) the solubility of calcite in pure cold water is 0.014 g/l; it is much higher in natural acidic cold waters. Let us assume a solubility 3 times greater, that is 0.042 g/l, a rainfall of 1000 mm/year (which is quite normal in tropical and sub-tropical countries) and a catchment area of 100 km^2. It is then easy to calculate that the quantity of limestone dissolved during a thousand years is 4.2 million tons equivalent to $1.6 \cdot 10^6$ m^3, in turn equivalent to a shaft of 40 X 40 m^2, and 1000 m deep.

The quantities of metals made available by the dissolution of supergene rocks can also be (roughly) estimated, on the basis of the above calculations. Assuming a Clarke value of 100 p.p.m (combined metals), which is a rather common figure for many carbonate formations exhibiting geochemical anomalies in ore-bearing regions, such as the Tri-State Mississippian and the Metalliferous Cambrian Limestones of Sardinia, it is easy to calculate that 420 tons of metal should be released in 1000 years, and, consequently, a metal stock of 10 million tons would be created by concentrating geologic events in about 24 million years, that is in a relatively short geological time interval.

FORMATION OF KARSTIC ACCUMULATIONS

Karsts can play a role in determining ore-mineral accumulation in different ways, namely:

(1) Karsts can act simply as surface water traps and filters. The simplest case is that of

surface waters flowing on a karstified surface; when they enter a karstic hollow, there is a decrease in turbulence and, consequently, the fluids deposit (at least a part of) the detrital particles. Many bauxites, clays, diamonds, and placer ores are accumulated in this way. In fact, the transportation capability of a stream is proportional to the square of its speed: should the cross-section of a constantly flowing stream increase three-fold when entering a hollow, its speed would decrease to 1/3 of the original value, and consequently the transportation capability would become 1/9. Moreover, waters entering a karst hollow sink away (at least partially if not completely) through fissures open in its bottom. This is another reason for the deposition of detrital or even ultra-detrital sediments in the same cavity.

(2) Karstification can yield local residual concentrations in ore-bearing complexes, with consequent partial to complete ore oxidation. The "circles with broken ground", from the Joplin District are typical examples (see Fig. 15).

(3) At the mature stage of a well-developed karst system, depositional possibilities are manifold, according to the different environmental conditions occurring within the system. The different sections of a mature karst system and their environmental conditions have been discussed above (pp. 183 ff.).

In a very broad scheme, it may be said that oxidized ore-mineral depositions (such as bauxites, laterites, barites) are prevalently formed in the percolating zone; chemical sulphidic ore deposition prevails in the imbibition zone, deposition in the permanent circulation zone having intermediate characteristics.

A consequence of the above situation is that substantial sulfide accumulations can occur only if and when the imbibition zone is well developed, that is during the senile stage of a karst. Therefore, the uplifting stage of the block of rock undergoing karstification is favourable to oxidized ore-mineral deposition and unfavourable to sulfide deposition; the opposite is the case in the subsequent erosional stage.

The, often rapid, variability of the conditions in the percolation and circulation zones has to be pointed out. As a consequence of this rapid cyclic changes, changes in ore-mineral deposition and in erosion versus deposition occur.

Let us consider a water stream flowing along a large vertical duct, as may be the case in the percolation zone. If the flow rate is low in comparison with the horizontal cross-

---

Fig. 8. Conceptual model of karstification and karstic accumulation. The events of karst formation and of karstic accumulation are schematically summarized and correlated.

The three possible controlling fluids, responsible for the above-mentioned phenomena (A), individually or mixed, enter a permeable, soluble rock complex, producing: (1) widening of its fissures, (the stages of which are summarized in B), the result being a karst system; (2) collection of soluble salts along their ways to (and/or into) the fissure system (C), thus becoming ore-metal-bearing fluids, which circulate within the karst system.

Intra-karstic ore-mineral accumulation and subsequent evolutions to form the end-product, that is, a karstic deposit with its present characteristics are summarized in D.

The controlling factors, in the different parts of the karst system, are schematized in E.

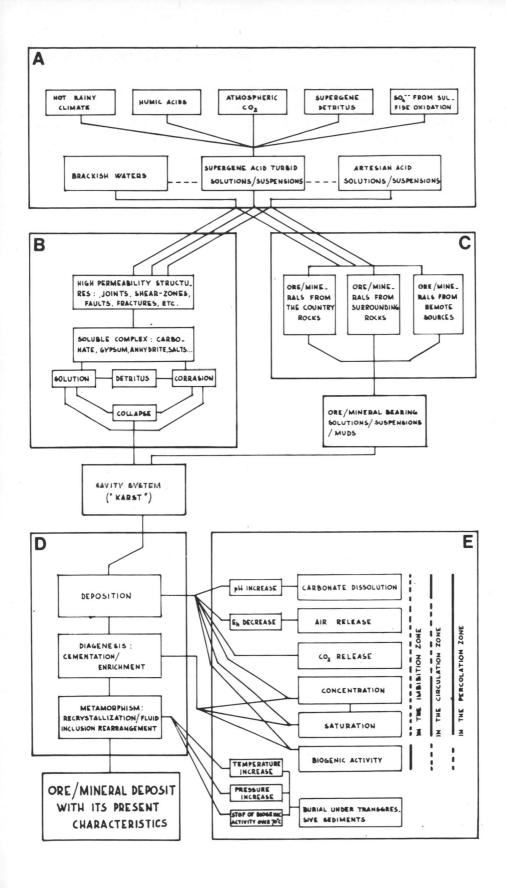

section of the passage ways, water laps on the walls and the confining pressure inside the duct equals the atmospheric pressure. Should the section decrease under the value $Q : v$ (Q being the water flow rate and v the water speed), water would fill the duct and flow conditions would become those of a penstock. Pressure would consequently increase with depth, equalling that of the atmosphere on the free water surface plus the hydraulic pressure. Solubility of gases (CO_2, H_2S) would thus be increased.

Should a narrow duct lead the water to a large cave in connection with the free atmosphere, penstock conditions would change to free-flow circulation and an abrupt pressure decrease would take place, with release of dissolved gases and consequent change in pH, Eh and of chemical equilibria of the dissolved components.

Another controlling factor on the gas content in karstic waters is *seasonal change in atmospheric temperature*: cold air, descending along the percolation zone in winter, causes partial pressure of gases to rise, thus facilitating their dissolution in water. The contrary takes place in summer.

In conclusion, one should not expect detail regularity in the distribution of intra-karstic dissolution versus deposition, and in the paragenesis of the deposits. This is particularly true in the percolation and circulation zones, in which rapid alternations are explained on the basis of different physicochemical equilibria in the micro-environments. This is the concept of *mozaic equilibrium*.

A gradual *pH increase* in circulating waters is caused by the increasing quantities of CaO dissolved in them as they flow along a karst system into carbonate host rock. This fact involves gradual change in ore-mineral depositions: e.g., $Fe(OH)_3$ is deposited at pH = 2.3, $Fe(OH)_2$ at pH = 5.5, and $Ni(OH)_2$ at pH = 6.7. As a consequence, when both Fe and Ni are present in the same karst deposit, they show mutual transition, Fe occurring in the upper and Ni in the lower portion of the karst.

Empty caves are sometimes utilized as shelters by animals, such as birds, bats, etc. Accumulations of bones, faeces and other *organic residues* are, therefore, possible and (if those do not become washed away, for example as under arid conditions) they can be transformed into phosphates and nitrates. Nitro-bacteria can intervene in the last case.

When sediments undergo *intra-karstic diagenesis,* the main controlling factors are:

(1) Evaporation of impregnating waters and consequent salt concentration in the sediments.

(2) Degasing of interstitial waters. Both these factors are particularly effective in the percolation zone.

(3) Biogenic activity, decay of organic matter, and development of H_2S by biogenic sulfate reduction: this is effective in the imbibition zone.

(4) Gradual pH change: this is effective in every section of the karst system.

Rather high homogeneization temperatures in paleo-karst ore-mineral deposits (see above, pp. 176, 182) are stressed by the supporters of hydrothermal geneses. Those temperatures, however, can be easily and reasonably explained by taking into account the degree of regional metamorphism of the deposits. In fact, burial to a depth of 4000 m, under conditions of a normal thermal gradient, would establish a temperature of $140°-150°C$, with consequent possible recrystallization and modification of the fluid inclusions.

The possibility of *sulfide recrystallization* is explainable, according to Bernard (1972), on the basis of temperature-controlled changes in equilibrium between bisulphide-metal complexing, chloride-metal complexing, both components being present in fluid inclusions. On the other hand, the often-described perfectly formed, fluid-inclusion-rich, clear-coloured sphalerite around dull sphalerite, may be considered as evidence of recrystallization.

The events controlling karstification and karstic accumulation can be summarized in the conceptual model of Fig. 8.

EXAMPLES OF ECONOMIC DEPOSITS IN KARSTS

Many examples are available of karstic ore-mineral deposition. Apart from the well-known carbonate foot-wall bauxite deposits, some recently described ones deserve attention and several of them are mentioned and briefly described below.

The lead—zinc (barite-fluorite-silica) deposits of Laurium, Greece (after Leleu, 1969; Bernard et al., 1972)

A Paleozoic series, called the Kamareza (Fig. 9) is composed of three members (K3 = calcitic-dolomitic marbles, overlain by calc-schist; K2 = micaschists with interbedded basic metavolcanics; and K1 = massive limestones), and is unconformably covered by a Mesozoic series. The latter is called the Plaka series and includes three members (P1 = limestones, underlain by conglomerates; P2 = sandy slates with interbedded limestones and metavolcanics; and P3 = limestones).

An emersion and erosion stage occurred between P1 and P2. The whole P series glided tectonically over the K series along the unconformity surface during pre-Miocene times. Acidic plutons intruded the series during the deposition of the P1 member.

Ore-deposits are of two different types: (1) volcano-sedimentary mixed-sulfide (Pb, Zn, Fe) deposits along the transition between K3 and K2; but dispersed, non-commercial, ores are also present along this horizon; (2) karst deposits of (oxidized and/or sulfidized) Pb, Zn and CaF_2 are related to the two emersion surfaces (E1, between P1 and K1; E2, between the P2 and P1).

The ore occurrences of type (1) are considered to be the source of metals for the type (2) deposits. These latter deposits have various shapes, as schematized in Fig. 9, namely:

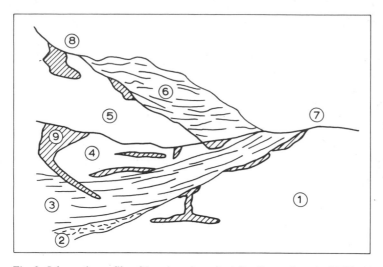

Fig. 9. Schematic profile of Laurium deposits (after Bernard et al., 1972).
   1 = K3 marbles; *2* = K3 calc-schists; *3* = K2 micaschists and interbedded metavolcanics; *4* = K1 limestones; *5* = P1 limestones and basal conglomerate; *6* = P2 sandy slates and metavolcanics; *7* = unconformity between P and K series; *8* = erosion surface between P1 and P2; *9* = vein no. 80.

(1) small pockets, irregular lenses along and below an erosion surface, or along the contact between two horizons with different permeability (e.g., marble/schists; calcitic/dolomitic); (2) veins along enlarged fracture or shear zones, and/or along the sides of

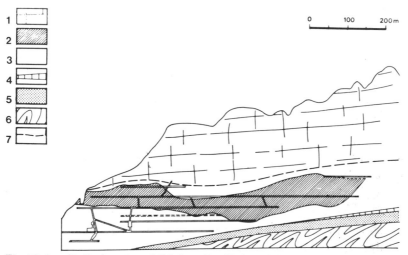

Fig. 10. Longitudinal section of Salafossa Zn–Pb deposit (after Bernard et al., 1972).
   1 = Ladinian–Anisian dolomites; *2* = ore-bearing breccia; *3* = dolomite; *1 + 2 + 3* = Mid-Triassic; *4* = Bellerophon Formation; *5* = Val Gardena Formation; *4 + 5* = Permian; *6* = undifferentiated Paleozoic basement; *7* = probable boundary between Anisian and Ladinian.

dykes; fractures in mica-schists can also be commercially mineralized (e.g., "vein no. 80");
(3) caves.

The sizes of the ore bodies are also variable: pockets and lenses rarely exceed 1 m in
depth, whereas the extent of the surface along which they are scattered can exceed some
tens of hectars. Vein deposits have thicknesses of some meters (up to 10), vertical extents
of 10—50 m, longitudinal extents (of each single vein) up to 100 m and total length of
the vein system can be several kilometers. The size of a cave deposit is about 5 X 5 X
10 m.

The zinc—lead deposit of Salafossa, central Italian Alps (Lagny, 1969; Bernard et al.,
1972)

This is a huge stratiform ore deposit (700 m long, 100 m thick and 200 m wide; see
Figs. 10 and 11). The structure is brecciated and the average size of individual fragments
ranges from some centimeters up to some tens of centimeters. The cement is ZnS and
calcite; PbS and FeS$_2$ are accessory. The breccia ore body is surrounded by an aureole of
sulfide veinlets which represents the initial stage of brecciation and mineralization. Some
cavities, filled up with ore crusts, ore stalactites and ore-bearing intra-karstic sediments,
lie beneath the main ore body.

Salafossa is an example of the mature stage of karstification, with a well developed
imbibition zone.

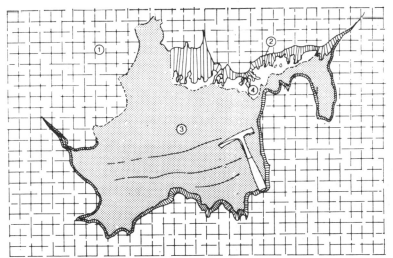

Fig. 11. Scheme of a filled cavity at the bottom of the main brecciated deposit of Sakafossa (after
Bernard et al., 1972).
    1 = limestone; *2* = stalactites (mainly pyrite); *3* = fine-bedded intra-karstic sediments (mainly
sphalerite); *4* = crust of "terra rossa".

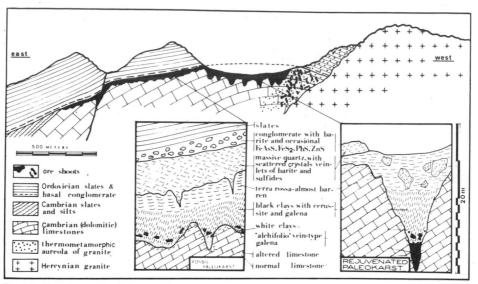

Fig. 12. Sketch profile of Arenas (Sardinia) lead–barite deposits. (After Padalino et al., 1972; courtesy Springer Verlag, Berlin.)

The galena-cerussite-barite deposit of Arenas, Sardinia (Benz, 1964; Padalino et al., 1972)

This deposit is shown diagrammatically in Fig. 12, and factual observations are described in the same figure. Its genesis involved three stages: (1) Karstification and low-grade, karstic ore accumulations (3–4% Pb) during the Cambro-Ordovician emersion. Their ore-mineral source was the metal contents of the same Cambrian carbonate complex, that contains up to 100 p.p.m. of Pb + Zn and up to 1000 p.p.m. and even more of Ba. (2) The Hercynian granite intrusion partially destroyed the above-mentioned carbonate complex and remobilized the elements, giving rise to high-grade ore shoots in the metamorphic aureole (up to 10% Pb). (3) The post-Hercynian peneplanation partially re-exhumed the Cambrian-Ordovician unconformity and karstification resumed with consequent ore-mineral redistribution; small but massive concentrations of PbS (the "alchifolio" of miners) thus originated at the very bottom of some rejuvenated karsts.

Arenas shows examples of (1) fossil (see p. 185), (2) rejuvenated, and (3) thermometamorphosed karsts. The presence of both fossil and rejuvenated karsts over many hundreds of square kilometers, even in areas very far from granites, is evidence that the correlations between granites and ore deposits are merely casual.

The barite deposits of Barega, Sardinia (Brusca et al., 1967; Tamburrini-Zuffardi, 1969; Padalino et al., 1972)

Figure 13 shows a sketch profile and some details of these deposits. In short, a Cambrian carbonate block, with a horst-and-graben structure, and having high Ba contents (up to 1000 p.p.m.) and, in places, small interbedded $BaSO_4$ seams and lenses, underwent

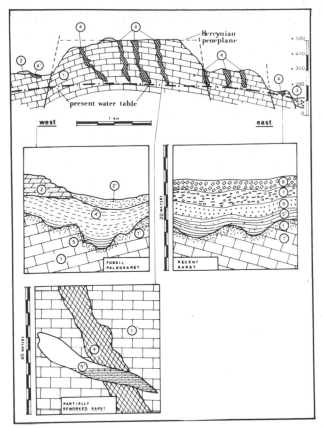

Fig. 13. Sketch profile of Barega (Sardinia) barite deposits. (After Padalino et al., 1972; courtesy Springer Verlag, Berlin.)

1 = Cambrian limestone; 1' = silicified, rubefied Cambrian limestone; 2 = Triassic(?) sandy dolomite with silicic-baritic cement; 2' = loose baritic ground (partially from decay of 2); 3 = Recent alluvium; 4 = barite (almost massive); 4' = thin-bedded barite-silica (+ occasional sulfides); 5 = white mudstone, sometimes with barite crystals (5') or pisolitic iron (5''); 6 = pisolitic iron with barite; 7 = fine-grained breccia with ferrous cement and barite pebbles; 8 = macro-breccia with limestone-silica-barite pebbles.

erosion and leaching from the beginning of the post-Hercynian peneplanation. This phenomenon continued, with varying intensity (according to the different climatic conditions and the possible impermeable covers deposited upon the block) up to the present, as a result of the more or less continuous uplifting of the structure.

A variety of karst features were thus formed: (1) shallow-depth fossil karsts with thin-bedded silicatic-baritic-galena depositions (see Fig. 4); (2) deep, pipe-like, rich baritiferous bodies, at places partially reworked in connection with changing hydrologic conditions; (3) shallow-depth karsts, with bedded baritic-limonitic deposition.

The depth of karstification is controlled by the present water table in cases (2) and (3), which are examples of juvenile stages of karstification, still proceeding at present.

The zinc deposits of the Lower Ordovician Kingsport Formation and Mascot Dolomite of East Tennessee

The genesis of these deposits is, according to the most recent ideas (Soc. Econ. Geologists, 1971) and also to some earlier authors (e.g., Ulrich, 1931) related to karstic breccias produced by collapse and fragmentation resulting from the solution of the Kingsport-Mascot carbonate complex. According to these authors, the present fluid-inclusions indicate that the most likely ore fluids were slightly warmed (90°–150°C) brines. The source of the zinc is thought to be elusive, but the hypothesis is often put forward that the diagenesis of regionally adjacent basins of thick argillaceous accumulations could be the source of those zinc-bearing brines (e.g., compaction waters, as advocated by numerous recent researchers; for a synopsis, cf., Wolf, 1976).

It has, however, to be pointed out that the possibility that the source of the zinc could be the same low-grade zinciferous, eroded Mascot Formation covering the Kingsport, should also be taken into consideration. The thermal characteristics of the present fluid inclusions could be related to recrystallization during post-ore metamorphism, rather than to the original temperature of the ore-forming brines.

Fluorite deposits in Triassic complexes of the Central Alps (Jadoul, 1973; Rodeghiero, 1974)

Irregular lenticular deposits of fluorite, in places associated with barite and Pb–Zn ores are scattered along the Carnian and Ladinian (Upper Trias) strata. They always occur: (1) in carbonate country rocks, and (2) beneath emersion surfaces, made evident by: (a) angular unconformities, (b) silicification/ferrugination, and (c) clastic overburden. They exhibit sedimentary textures (slumping, gravity moulds, graded bedding, stylolites, etc.). The accumulation of the deposits in karsts during recurrent upheavals of the Triassic series is generally accepted.

The sources of fluor and of the other admixed ore minerals are not precisely defined. Maybe different ore minerals have different sources, e.g., (1) syn-sedimentary volcanism, that fed the surface waters with them; (2) leaching of sandy/clayey interbedded ore-mineral-bearing sediments and/or of the same volcanics (both well known in this same area); and (3) erosion and leaching of Hercynian ore-mineral-bearing porphyries and/or granitoids. This could apply particularly to fluorite, because of the high contents of fluor in some of these rocks.

The similarity between the above-mentioned fluoritic deposits and those of Hamman Zriba (Tunisia) and of Illinois-Kentucky is striking.

The lead–zinc deposit of Ali-ou-Daoud, Morocco (Emberger, 1969)

This is a stratiform deposit within the lower member of a Liassic carbonate series. An emersion surface, marked by initial karstification and general silicification features divides

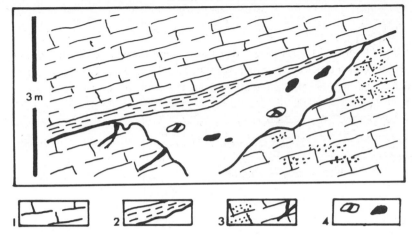

Fig. 14. Ali-ou-Daoud (Morocco) lead–zinc deposit. (After Emberger, 1969; courtesy Ente Minerario Sardo, Cagliari.)
  1 = barren limestones and dolomites; *2* = clayey seam; *3* = low-grade, disseminated, and veinlets of ore-bearing limestones and dolomites, silicified at the top; *4* = karstic clays with boulders of member under 3, with small masses of PbS, $PbCO_3$, $ZnCO_3$.

the ore-bearing member from the overlying barren one (see Fig. 14). Karstic accumulations have higher metal grades than the sheeted occurrences. The conceptual similarity with "broken ground" and "circles" near surface and "sheet-ground" deposits of the Joplin district (see Fig. 15) is notable.

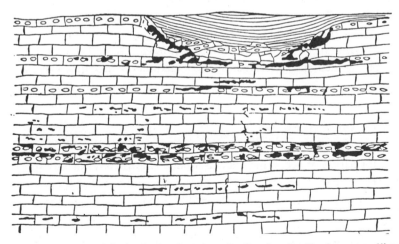

Fig. 15. Diagram of the lead–zinc deposits at Joplin, showing "broken ground" covered by Pennsylvanian Shale, around "circles" near surface and "sheet ground" in Mississippian Limestone. The total length of the section is about 100 meters. (From Lindgren, 1933; courtesy McGraw-Hill, New York, N.Y.)

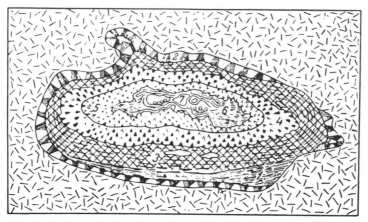

Fig. 16. Horizontal schematic section of the "main vein" of Tyuya Mayun. (After Routhier, 1969; courtesy Masson, Paris.)

1 = limestone crust; 2 = limestones, stalactites; 3 = "ore-bearing marble", i.e., coarse-grained calcite admixed with tyuyamanite (CaO · 2UO$_3$ · V$_2$O$_5$ · 2H$_2$O), and other vanadate of Cu, radiobarite, radiocalcite; 4 = ruby barite; 5 = transparent honey-coloured barite; 6 = "terra rossa"; 7 = empty space.

The total tickness of the crust can reach 1.5 m; their distribution along the orebody (that has the shape of a sub-vertical winding pipe, varved in Devonian limestones) is very variable.

The uranium-vanadate ("Tyuyamanite") deposit of Tyuya Mayun (Fig. 16), Fergana, U.S.S.R. (Routhier, 1963)

The genesis of this deposit, one of the most important in the U.S.S.R., is generally considered to be karstic, but the source of metals is unknown or, at least is controversial. It has been suggested that U and V were present as dispersions in the overlying graphitic shales, which later on supplied the elements through remobilization. As an alternative, a completely eroded and, consequently, now unknown source rock may have existed.

COMPARISON WITH SIMILAR DEPOSITS

The possibility of formation of economic mineral accumulations controlled by supergene descending waters in rocks different from the ones dealt with in the above sections (carbonates, gypsum and anhydrite, halites, i.e. the readily soluble rocks) has to be taken into consideration.

The main characteristic of karstification is indeed the widening of fissures, caused by aggressive descending waters; large open spaces suitable for the accumulation can in this manner originate. Apart from this, many other phenomena, prerequisites for the origin of economic deposits (ore-mineral-bearing solutions, pH and Eh variations, biogenic activity at depth, etc.) could also be established in *non*-carbonate rocks.

The chief problem, in the origin of ores within karst systems, is, therefore, related to the mechanisms that lead to the formation of sufficient open spaces to permit the accumulation of large tonnages of economic deposits. From this point of view, it has to be born in mind that *mechanical erosion* can cause large gulches and caves in loesses and in clayey complexes.

The important role played by CO_2 and oxygen in the atmospheric alteration of rocks must also be remembered. As was mentioned above (p. 189 ff.), alkali–Ca–silicate can be dissolved in this way. The consequent formation of alkali-carbonate and silicate solutions enhances their ability to corrode and dissolve; Lukashev (1970) describes rapid alteration of andalusite, kyanite and topaz under such conditions. Moreau et al. (1966) described supergene dequartzification of granites in the Plateau Central (France) under (Triassic) tropical/subtropical paleoclimatic conditions.

The most evident examples of accumulations controlled by descending waters, independent of type of country rocks, are those formed by supergene oxidation-reduction processes. The geochemical aspects are rather well known, but, on the other hand, the quantitative factors (i.e., tonnages involved, grade increase) have probably been overlooked, at least in some cases. For example, Padalino et al. (1972) described noteworthy accumulations of pure CaF_2 occurring at the base of gossans derived from oxidation of CaF_2—mixed sulfide-bearing skarns overlying Hercynian granite in Sardinia.

Zuffardi (1968, 1969, 1970) decribed huge, high-grade, galena shoots, related to ancient water-tables in the Pb–Zn district of Iglesias (Sardinia) in carbonate country rocks. The same author (1962) described a (possibly) similar phenomenon in a quartz—carbonate—PbS–ZnS–FeS_2 vein in slaty country rocks of the same region (Montevecchio mine).

Aside from these supergene accumulations, other good examples of non-karstic deposits, made up "per descensum" are: (1) the uraniferous veins of the Plateau Central (Moreau et al., 1966), and (2) the fluorite deposits on an eroded granitic surface of the Morvan district, (France) (Soulé de la Fonte, 1967). In the first case, it is a matter of economic-grade pitchblende—calcite—red chalcedony—earthy hematite accumulations along sheared, fractured and leached zones of Hercynian granites, with high U contents (5–20 p.p.m.). Ore shoots occur down to depth of some tens of meters (exceptionally 200–300 m), beneath a partially re-exhumed Triassic erosion surface, and are held in minor fractures of horsts in a horst-and-graben structure. The absolute model age of pitchblende is Triassic (i.e., younger than granites).

Moreau et al. (1968) explained these factual observations, by proposing the hypothesis that these vein accumulations of uranium are the result of erosion and supergene leaching of uraniferous granites under tropical/subtropical paleo-climates.

The Morvan fluorite deposits occur as irregular beds and crusts (rarely as short veins) up to several meters thick, along the same partially re-exhumed Triassic erosion surface of Hercynian granites, which, in this area, holds minor accessory quantities of CaF_2 and

$BaSO_4$, CaF_2 grade in these beds and crusts ranges from a few percent up to 25–30%, whereas $BaSO_4$ reaches, at the most, 8–10% (and, therefore, is not of economic grade). Silica (chalcedony and fine-grained quartz) is widespread. The accumulations, according to Soulé de la Fonte (1967), were the result of residual (partially mechanical, but mostly chemical) concentrations, under tropical/subtropical Triassic paleoclimatic conditions.

CONCLUSION

As a conclusion of this review one may say that it seems probable that ore-mineral accumulations controlled by descending supergene solutions play a more important role than was believed until comparatively recently. Bearing in mind this possibility, probably more of the so-called "telethermal" and/or "cryptothermal" deposits could be re-interpreted. Accordingly, up-dated conceptual guides for ore prospecting should be established. More thorough investigations in peneplaned regions, especially (but not only), where they are composed of carbonate and other readily soluble rocks, should also be carried out to find more and new natural resources.

BIBLIOGRAPHY

I. KARSTIFICATION (apart from economic accumulation in karsts)

Abrami, G., 1966. Hypothèses sur l'évolution de la morphologie et de l'hydrologie karstique. *Soc. Ital. Sci. Nat., Mus. Civ.*, 105: 61–90.
Birot, P., 1954. Problèmes de morphologie karstique. *Ann. Géogr.*, 63: 161–162.
Bogli, A., 1964. Corrosion par mélange des eaux. *Int. J. Speleol.*, 1 (1/2): 61–70.
Bonte, A., 1963. Les remplissages karstiques. *Sedimentology*, 2: 333–340.
Caumartin, V. and Renault, Ph., 1958. La corrosion biochimique dans un réseau karstique. *Notes Biospeleol.*, 13: 87–109.
Cavaille, A., 1962. Le système karstique et l'évolution des grottes. *Spelunca–Ann. 4e Congr. Natl. Speleol.*, pp. 9–28.
Chevalier, P., 1953. Erosion ou corrasion (essai de contrôle du mode de creusement des réseaux souterrains). *Publ. 1er Congr. Int. Speleol.*, 2: 35–39.
Cvijic, J., 1918. Hydrographie souterraine et évolution morphologique du karst. *Trav. Inst. Géogr. Alp., Univ. Grenoble*, 6 (4): 56 pp.
Dublanskij, N.V., 1963. Sur le rôle de la neige dans la karstification et l'alimentation des eaux karstiques. *Izv. Akad. Nauk. S.S.S.R., Serv. Géol.*, 2: 69–74.
Elliot, J.K., 1964. Carbonate characteristics and its relationship to cavern development. *Bull. Natl. Speleol. Soc.*, 26 (2): 76.
Ford, D.C., 1971a. Research methods in karst geomorphology. *Symp. Geomorphol.*, 1st, 1969: 23–47.
Ford, D.C., 1971b. Geologic structure and a new explanation of limestone cavern genesis. *Cave Res. Group., Gr. Br., Trans.*, 13: 91–94.
Ford, T.D., 1969a. The stratiform ore deposits of Derbyshire. In: Sedimentary Ores - Ancient and Modern - Revised. *Inter-Univ. Geol. Congr., 15th, Leicester, 1967*: 73–96.

Gèze, B., 1955. La genèse des gouffres. *Congr. Int. Speleol., Paris, 1953*, 2: 11–23.

Gèze, B., 1961. L'évolution karstique dans ses rapports avec les alternances climatiques quaternaires. *Rass. Speleol. Ital., Mem.*, 5 (1): 111–117.

Gèze, B., 1962. Sur quelques caractères fondamentaux des circulations karstiques. *Ann. Speleol.*, 18 (4): 5–22.

Ginzburg, I.I., 1952a. Rol' mikroorganizmov vyvetrivanii porod i obrazovarii mineralov. (Role of micro-organisms in rock weathering and mineral formation.) *Sb. Kora Vyv.*, 1.

Howard, A.D., 1964. Processes of limestone cave development. *Int. J. Speleol.*, 1 (1/2): 47–60.

Kaye, C.A., 1957. The effect of solvent motion on limestone solution. *J. Geol.*, 65 (1) 35–46.

Kiersch, G.A. and Huches, P.W., 1952. Structural localization of ground water in limestone of Big Bend District, Texas. *Econ. Geol.*, 47: 794–806.

Kukla, J. and Lozek, V., 1958. Problematice vyzkumus jeskynnich vyplni. (The problems of investigation of cave deposits.) *Cesk. Kras*, 11: 19–83.

Leleu, M., 1966. Le karst et ses incidences métallogeniques. *Sci. Terre*, 11 (4): 385–413.

Llopis Llado, N., 1953. Karst holofossile et mérofossile. *1er Congr. Int. Speleol., Paris*, 2: 41–50.

Maiklem, W.R., 1971; Evaporative drawdown – a mechanism for water-level lowering and diagenesis in the Elk Point Basin. *Bull. Can. Pet. Geol.*, 19: 487–503.

Maksimovitch, G.A., 1957. Principaux types de profils hydrodynamiques des régions de karst dans les formations carbonatées et sulfatées. *Dokl. Akad. Nauk. S.S.S.R.*, 112(3): 501. (Trad. BRGM, 1525.)

Meinzer, O.E., 1923. Outline of ground water hydrology, with definitions. *U.S. Geol. Surv., Water Supply Pap.*, 494.

Muore, G.W., 1964. Abrupt change in cave history when ventilation begins. *Bull. Speleol. Soc.*, 26 (2): 76.

Nishihara, G.S., 1914. The rate of reduction of acidity of descending waters by certain ore and gangue minerals and its bearing upon secondary sulphide enrichments. *Econ. Geol.*, 9 (8): 743–757.

Quinlan, J.F., 1970. Review of problems of karst denudation. *Caves Karst*, 12: 17–20.

Quinlan, J.F., 1972a. *Karst, Pseudokarst, and Dolines: Classification and a Review*. Thesis Univ. Texas, Austin.

Quinlan, J.F., 1972b. Karst-related mineral deposits and possible criteria for recognition of paleo-karsts: a review of preservable character of Holocene and older karst terranes. *24th Int. Geol. Congr., Sect 6*: 156–168.

Quinlan, J.F. and Smith, A.R., 1967. Drip pockets and related karst cavities in gypsum (abs.). *Cave Karst*, 9: 50.

Renault, Ph., 1957. Sur deux processus d'effondrement karstique. *Ann. Speleol.*, 12 (1/4): 19–46.

Renault, Ph., 1960. Rôle de l'érosion et de la corrasion dans le creusement d'un réseau karstique. *Ann. Speleol.*, 13 (1/4): 23–49.

Renault, Ph., 1964. Remarques sur la signification des expériences en géodynamique karstique. *Int. J. Speleol.*, 1 (1/2): 109–152.

Reparaz, A. and Weydert, P., 1966. Etudes de climatologie souterraine et de karstologie dans le massif de Sainte-Victoire. *Spelunca, Bull. Fr.*, 6 (1): 52.

Rhoades, R. and Sinacori, M.N., 1941. The pattern of ground water flow and solution. *J. Geol.*, 49: 785–794.

Roques, H., 1959. Sur la répartition du CO_2 dans les karst. *Ann. Speleol.*, 14 (2): 1–22.

Roques, H., 1963a. Sur la répartition du CO_2 dans les karst. *Ann. Speleol.*, 18 (2): 142–184.

Swinnerton, A.C., 1932. Origin of limestone caverns. *Geol. Soc. Am. Bull.*, 43: 366–694.

Thrailkill, J.V., 1966. Studies in the excavation of limestone cave and the deposition of speleothems. *Diss. Abstr.*, 26 (7): 2872.

White, W.B., 1964. Sedimentation in caves–a review. *Bull. Speleol. Soc.*, 26 (2): 77–78.

II. CHEMICO-PHYSICAL EQUILIBRIA of water-gases-soluble salts systems, which can have a bearing on karstification and ore-mineral accumulation in karst

Cigna, A.A., Cigna, L.R. and Vido, L.L., 1963. Quelques considérations sur l'effet du sel dans la solubilité des calcaires. *Ann. Speleol.*, 18: 185–191.

Devigne, J.P., 1968. Une bactérie saturnophile, *Sarcina flava* Bary 1887. *Arch. Inst. Pasteur Tunis, A*, 45: 341–358.

Garrels, R.M., 1954. Mineral species as functions of pH and oxydation-reduction potentials, with special reference to the zone of oxidation and secondary enrichment of sulphide ore deposits. *Geochim. Cosmochim. Acta*, 5 (4): 153–168.

Garrels, R.M., 1960. *Mineral Equilibria at Low Temperature and Pressure*. Harper and Row, New-York, N.Y.

Garrels, R.M. and Christ, Ch.L., 1965. *Solutions, Minerals and Equilibria*. Harper and Row, London, New York, Tokyo, 450 pp.

Garrels, R.M. and Dreyer, R.M., 1952. Mechanism of limestone replacliment at low temperatures and pressures. *Bull. Geol. Soc. Am.*, 63 (4): 325–379.

Garrels, R.M. and Richter, D.H., 1955. Is carbon dioxide an ore-forming fluid under shallow earth conditions? *Econ. Geol.*, 50: 446–458.

Helgeson, H.C., 1964. *Complexing and hydrothermal ore deposition*. Pergamon Press, London, 128 pp.

Kielland, J., 1937. Individual ion activity coefficients in aqueous solutions. *J. Am. Chem. Soc.*, 59: 1675–1678.

Konstantinov, M.M. and Rafal'skij, R.P., 1960. Dissolving of galena and the migration of lead under conditions prevailing near the surface. *Geokhim. S.S.S.R.*, 3: 280–281.

Leleu, M., 1969. Essai d'interprétation thermodynamique en métallogénie: les minéralisations karstiques de Laurium (Grèce). *Bull. B.R.G.M., 2. Sér.*, 4: 1–62.

Michard, G., 1967. Signification du potentiel redox dans les eaux naturelles. Conditions d'utilisation des diagrammes (Eh, pH). *Miner. Deposita*, 2 (1): 34–37.

Nancollas, G.H., 1960. Thermodynamics of ion association in aqueous solution. *Chem. Soc. Lond., Spec. Publ.*, 14 (4): 402–427.

O'Connor, J.T. and Renn, L.E., 1964. Soluble adsorbed zinc equilibrium in natural waters. *J. Water Works Assoc.*, 56 (8).

Pourbaix, M.J.N., 1945. *Thermodynamique des Solutions Aqueuses Diluées*. Thesis Delft, 122 pp.

Razzell, W.E., 1962. Lessivage bactérien des sulfures métalliques. *Can. Min. Metall. Bull.*, 55: 190–191.

Schoeller, H., 1955. Géochimie des eaux souterraines. *Rev. Inst. Fr. Petr.*, 10 (3,4,5).

Schoeller, H., 1962. *Les Eaux Souterraines*. Masson, Paris, 642 pp.

Shcherbakov, A.V., 1956. Critères géochimiques des milieux oxydoréducteurs dans l'hydrosphère souterraine. *Sov. Geol.*, 56: 72–82. (Trad. B.R.G.M., 2289.)

Sokolova, G.A. and Karavaiko., 1968. Physiology and geochemical activity of Thiobacilli. *Israel Progr. Sci. Transl.*, 74 pp.

Strakhov, N.M., 1967. *Principles of Lithogenesis, 1*, Consultants Bureau, New York, N.Y., 245 pp.

Verhoogen, J., 1938. Thermodynamical calculation of the solubility of some important sulphides up to 400°C. *Econ. Geol.*, 33: 34–51.

White, D.E., Hem, J.D. and Waring, G.A., 1963. Chemical composition of sub-surface waters. (Data of geochemistry, 6th ed.) *U.S. Geol. Surv., Prof. Pap.*, 440-F: 67 pp.

III. GENERAL PROBLEMS OF KARSTS AND ORE-MINERAL DEPOSITS

Altschuler, Z.S. and Young, E.J., 1960. Residual origin of the "Pleistocene" sand mantle in Central Florida uplands and its bearing on marine terraces and Cenozoic uplift. *U.S. Geol. Surv., Prof. Pap.*, 400-B: 202–207.

Altschuler, Z.S., Cathcart, J.B. and Young, E.J., 1964. Geology and geochemistry of the Bone Valley Formation and its phosphate deposits, west-central Florida. *Geol. Soc. Am., Guideb., Annual Meet., Miami Beach*, 68 pp.

Andrieux, C., 1965a. Sur la mesure précise des caractéristiques météoclimatiques souterraines. *Ann. Speleol.*, 20 (3): 319–340.

Andrieux, C., 1965b. Etude des stalactites tubiformes monocristallines. Mécanisme de leur formation et conditionnement de leur dimension transversale. *Bull. Soc. Fr. Mineral. Cristallogr.*, 88: 53–58.

Anelli, F., 1958. Le cavità di riempimento bauxitico di Spinazola (Bari). Forme paleocarsiche bicicliche nelle Murge Nordoccidentali. *Atti Congr. Int. Speleol., Bari-Lecce-Salerno*, 1 (1): 201–215.

Aubert, D., 1966. Structure, activité et évolution d'une doline. *Bull. Soc. Neuchatel. Sci. Nat.*, 89: 113–120.

Baranova, N.N. and Barsukov, V.L., 1965. Transport of lead by hydrothermal solutions in the form of carbonate complexes. *Geochem. Int.*, 2 (5): 802–809.

Bardossy, G. (Editor), 1970. *International Conference on Bauxite Geology, Budapest, 1969, Proceedings–Inst. Geol. Pubbl. Hung., Ann.*, 54 (3): 485 pp.

Barnes, H.L., 1966. Sphalerite solubility in ore solutions. *Econ. Geol.* Monogr., 3: 326–332.

Barnes, H.L. and Czamanske, G.K., 1967. Solubilities and transport of ore minerals. In: H.L. Barnes (Editor), *Geochemistry of Hydrothermal Ore Deposits*. Holt, Rinehart and Winston, New York, N.Y., pp. 334–381.

Beales, F.W. and Jackson, S.A., 1968. Pine Point - a stratigraphical approach. *Can. Inst. Min. Metall. Trans.*, 61: 867–878.

Beales, F.W. and Oldershaw, A.E., 1969. Evaporite-solution brecciation and Devonian carbonate reservoir porosity in western Canada. *Am. Assoc. Pet. Geol., Bull.*, 53: 503–512.

Benz, J.P., 1964. *Le Gisement Plombo-zincifère d'Arenas (Sardaigne)*. Thesis Univ. Nancy. 126 pp.

Behre, C.H. and Garrels, R.M., 1953. Groundwater and hydrothermal deposits. *Econ. Geol.*, 38: 65–69.

Bernard, A., 1972. Metallogenic processes of intra-karstic sedimentation. In: G.C. Amstutz and A.J. Bernard (Editors), *Ores in Sediments*. Springer, Berlin, pp. 43–56.

Bernard, A. and Leleu, M., 1967. A propos de la concentration résiduelle de la blende et de la galène. *C.R. Acad. Sci. Fr.*, 265 (D): 279.

Bernard, A., Lagnay, Ph. and Leleu, M., 1972. A propos du rôle métallogénique du karst. *Proc. 24th Int. Geol. Congr., 1972*, Sect. 4: 411–422.

Bleifuss, R.L., 1966. The origin of the iron ores of southeastern Minnesota. *Diss. Abstr., B*, 28: 2475–2476.

Blount, D.N. and Moore, C.H., Jr., 1969. Depositional and non-depositional carbonate breccias. Chiantla Quadrangle, Guatemala. *Geol. Soc. Am. Bull.*, 80: 429–442.

Bogacz, K., Dzulynsky, S. and Haranczyk, C., 1970. Ore-filled hydrothermal karst features in the Triassic rocks of the Cracow-Silesian region. *Acta Geol. Pol.*, 20: 247–267.

Borchert, H., 1970. On the ore-deposition and geochemistry of manganese. *Miner. Deposita*, 5: 300–314.

Bosellini, A., Carraro, F., Corsi, M., De Vecchi, G.P., Gatto, G.O., Malaroda, R., Sturani, C., Ungaro, S. and Zanettin, B., 1967. Note illustrative della Carta Geologica d'Italia alla scala 1.100.000, 49. Verona. *Serv. Geol. Ital.*, 61 pp.

Bottke, H., 1969. Die Eisenmanganerze der Grube Geier bei Bingen/Thein als Verwitterungsbildungen des Mangans vom Typ Lindener Mark. *Miner. Deposita*, 4: 355–367.

Bourgin, A., 1945. Hydrographie karstique. La question du niveau de base. *Rev. Geogr. Alp.*, 33: 99–107.

Bowles, C.G., 1968. Theory of uranium deposition from artesian water in the Edgemont District, southern Black Hills. *Wyo. Geol. Assoc., 20th Field Conf. Guideb.*, pp. 125–130.

Bowles, C.G. and Braddock, W.A., 1963. Solution breccias of the Minnelusa Formation in the Black Hills, South Dakota and Wyoming. *U.S. Geol. Surv., Prof. Pap.*, 475-C: 91–95.

Braddock, W.A. and Bowles, C.G., 1963. Calcitization of dolomite by calcium sulphate solutions in the Minnelusa Formation, Black Hills, South Dakota and Wyoming. *U.S. Geol. Surv., Prof. Pap.*, 475-C: 96–99.

Bretz, J.H., 1950. Origin of the filled sink structures and circle deposits of Missouri. *Geol. Soc. Am. Bull.*, 61: 789–834.

Brown, J.S. (Editor), 1967. Genesis of stratiform lead–zinc–barite deposits. *Soc. Econ. Geol., Monogr.*, 3: 443 pp.

Brusca, C. and Dessau, G., 1968. I giacimenti piombo-zinciferi di S. Giovanni (Iglesias) nel quadro della geologia del Cambrico Sardo. *Ind. Min.*, 14: 470–489; 533–552; 597–609.

Brusca, C., Pretti, S. and Tamburrini, D., 1967. Le mineralizzazioni delle coperture di Mte Sa Bagattu (Iglesiente-Sardegna). *Rend. Assoc. Min. Sarda*, 72 (7): 89–106.

Burchard, E.F., 1960. Russellville brown iron ore district, Franklin Co., Alabama. *Ala. Geol. Surv. Bull.*, 70: 96 pp.

Burdon, D.J. and Al-Sharhan, A., 1968. The problem of the paleokarstic Damman limestone aquifer in Kuwait. *J. Hydrol.*, 6: 385–404.

Cadek, J. and Malkovsky, M., 1965. Contribution to the problems of transport of fluorine at low temperatures. In: *Symposium on Problems of Postmagmatic Ore Deposits, Prague, 1963*. Publ. House Czechosl. Acad. Sci., Prague, pp. 407–412.

Calembert, L., 1945. *Les Gisements de Terres Plastiques et Refractaires d'Andenne et du Condroz*. Vaillant-Carmanne, Liège, 204 pp.

Callahan, W.H., 1965. Paleophysiographic premises for prospecting for stratabound base-metal mineral deposits in carbonate rocks. *Symp. Min. Geol. Base Metals, Ankara, 1964:* 191–248.

Cavaille, A., 1960. Les argiles des grottes. *Ann. Speleol.*, 15: 383–400.

Chikishev, A.G., 1962. Importance économique nationale du karst de l'Oural moyen. *Bull. Mosk. Ova. Ispyt. Prir., Otd. Geol.*, 37 (6): 145–146.

Ciry, R., 1962. Le rôle du froid dans la spéléogenèse. *Spelunca Mem.*, 2: 29–34.

Clifton, H.E., 1967. Solution-collapse and cavity filling in the Windsor Group, Nova Scotia, Canada. *Geol. Soc. Am. Bull.*, 78: 819–832.

Collier, C.R., Krieger, R.A. and Whetstone, G.W., 1964. Hydrochemistry and sedimentation in Mammoth Cave. *Bull. Natl. Speleol. Soc.*, 26 (2): 83–84.

Collins, J.A. and Smith, L., 1972. Sphalerite as related to the tectonic movements, deposition, diagenesis and karstification of a carbonate platform. *Proc. 24th Int. Geol. Congr., Sect. 6*: 208–214.

Combes, P.J., 1965. Dissolution karstique sous une couche bauxitique. Remarques sur l'origine des gisements en poche. *C.R. Soc. Geol. Fr.*, 260: 123–124.

Comel, A., 1943. Appunti sulle terre rosse dell'Albania. *Boll. Soc. Geol. Ital.*, 61 (3): 400–404.

Corbel, J., 1959. Les grandes cavités de France et leurs relations avec les facteurs climatiques. *Ann. Speleol.*, 14: 31–47.

Cramer, H., 1941. Die Systematik der Karstdolinen. *Neues Jahrb. Mineral. Geol. Paleontol., Abt. B*, 85: 293–382.

Crawford, J. and Hoagland, A.H., 1968. The Mascot-Jefferson City zinc district, Tennessee. In: J.D. Ridge (Editor), *Ore Deposits in the United States 1933/1967*. AIME, New York, N.Y., pp. 242–256.

Cumming, L.M., 1968. Table head disconformity and Zn mineralization, western Newfoundland. *Can. Inst. Min. Metall. Bull.*, 61: 721–725.

Davis, J.B., and Kirkland, D.W., 1970. Native sulphur deposition in the Castille Formation, Culberson County, Texas. *Econ. Geol.* 65: 107–121.

Dean, B.G. and Brobst, D.A., 1955. Annotated bibliography and index map of barite deposits in United States. *U.S. Geol. Surv. Bull.*, 1019-C: 145–186.

Derry, D.R., 1969. Supergene remobilization at the Tynagh Mine, Ireland, of Northgate Exploration. In: P. Zuffardi (Editor), *Remobilization of Ores and Minerals*. Ente Minerario Sardo, Cagliari, pp. 205–209.

Dublanskij, V.N. and Slutov, Yu.L., 1967. Gas composition of air in karst cavities of the Crimean Highland. Am. Geol. Inst., Wash., D.C. (transl. from Russian), pp. 39–41.

Emberger, A., 1969. Problèmes des remodilisations dans les gites de plomb et de zinc. In: P. Zuffardi (Editor), *Remobilization of Ores and Minerals*, Ente Minerario Sardo, Cagliari, pp. 37–57.

Emery, O.K., 1946. Marine solution basins. *J. Geol.*, 54: 209–228.

Faucherre, J. and Bonnaire, Y., 1959. Sur la constitution des carbonates complexes de cuivre et de plomb. *C.R. Acad. Sci. Fr.*, 248: 3705–3707.

Feulner, A.J. and Hubble, J.H., 1960. Occurrence of strontium in the surface and ground waters of Champaign County, Ohio. *Econ. Geol.*, 55 (1): 176–186.

Ford, T.D., 1969b. The Blue John fluorspar deposits of Treak Cliff Cave, Derbyshire, in relation to the boulder bed. *Yorksh. Geol. Soc., Proc.*, 37 (7): 153–157.

Ford, T.D. and King, R.J., 1969. The origin of the silica sand pockets in the Derbyshire Limestone. *Mercian Geol.*, 3: 51–69.

Ford, T.D. and Sarjeant, W.A.S., 1964. The stalactitic barytes of Derbyshire. *Proc. Yorksh. Geol. Soc.*, 34 (4-19): 371–386.

Fuchs, Y., 1969. Quelques exemples de remobilisations dans le domaine épicontinental (Sud de Massif Central). In: P. Zuffardi (Editor), *Remobilization of Ores and Minerals*. Ente Minerario Sardo, Cagliari, pp. 161–183.

Gèze, B., 1949. Les gouffres à phosphate du Quercy. *Ann. Speleol.*, 4: 89–107.

Ginzburg, I.I., 1952b. Mezozoiskie karsty i svyazannye s nimi poleznye iskopaenye na Urale (Mesozoic karsts and related mineral resources in the Urals). *Sb. Kora Vyvetr.*, 1.

Ginzburg, I.I., 1966. Karst und Erzbildung. *Z. Angew. Geol.*, 12 (2): 67–71.

Glagoleva, N.A., 1958. Les formes de migration des éléments dans les eaux courantes. *Dokl. Akad. Nauk S.S.S.R.*, 121 (6): 1052–1055.

Goguel, J., 1953. Données techniques sur l'effondrement des cavités souterraines. *Ann. Speleol.*, 8: 1–8.

Goni, J., Pierrot, R. and Passaqui, B., 1965. Le problème du plomb dans certaines aragonites plombifères. *Bull. Soc. Fr. Mineral. Cristallogr.*, 88: 273–280.

Gorlich, E. and Gorlich, Z., 1958. Adsorption series of some cations on pure calcium carbonate and on natural limestone and dolomite. *Bull. Acad. Pol. Sci., Ser. Sci. Chim. Geol. Geogr.*, 6: 669–674.

Grossman, I.G., 1968. Origin of the sodium sulfate deposits of the northern Great Plains of Canada and the United States. *U.S. Geol. Surv., Prof. Pap.*, 600-B: 104–109.

Gruszcyk, H., 1967. The genesis of the Silesian-Cracow deposits of lead–zinc ores. *Econ. Geol. Monogr.*, 3: 169–177.

Habashi, F., 1966. The mechanism of oxidation of sulphide ore in nature. *Econ. Geol.*, 61 (3): 587–591.

Hagni, R.D., and Desai, A.A., 1966. Solution thinning of the M Bed host rock limestone in the Tri-State district, Missouri, Kansas, Oklahoma. *Econ. Geol.*, 61: 1436–1442.

Hart, O.M., 1958. Uranium deposits in the Pryor-Big-Horn-Mountains, Carbon County, Montana, and Big Horn County, Wyoming. *U.N. Int. Conf. Peaceful Uses At. Energy, Geneva, 1958, Proc.*, 2. 523–526.

Hirschmann, G. and Neuhof, G., 1964. Beziehungen zwischen Verwitterungsvorgänge und Lagerungsverhältnisse im Bereich des Ludwigsdorfer Unterkambriums. *Geologie*, 13 (5): 524–542.

Hitchon, B., Levinson, A.A. and Reeder, S.W., 1969. Regional variations of river water composition resulting from halite solution, Mackenzie River drainage basin. *Can. Water Resour. Rep.*, 5: 1395–1403.

Hoagland, A.D., Hill, W.T. and Fulweiler, R.E., 1965. Genesis if the Ordovician zinc deposits in east Tennessee. *Econ. Geology*, 60: 693–714.

Holland, H.D., 1964. The chemical evolution of some cave water. *Bull. Speleol. Soc.*, 26 (2): 69.

Holter, M.E., 1969. The Middle Devonian Prairie Evaporite of Saskatchewan. *Sask. Dep. Miner. Resour. Rep.*, 123: 49–56.

ICSOBA (Int. Comm. Stud. Bauxites, Oxides and Hydroxides of Aluminum), 1970/1971. *Int. Symp. ICSOBA, 2nd, Budapest 1969, Proc.*, 1, 2.

Ingham, F.T. and Bradford, E.F., 1960. The geology and mineral resources of the Kinta Valley, Perak. *Malaya, Geol. Surv. Dep., Mem.*, 9: 347 pp.

Ioannou, I., 1966. La grotte-mine de Lefki Kavalla. *Delt. Ellenikes Spelailog. Etat.*, 8 (5): 137–142.

Itkina, E.S., 1963. Fluorine geochemistry in sedimentary rocks. *Geokhim. Gidrokhim. Neft. Mestor. Akad. Nauk. S.S.S.R.*, 1963: 57–64.

Jadoul, F., 1973. *Il Giacimento di Fluorite di Paglio Pignolino (Bergamo)*. Thesis, Univ. Milano. 93 pp.

Jordan, R.H., 1950. An interpretation of Floridan karst. *J. Geol.*, 58: 261–268.

Jennings, J.N., 1968. Syngenetic karst in Australia. *Aust. Natl. Univ. Dep. Geogr. Publ.*, G/5: 41–110.

Kaisin, F., Jr., 1956. Le rôle de la substitution dans l'altération météorique des roches sédimentaires, spécialement des calcaires. *Univ. Louvain, Inst. Géol. Mem.*, 20: 47–164.

Keller, W.D., Westcott, J.F. and Bledsoe, A.O., 1954. The origin of Missouri fire-clays. *Natl. Clay Conf., 2nd, Proc.*, pp. 7–46.

Kesler, T.L., 1950. Geology and mineral deposits of the Cartersville District, Georgia. *U.S. Geol. Surv., Prof. Pap.*, 224: 97 pp.

Koritnig, S., 1951. Ein Beitrag zur Geochemie des Fluors. *Geochim. Cosmochim. Acta*, 2: 89–116.

Krumbein, W.C. and Garrels, R.M., 1952. Origin and classification of chemical sediments in terms of pH and oxidation-reduction potentials. *J. Geol.*, 60: 1–33.

Kunsky, J., 1957. Thermomineral karst and caves of Zbrasov, northern Moravia. *Sb. Cesk. Spolecn. Zemep.*, 62: 306–351.

Lagny, Ph., 1969. Minéralisations plombo-zincifères triasiques dans un paléo-karst (gisement de Sala-fossa, province de Belluno, Italia). *C.R. Acad. Sci. Ser D*, 268: 1178–1181.

Lagny, Ph., 1971. Les minéralisations plombo-zincifères de la région d'Auronzo (province de Belluno, Italie); remplissage d'un paléokarst d'âge Anisien supérieur. *C.R. Acad. Sci., Ser D*. 273: 1539–1542.

Landes, K.K., 1945. The Mackinac Breccia. *Mich. Geol. Surv. Publ.*, 44: 121–154.

Larson, L.T., 1970. Cobalt and nickel-bearing manganese oxides from the Fort Payne Formation, Tennessee. *Econ. Geol.*, 65: 952–962.

Lebedev, L.M., 1967. *Metacolloids in Endogenic Deposits*. Plenum Press, New York, N.Y., transl. from Russian 298 pp.

Leith, C.K., 1925. Silicification of erosion surfaces. *Econ. Geol.*, 20: 513–523.

Lindgren, W., 1933. *Mineral Deposits*. McGraw-Hill, New York, N.Y., 4th. ed., pp. 430–432.

Lukashev, K.I., 1970. *Lithology and Geochemistry of the Weathering Crust*. Israel Program for Scientific Translations, Jerusalem, translated from Russian, 368 pp.

Marcello, A., 1969. Déposition supergénique de galène: les examples de la Sardaigne. In: P. Zuffardi (Editor), *Remobilization of Ores and Minerals*. Ente Minerario Sardo, Cagliari, pp. 293–303.

Materikov, M.P., 1961. Specific features of tin ore deposits in areas of carbonate rocks. *Am. Geol. Inst., Transl.*, SG-61-9-4: 15 pp (transl. from *Sov. Geol.*, 9: 96–107.)

Mather, W.B., 1947. Barite deposits of Missouri. *Am. Inst. Metall. Eng., Tech. Publ.*, 2246, *Mineral. Technol.*, 11 (5): 15 pp.

Mills, J.W. and Eyrich, H.T., 1966. The role of unconformities in the lozalization of epigenetic mineral deposits in the United States and Canada. *Econ. Geol.*, 61: 1232–1257.

Monaco, A., 1964. Remplissage karstique du Djebel Sidi Amara (Cap Bon, Tunisie). *C.R. Soc. Geol. Fr.*, 9: 389.

Moore, D.P., 1971. The occurrence of biogenetic sulphur deposits in west Texas and their implications for future explorations activity. *Soc. Min. Eng.*, preprint.

Moore, G.K., Burchett, C.R. and Bingham, R.H., 1969. Limestone hydrology in the upper Stones River basin, central Tennessee. *Tenn. Div. Water Resour.*, 58 pp.

Moreau, M., Poughon, A., Puibaraud, Y. and Sanselme, H., 1966. L'Uranium et les granites. *Chron. Mines Rech. Minière*, 350: 47–51.

Morgan, A.M., 1942. Solution-phenomena in the Pecos basin in New Mexico. *Am. Geophys. Union Trans.*, 1: 27–35.

Morrissey, C.J. and Whitehead, D., 1970. Origin of the Tynagh residual orebody, Ireland. *9th Commonw. Min. Metall. Congr., London, 1969, Proc.*, 2: 131–145.

Munch, W., 1960. Ricerche geo-giacimentologiche in Sardegna. In: *Giornate di studio sulle ricerche geo-giacimentologiche, Rome, 1960*, 1 (2): 8.

Nicolas, J., 1968. Nouvelles données sur la genèse des bauxites à mur karstique du sud-est de la France. *Miner. Deposita*, 3: 18–33.

Obellianne, J.M., 1963. Le gisement de phosphate tricalcique de Makatéa (Polynésie Française-Pacifique Sud). *Sci. Terre*, 9: 5–60.

Ozoray, G., 1961. The mineral filling of thermal spring caves. *Rass. Speleol. Ital., Mem.*, 5 (2): 152–170.

Padalino, G., Pretti, S., Tocco, S. and Violo, M., 1971. Some examples of lead–zinc–barite depositions in karstic environment. *2nd. Int. Symp. Miner. Deposits Alps, Bled*, pp. 109–113.

Padalino, G., Pretti, S., Tamburrini, D., Tocco, S., Uras, I., Violo, M. and Zuffardi, P., 1972. Ore deposition in karst formation with examples from Sardinia. In: G.C. Amstutz and A.J. Bernard (Editors), *Ores in Sediments*. Springer, Berlin, pp. 209–220.

Panos, V. and Stelcl, O., 1968. Physiographic and geologic control in development of Cuban Mogotes. *Z. Geomorphol.*, 12: 117–173.

Perna, G., 1972. Fenomeni carsici e giacimenti minerari. (Proc. Semin. Speleogenesi, 1972.). *Grotte Ital., Ser. 4*, 4: 34 pp.

Pittman, J.S. and Folk, R.L., 1971. Length-slow chalcedony after sulphate evaporite mineral in sedimentary rocks. *Nat. Phys. Sci.*, 230 (11): 64–65.

Poliak, J., 1952. Über die Erscheinung von fossilen Karstformen und ihren Zusammenhang mit der Erzlagerstätte am Debeljak im nordlichen Velebit. *Geol. Vjesn.*, 2 (4): 99–110.

Renault, Ph., 1963. Quelques réalisations de spéléologie expérimentale: vermiculation argileuse et corrosion sous remplissage. *Spelunca, 4, Mem., 3–5th Congr. Speleol., Millau*, pp. 48–54.

Roberst, A.E., 1966. Stratigraphy of the Madison Group near Livingston, Montana, and discussion of karst and solution-breccia features. *U.S. Geol. Surv. Prof. Pap.*, 526-B: 23 pp.

Roch, E., 1966. A comparison of some European bauxites with those of the Caribbean. *Geol. Soc. Jamaica J.*, 7: 1–23.

Rodeghiero, F., 1974. *Le Mineralizzazioni a Pb–Zn, Fluorite, Barite della Zona del Pizzo Presolana (Prealpi Bergamasche)*. Thesis Univ. Milan, 166 pp.

Roedder, E., 1966. Environment of deposition of stratiform Mississippi Valley-type ore deposits, from studies of fluid inclusions. *Econ. Geol. Monogr.*, 3: 349–362.

Roques, H., 1963b. Observations physiochimiques sur les eaux d'alimentation de concrétion. *Ann. Speleol.*, 18 (4): 337–404.

Roques, H., 1965. Sur la genèse des formations aragonitiques naturelles. *Ann. Speleol.*, 20 (1): 47–54.

Routhier, P., 1963. *Les Gisements Métallifères*. Masson, Paris, 1282 pp.

Salvadori, I. and Zuffardi, P., 1964. Supergene sulfides and sulfates in the supergene zones of sulfide ore deposits. In: C.G. Amstutz (Editor), *Sedimentology and Ore Genesis (Developments in Sedimentology, 2)*. Elsevier, Amsterdam, pp. 91–99.

Sawkins, F.J., 1969. Chemical brecciation, an unrecognized mechanism for breccia formation. *Econ. Geol.*, 64: 613–617.

Searls, F.Jr., 1952. Karst ore in Yunnan. *Econ. Geol.*, 47: 339–346.

Siegel, F.R., 1965. Aspects of calcium carbonate deposition in Great Onyx Cave, Kentucky. *Sedimentology*, 4 (4): 285–299.

Smith, L. and Collins, J.A., 1971. Ordovician karst and cave deposits, northwestern Newfoundland (abs.), *Caves Karst*, 13: 49–50.

Smith, D.I. and Nicholson, F.H., 1963/64. A study of limestone solution in northwest Co-Clare, Eire. *Proc. Speleol. Soc. Univ. Bristol*, 10 (2): 119–138.

Society of Economic Geologists, 1971. A paleoaquifer and its relation to economic mineral deposits. The Lower Ordovician Kingsport Formation and Mascot Dolomite—a symposium. *Econ. Geol.*, 66: 695–810.

Soulé de la Fonte, F., 1967. Les gîtes de fluorine stratiform de la bordure Nord du Morvan. *Chron. Mines Rech. Minière, Fr.*, 361: 85–108.

Stecl, O., 1964. Composition chimique de l'eau s'égouttant de stalactites dans quelques cavernes du karst de Moravie. *Cesk. Kras*, 16: 23–32.

Steinbreker, B., 1959. Die Suberosion des Zechsteingebirges im östlichen und nordöstlichen Harzvorland mit besonderer Berücksichtigung der Edderitzer Mulde. *Geologie*, 8: 489–522.

Takahashi, T., 1960. Supergene alteration of zinc and lead deposits in limestone. *Econ. Geol.*, 55 (6): 1083–1115.

Tamburrini, D. and Zuffardi, P., 1967. Ulteriori sviluppi delle conoscenze e delle ipotesi sulla metallogenesi sarda. In: *Giornata di Studi Geominerari, Agordo, October 7, 1967*, pp. 149–168. (A seminar organized by the Istituto Tecnico-Industriale Minerario Statale U. Follador and by the Istituti di Geologia e Mineralogia dell'Università di Padova.)

Tamburrini, D. and Zuffardi, P., 1969. Field evidences of supergene remobilization of barium (and possibly of barite) in Sardinia. In: P. Zuffardi (Editor), *Remobilization of Ores and Minerals*. Ente Minerario Sardo, Cagliari, pp. 305–314.

Thomas, T.M., 1963. Solution subsidence in southeast Carmarthenshire and southwest Breconshire. *Inst. Br. Geogr., Trans. Pap.*, 33: 14–60.

Toit, A.L. du., 1951. The diamondiferous gravels of Lichtenburg. *S. Afr. Geol. Surv., Mem.*, 44: 50 pp.

Ulrich, E.O., 1931. Origin and stratigrafical horizon of the zinc ores of the Mascot District of east Tennessee. *Wash. Acad. Sci. J.*, 21: 30–31.

Valeton, I., 1966. Sur la genèse des gisements de bauxite du Sud-Est de la France. *Bull. Soc. Geol.*, 8: 685–701.

Verwoerd, W.J., 1957. The mineralogy and genesis of the lead–zinc–vanadium deposit of Abenab West in the Otavi Mountains, South West Africa. *Univ. Stellenbosch Ann., Sect. A*, 33: 235–319.

Villiers, J. de, 1960. The manganese deposits of the Union of South Africa. *S. Afr. Geol. Surv., Handb.*, 2: 271 pp.

Walker, R.T., 1928. Deposition of ore in pre-existing limestone caves. *Am. Inst. Min. Metall. Eng., Tech. Publ.*, 154: 43 pp.

Wash, P.T., and Brown, E.H., 1971. Solution subsidence outliers containing probable Tertiary sediments in northeast Wales. *Geol. J.*, 7: 299–320.

Wedepohl, K.H., 1956. Untersuchungen zur Geochemie des Bleis. *Geochim. Cosmochim. Acta*, 10: 69–148.

White, D.E., 1966. Outline of thermal and mineral waters as related to origin of Mississippi Valley ore deposits. *Econ. Geol.*, 13: 379–382.

White, D.E., 1968. Environments of generation of some base-metal ore deposits. *Econ. Geol.*, 63: 301–335.

Wolf, K.H., 1976 Ore genesis influenced by compaction. In: G.V. Chilingar and K.H. Wolf (Editors), *Compaction of Coarde-Grained Sediments, 2*. Elsevier, Amsterdam, in press.

Wolfenden, E.B., 1965. Bau mining district, west Sarawak, Malaysia, 1. Bau. *Malaysia Geol. Surv., Borneo Region, Bull.*, 7: 147 pp.

Yates, R.G. and Thompson, G.A., 1959. Geology and quicksilver deposits of the Terlingua District, Texas. *U.S. Geol. Surv., Prof. Pap.*, 312: 111 pp.

Zans, V.A., 1959. Recent views on the origin of bauxite. *Geonotes*, 1: 123–132.

Zuffardi, P., 1962. Fenomeni di ricircolazione nel giacimento di Montevecchio e l'evoluzione in profondità della sua mineralizzazione. *Res. Assoc. Min. Sarda*, 56 (1/2): 17–73.

Zuffardi, P., 1968. Transformism in the genesis of ore deposits: examples from Sardinian lead–zinc deposits. *23th Int. Geol. Congr.*, 7: 137–149.

Zuffardi, P., 1969. Remobilization in Sardinian lead–zinc deposits. In: *Remobilization of Ores and Minerals*. Ente Minerario Sardo, Cagliari, pp. 283–292.

Zuffardi, P., 1970. La métallogenèse du plomb, du zinc et du barium en Sardaigne: un exemple de permanence de polygénétisme et de transformisme. *Ann. Soc. Geol. Belg.*, 92: 321–344.

Chapter 5

PLACER DEPOSITS

JOHN R. HAILS[1]

INTRODUCTION

Placers may be defined as surficial mineral deposits formed by the mechanical concentration of mineral particles from weathered debris. The mechanical agent is usually
alluvial but can also be marine, aeolian, lacustrine, or glacial, and the mineral is usually
a heavy metal such as gold (*Glossary of Geology,* 1972; see also Tables I and II).

Various classifications have been presented in both mining and geological texts, but
in this chapter placers are divided into marine (offshore), alluvial (stream), including
river-terrace, beach, eluvial (slope), residual and fossil. The last type is often referred
to as a deep-lead or buried placer which is an ancient deposit that has been buried under
overburden varying in thickness from 30 m to 305 m and more (Griffith, 1960). Emery
and Noakes (1968) have usefully divided placer deposits into three groups, each one being

TABLE I

Physical properties of the more common placer deposits

Mineral	Hardness (Mohs scale number)	Specific gravity	Resistance to weathering*
Cassiterite (tin)	6–7	6.8–7.1	high
Diamond	10	3.52	very high
Garnet	6.5–7.5	3.5–4.3	moderate
Gold	2.5–3	19.3	very high
Ilmenite	5–6	4.5–5	high
Magnetite	5.5–6.5	5.1–5.18	high
Monazite	5	4.9–5.3	high
Platinum	4–4.5	21.46 (chemically pure); 14–19 (native)	very high
Ruby and sapphire	9	3.95–4.10	very high
Zircon	7.5	4.5–4.7	very high

* According to Kukharenko (see Lukashev, 1970).

---

[1] Contribution prepared at Institute of Oceanographic Sciences, Taunton, Great Britain.

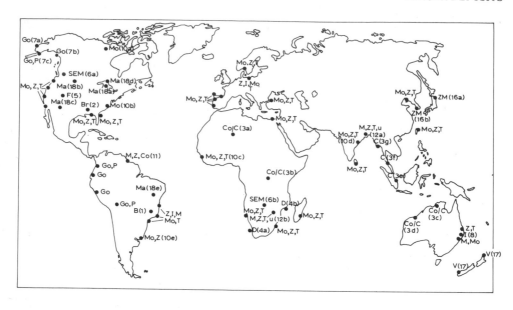

Fig. 1. World map to show the distribution of diamonds, platinum, monazite, ilmenite, magnetite, rutile, zircon and other placers.

1 = baddeleyite placers (B): Pocos de Caldas

2 = placers containing niobium-bearing brucite (Br): Arkansas

3 = deposits and ore manifestations containing columbite and cassiterite (Co/C), for example: (a) Ahaggar; (b) Zaire; (c) Northern Territory of Australia; (d) Pilbara – cassiterite (C): (e) Indonesia; (f) Malaysia; (g) Thailand.

4 = diamonds (D), for example: (a) Southwest Africa; (b) Mozambique channel, southeast Africa.

5 = placers containing fergusonite (F): Colorado.

6 = placers containing samarskite, euxenite and monazite (SEM): (a) British Columbia; (b) Zambia.

7 = gold (Go), for example: (a) Norton Sound, Alaska; (b) Stephens Passage, Alaska; (c) Nome, Alaska.

8 = ilmenite (I): see Table II

9 = magnetite (M): see Table II

10 = monazite (Mo), for example: (a) Lake Yamba; (b) North and South Carolina; (c) Sierra Leone; (d) India (east coast); (e) Uruguay; see also Table II

11 = placers containing monazite, zircon and columbite (MZCo): (a) West Bengal; (b) Guyana.

12 = placers containing monazite, zircon, thorite and uraninite (MZTU): (a) Pakistan; (b) Rhodesia.

13 = platinum (P): Colombia, Bolivia; see also Table II

14 = ruby and sapphire (RS): see Table II

15 = zircon (Z): see Table II

16 = placers containing zircon, monazite, fergusonite, xenotite and samarskite (ZM): (a) Japan, (b) South Korea.

17 = deposits containing vanadium-bearing titano-magnetite (V).

18 = metamorphosed placers containing monazite, zircon and uranium minerals (Ma), for example: (a) Palmer; (b) Little Bighorn; (c) Gallup; (d) Blind River (or Elliot Lake) Ontario; (e) Serra de Jacobina.

19 = titanium minerals (T), for example: Australia (New South Wales and Queensland), Brazil, Chinese People's Republic, East Germany, Iberian Peninsula, India (east and west coasts), Malagasy, Natal, Norway, Poland, Sri Lanka, Southwest Africa, Taiwan, Turkey, United Arab Republic (Egypt), U.S.A. (California, Florida, Oregon, Washington).

TABLE II

General classification and distribution of placers (to be used in conjunction with Fig. 1). (Based on Cheraskov et al., see Vlasov, 1968, his fig. 133 and table 176)

Classification	Genetic group	Type of placer (also rare-metal placer and other economic minerals)	Geographical distribution
Paragenetic (geological complexes) near-source placers	eluvial, talus, proluvial, fluvio-glacial and moraine placers; placers of small local glaciers; alluvial placers of river head-waters; lacustrine placers of small lakes	cassiterite(tin); monazite; zircon (baddeleyite; cinnabar; columbite-tantalite; corundum; euxenite; hessite; loparite; native copper; phosphorite; pyrochlore; silver; smarskite; thorite; uranothorite; xenotite)	Ahaggar; Air; Australia (Northern Territory, Pilbara); Brazil (Pocos de Caldas); Canada (British Columbia; Lake Yamba); Nigeria, area of Lake Kivu; Sierra Leone; South-east Asian Tin Belt (Burma, Thailand, Indonesia, Malaysia); Transvaal; U.S.A. (Arkansas, Colorado)
Alluvial (stream) placers	alluvial, aeolian; lacustrine (cut-off/oxbow lake) placers	cassiterite; diamonds; gems; gold; monazite; platinum; scheelite; wolframite (tungsten); zircon	Argentina; Australia; Bolivia; Brazil; Colombia; Ecuador; Ghana; Guyana; Japan; Mozambique; Pakistan; Rhodesia; Romania; Siberia (Amur, Lena and Yenisly Rivers); South Africa; South Korea; U.S.A. (Alaska, California, Montana, North and South Carolina, Wyoming); Uruguay; West Bengal; Yukon
Placers of water bodies	marine, lacustrine (large lakes), aeolian placers	chromite; garnet, ilmenite; monazite; rutile; sillimanite; vanadium-bearing titano-magnetite; zircon	Albania; Australia (east coast); Brazil (coast); Chinese People's Republic; East Germany; Gambia; Guinea (coast); India (east and west coasts); Ivory Coast (coast); Malagasy Republic; Mauritania (coast); Mozambique (coast); Norway; Poland; Portugal; Senegal (coast); Sierra Leone (coast); Somali Republic; Spain; Sri Lanka; South West Africa; Taiwan; Tanzania (coast); Turkey; United Arab Republic (Egypt); U.S.A. (States of California, Florida, Maryland, Oregon, Washington); Yugoslavia
Metamorphosed placers		monazite; uranium minerals; zircon	Brazil; Canada; U.S.A.

characterized by its own unique properties and particular environment of deposition. These are heavy heavy minerals (gold, tin and platinum) which occur mainly in streams, light heavy minerals (ilmenite, rutile, zircon and monazite) occurring on beaches, and gems (mainly diamonds) which are chiefly alluvial placers (see also Table II and Fig. 1).

Fossil placer deposits of economic size are relatively scarce (e.g., Witwatersrand[1], Blind River and Serra de Jacobina, see below, p. 221) and as Dunham (1969) has stated: "It is a curious fact that placer concentrations are virtually unknown (or perhaps unrecognized) in the older, consolidated rocks" (see Zimmerle, 1973). The majority of minerals recovered from placers are near primary sources, as cassiterite, gold, platinum and diamonds, for example. But, because there may be more than one primary source, it is often difficult to determine the exact distance of transportation. Zeschke (1961), prospecting for ore deposits by panning heavy minerals from river sands, has found that some minerals are transported much farther by streams than had been thought earlier.

The need for additional minerals to meet industrial demand at a time when high-grade placer reserves are rapidly becoming exhausted in several parts of the world, has been reflected during the past decade by increased interest in methods of recovering metals from low-grade deposits on land and in the intensive search for, hitherto, unexploited deposits on the seafloor (Emery and Noakes, 1968). In the late 1960s, for example, more than seventy dredging operations were active, exploiting such diverse products as diamonds, gold, heavy-mineral and tin sands, as well as sand and gravel (Cruickshank, 1969). The purpose of this chapter is to examine briefly the general distribution of placers and their exploitation, past and present, in the light of engineering technology and demand. Each type of placer deposit will be considered, in turn, although not necessarily in a separate section. At the time of writing, the author has endeavoured to cite the most recent literature although he acknowledges that some references may have been inadvertently omitted. The reader is referred not only to the bibliography at the end of this paper, but also to many of the references contained in the papers cited here.

Inevitably, a slight degree of overlap occurs between the discussions on the different types of placers and those metals/minerals mentioned individually. Nevertheless, every attempt has been made to present a comprehensive review of such deposits and to highlight their importance in the present-day economy.

CASSITERITE (SnO$_2$)

In general, tin placer deposits are more productive and more economic to mine than lode deposits. Cassiterite, from which most tin is smelted, is contained in placer concentrates that also yield valuable amounts of niobium, tantalum, and rare-earth minerals. A comprehensive description of the types of tin deposits and main tin-producing areas

---

[1] See Chapters 1 and 2 by Pretorius, Vol. 7.

of the world has been compiled by Sainsbury (1969) from whose report part of the following account has been abstracted.

The main source of cassiterite is the Southeast Asian tin belt, extending more than 2400 km from northern Burma through western Thailand and Malaysia to Billiton (Belitung) Island in Indonesia (Figs. 1, 2). Geologic factors considered to have contributed to the unusual size and value of this belt include: high concentration of tin granite, deep and rapid tropical weathering which has released large quantities of primary tin from lode sources, the formation of marine and alluvial placers many of which are now seaward of the present coastline, and the preservation of placers resulting from relatively low terrain drained by low-velocity streams.

Cassiterite is concentrated by the selective chemical weathering and removal downslope of feldspar and, to a lesser degree, of quartz. Intense tropical weathering is by far the most important geologic process throughout the Southeast Asian tin belt with the result that important concentrates are derived from primary low-grade deposits. According to Aleva et al. (1973a), a large bucket dredge can operate economically in areas with 0.2 kg tin metal per m^3, which corresponds to approximately 100 ppm. Also, primary mineralization with a cassiterite content of as little as 1—10 ppm may lead to tin deposits of economic importance, provided the process of concentration is favourable.

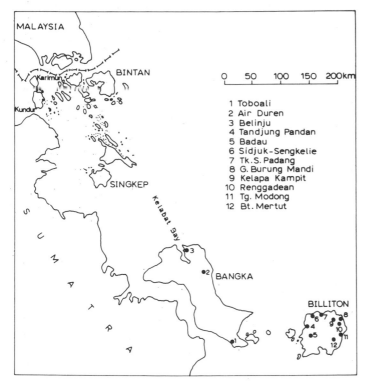

Fig. 2. Map to show the distribution of tin in the Indonesian (Southeast Asia) tin belt.

Aleva (1973), together with his co-workers (Aleva et al., 1973a,b) have concluded from detailed studies of the environmental influence on secondary tin deposits that cassiterite deposits are of a residual-elutriational origin and not alluvial sensu stricto. Thus, the concentration of cassiterite within the Thailand—Malaysia—Indonesia tin belt is entirely dependent upon secondary processes although only a few primary de-

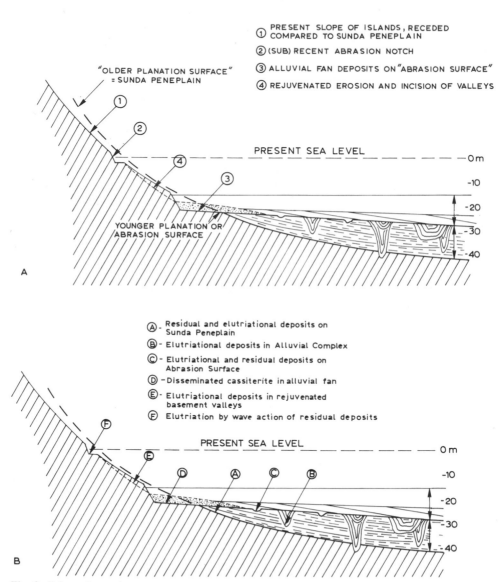

Fig. 3. Schematic section perpendicular to the coast of Tjebia or Bangka, (A) summarizing the geologic history; (B) indicating the placer possibilities. (After Aleva, 1973.)

posits of marginal grade would exist. These secondary processes are chemical weathering as just mentioned, trap-forming by the occurrence of a coarse gravel layer of weathering-resistant vein quartz (kaksa) on a sedimentary or granitic bedrock, or of sinkholes in a limestone bedrock, and the removal of fine-grained and light minerals by groundwater, or fluvial, or wave action, resulting in a relative enrichment in the cassiterite and other heavy minerals. Type localities (Fig. 3) for such deposits are the islands of Bangka and Billiton (Adam, 1960; Osberger, 1967, 1968; and references cited therein).

A brief word is pertinent here about the origin of the cassiterite deposits on these two islands, in particular about the "kaksa" hypothesis advanced by Adam (1932, 1933). He proposed that the cassiterite placer deposit is, in fact, the normal weathering residue found on erosional surfaces within the humid tropics, and not entirely the result of chemical weathering, erosion and transport. Therefore, it mainly consists of coarse grains and angular fragments of vein quartz derived from the in situ bedrock, and all insoluble heavy minerals contained within it.

Between 30% and 55% of the world's annual tin production comes from Malaysia and about 10% from Thailand and Indonesia. In the mid-1960s more than 1000 gravel pump and other mines were in operation in Malaysia where placer cassiterite deposits of various origins are associated with widespread unconsolidated clays and gravels. These are mined throughout the country in two belts, almost parallel to the coastline and coinciding with two mountain ranges which consist of batholiths, stocks of biotite—muscovite granite, and related rocks of probable Cretaceous age. The igneous strata intrude a thick series of sedimentary and metamorphic rocks of Cambrian to Cretaceous age.

The Kinta Valley, the geology of which has been described in detail by Ingham and Bradford (1960), is the most productive area in Malaysia. Crystalline limestone, encircled by hill-forming biotite-granite, and characterized by numerous deep solution hollows, is the predominant bedrock in the valley. These hollows have trapped the alluvial cassiterite which is recovered by dredging, gravel pump mining and hydraulicking. Trace amounts of columbite—tantalite, monazite, ilmenite, cobaltite, rutile and struverite occur in the placer concentrates. The Selangor River and the Frasers Hill area are also important for placer production in Malaysia (Roe, 1953). The deposits in Thailand and Burma have an almost identical origin in that they are closely associated with biotite granite intrusive rocks of Late Cretaceous age (Brown et al., 1951).

Schuiling (1967) has plotted all the known economic and uneconomic occurrences of tin on the continents around the Atlantic Ocean and has shown that relatively narrow zones, referred to as "tin belts", extended unbroken from one continent to another before the advent of continental drift. He speculates that concentrations of workable tin deposits occur at the intersection of orogenic belts with zones of primitive enrichment of tin. The data derived from the investigation may be useful for planning future prospecting for tin deposits. For example, Schuiling argues that the rich Bolivian tin deposits are located where the Rondonia—Guyana belt strikes into the Andes, whereas

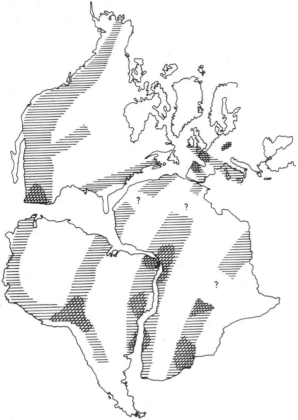

Fig. 4. Continental drift and tin belts on the continents around the Atlantic Ocean. (After Schuiling, 1967.) Position of tin belts on a reconstruction of the continents. Darker areas in tin belts indicate concentration of workable deposits.

elsewhere in the Andes the same apparent geological processes produced only insignificant concentrations of tin minerals (Fig. 4).

GOLD

Gold occurs in placers of varying geological ages. Young placer deposits are composed of unconsolidated or semi-consolidated sand and gravel containing small quantities of native gold and other heavy minerals. Most are stream deposits which occur within present-day valleys, or on benches or terraces of pre-existing rivers. In the United States, for example, gold is found along such rivers as the American, Feather and Yuba in the Sierra Nevada of California; the Alder Gulch at Virginia City, Montana; on the Yukon River at Fairbanks, Alaska, and on or near the beach at Nome, Alaska. Elsewhere, gold has been located in young placer deposits at Klondike in the Yukon; the upper Lena, Yenisey and Amur Rivers in Siberia, and along the east flank of the Andes Mountains in Bolivia, Peru, Ecuador and Colombia (Brobst and Platt, 1973).

The Tertiary placers of the Sierra Nevada and geologically similar placers at Ballarat, Australia, as well as the Upper Cretaceous to Lower Tertiary auriferous conglomerates of northwestern Wyoming are also placed in this group. However, they are older and more consolidated than the typical young placers, but nevertheless they are more akin to young than to fossil placers.

Notable ancient placers occur in the Witwatersrand District, Republic of South Africa; Tarkwa, Ghana, and the Serra de Jacobina, Bahia, Brazil. These are of Precambrian age and have been lithified to conglomerate to become part of the bedrock and may be termed "fossil placers" (Brobst and Pratt, 1973; Koppel and Saager, 1973; cf. also Chapters 1 and 2 by Pretorius, Vol. 7).

Production from placer mines has been declining worldwide for many years and significant quantities of gold, probably accounting for about 5% to 10% of the world production, only occur now in the U.S.S.R. and Colombia. Brobst and Platt (1973) state that little is known about production of placer gold from the U.S.S.R. lately, although it is estimated that 50% of the total production of about 290–350 million ounces has come from placers. Formerly productive placers in Egypt, Spain and Australia are now virtually exhausted while those in Brazil, Peru and Bolivia are being worked on a much reduced scale.

In 1969, placer gold production accounted for little more than 1% of the total U.S. output. According to Brobst and Platt (1973), there are various reasons for this marked decline, including environmental problems such as water pollution and siltation; increasing production costs, and ease of exploitation resulting in rapid exhaustion. Consequently, placer deposits are not expected to be a significant factor in future U.S. gold production.

The large reserves of the Witwatersrand, estimated to be about 600 million ounces, accentuate the future importance of fossil placer deposits in this region (Gold Producers Committee, 1967; Kavanagh, 1968; U.S. Bureau of Mines, 1972). Whether or not other ancient (fossil) placer deposits, on the same scale as the Rand, are likely to be discovered elsewhere is a matter for conjecture, although Brobst and Pratt (1973) point out that it seems possible that some of the ancient shield areas — Canadian, Brazil-Guyanan and African—Arabian — may contain extensive gold-bearing conglomerates. The uranium-bearing conglomerates discovered at Blind River (or Elliot Lake), Ontario, Canada, in 1949, only contain minor amounts of gold (Roscoe, 1957; Bateman, 1958).

By contrast, marine placers are virtually impoverished of economically viable gold deposits mainly because such placers are polygenetic and polycyclic in origin as a result of gold-bearing sediments being reworked in response to eustatic changes of sea level (Clifton et al., 1967; Hails, 1972). The only reported production of gold from a marine placer is a small amount recovered as a by-product of tin dredging in Malaysia (Brobst and Platt, 1973), but, according to McKelvey et al. (1968), fine gold deposits are thought to be widespread in the Bering Sea. Only negligible quantities of gold are contained in the marine placer deposits of tin exploited in Malaysia, Thailand and Indonesia.

Plant and Coleman (1972) have described the application of a neutron activation method of gold analysis to the evaluation of placer gold occurrences with particular reference to the Strath of Kildonan, Sutherland, Scotland. The material investigated consists of alluvium and fluvioglacial detritus, which infills a glacially overdeepened section of the Helmsdale River Valley. The concentration of gold in the alluvium is about 0.1 ppm. Granite veins associated with a Caledonian injection complex are the primary source of the gold, and it seems likely that natural beneficiation has occurred as a result of weathering processes during an earlier climatic regime. According to Plant and Coleman further reworking and concentration of the gold is attributable to the melting and retreat of the Pleistocene ice sheet. This method permits large samples to be analysed and affords more representative results than can be obtained by atomic absorption of fire assay methods, which involve the analysis of only 20–100 g of material.

MONAZITE

The major source of thorium is monazite, one of two principal ores for the cerium group of the rare earths. In industry, thorium and the rare earths are needed in gas mantles, cores of carbon electrodes, optical glass, coloured glass, ceramics, glazes, glass polishing, pytophoric alloys, metallurgical processes, printing and dyeing, magnesium alloys and radioactive energy applications.

The geological occurrence of monazite in general and of fluvial monazite deposits in the southeastern United States in particular have been thoroughly investigated by Mertie, 1953; Overstreet et al., 1959, 1963, 1968, 1969, 1970a, 1970b and 1971; Overstreet, 1967; Dirkle and Yoho, 1970; Davis and Sullivan, 1971.

In brief, the occurrence of monazite in sedimentary rocks is related to both its relative stability in the weathering profile, as evidenced by resistance of fossil placers to intrastratal solution, and its high specific gravity compared with ordinary rock-forming minerals. Also, because of its resistance to mechanical abrasion, monazite is concentrated at the site of weathering and in streams, but the enriched coastal sand barrier and beach deposits have been concentrated by successive cycles of erosion and deposition.

North America was the first important source of monazite in world commerce when the mineral was mined from fluviatile placers in the Piedmont province of North and South Carolina from 1887 through 1917. Other exploited sources in North America are stream deposits in Idaho and beach placers in Florida. Also, large resources of monazite have been discovered in fossil placers that range in age from Precambrian to Late Cretaceous. It is believed, although not unequivocally proven, that large low-grade resources of monazite doubtless exist with ilmenite in the sedimentary rocks of the Atlantic and Gulf Coastal Plain, and in offshore deposits of the southern U.S. and Gulf Coast of Mexico (Overstreet, 1967, and other references cited therein).

A pertinent point to bear in mind is that the dependence of the United States on monazite as a source for the rare earths and thorium has been greatly lessened by the discovery of very large deposits of bastnaesite in California and thorite in the Rocky Mountains.

Overstreet et al. (1968) have found that heavy resistate accessory minerals derived through weathering and erosion of source rocks can be preferentially accumulated in stream (alluvial) sediments so that the coarse and heavy grains are proportionately more abundant in gravel than in sand, silt or clay. Thus, coarse grains of monazite, which originated mainly in rocks, such as pegmatites that represent only a small percentage of the volume of rock in a drainage basin, tend to be disproportionately concentrated in alluvial gravel. Such gravel is the medium commonly used as a source for alluvial concentrates in prospecting.

Chemical analyses for ThO_2 and U_3O_8 content in 81 sized fractions of detrital monazite from placer concentrates of 33 drainage basins in North and South Carolina show that the percentage of ThO_2 tends to vary according to particle sizes of monazite at each locality. Higher percentages of ThO_2 are associated with coarse-grained rather than with fine-grained monazite. Uranium-rich monazites in the United States, defined as being 0.95% or more U_3O_8, have been located so far only in parts of the Inner Piedmont province of North and South Carolina (Overstreet et al., 1970b). The principal sources of these monazites are granitic rocks and pegmatites that range in age from Ordovician to Permian. As Overstreet et al. (1970a) point out, investigations of regional relations of the chemical composition of selected detrital mineral grains to the geology may require particular effort to overcome the bias introduced by grain size.

Most of the world output of monazite has come from Brazil, India and South Africa but twelve other countries, including Malaysia, Thailand, Indonesia, Australia and Argentina, have produced this mineral more recently. The Brazilian ilmenite—monazite beach and barrier placers still constitute one of the largest known sources of monazite, but the value of stream placers in that country have not been realised because the interior of Brazil is virtually unexplored. Although the placers from the southern coast of Brazil were the main source of commercial monazite from 1895 until 1913, the international importance of the Brazilian production has been eroded during the past decade or so by new discoveries in Africa, Asia, Australia and North America.

MARINE (OFFSHORE) PLACERS

Although some sea-floor deposits have been exploited, they have received little systematic study in terms of cost-effectiveness (Aksenov et al., 1965; Emery and Noakes, 1968; Wilcox et al., 1972; Cruickshank, 1974, and references contained therein). The problems of mining offshore have been reviewed by Cruickshank et al. (1969). Although knowledge has been gained from observing contemporary continental shelf mining operations in various areas of the world, such as tin dredging in Thailand and Indonesia, and diamond dredging in Southwest Africa, there is still much to be learned about locating offshore minerals and the degree to which the marine environment can influence exploration techniques (see Noakes and Harding, 1971; Noakes et al., 1974). Thus, it is essential to monitor and to evaluate critically the frequency of occurrence of different wind speeds, swell conditions, wave heights, tidal fluctuations and other parameters in order to predict the economic feasibility of proposed offshore placer programmes.

There are differences of opinion as to the economic viability of recovering placer deposits from the sea floor. In some areas the development of marine placers is restricted by both the cost of exploitation and by the difficulties of exploration. Nevertheless, some authorities think that offshore placer and other types of unconsolidated deposits are considered to have the greatest potential for low capital investment, quick returns, and high profit margins (Cruickshank et al., 1968; Tooms, 1970). But, as Cruickshank and co-workers recognize, as exploration methods and mining technology become more sophisticated the emphasis may change. The key question, surely, is how soon, when so little is known about the potential source of the sea floor?

Some results of recent and current research pertaining to detrital heavy-mineral placers are summarized in this paper. As this writer has indicated elsewhere, such deposits are basically *polygenetic* because their origin is complex in relation to time, process and place (Hails, 1969, 1972). During the Pleistocene and Holocene epochs heavy-mineral deposits have been reworked on, and transported considerable distances across, the continental shelves of the world in response to eustatic changes of sea level with the result that the degree of concentration of some economic placers varies markedly from one locality to another within relatively short distances, sometimes little more than a few hundred metres. This concentration invariably reflects the interaction between current movement and velocities (as in the case of cassiterite, for example), wave action and the relation of grain diameter to settling velocity, threshold velocity and roughness velocity (see Inman, 1949; Hill and Parker, 1970, 1971).

Tooms (1970) has classified marine placers into four main types: (1) submerged alluvial placers; (2) eluvial deposits resulting from the weathering of underlying bedrock lodes; (3) drowned beach and dune heavy mineral deposits; (4) marine placer deposits.

Each of these types will be discussed later with respect to the placer minerals mentioned in this paper. Briefly, drowned alluvial (stream) placers can be buried beneath varying thicknesses of marine sediments and consequently their presence will not usually be reflected in the surface sediment (see, for example, Van Overeem, 1960a,b; Tooms, 1970; Aleva et al., 1973a,b). However, continuous seismic profiling in association with reconnaissance drilling and profile sampling has proved a relatively successful technique for locating such deposits in buried valleys. So far, most offshore placer mining has essentially been a continuation of onshore placer dredging operations, as exemplified by the exploitation of tin off Thailand, and the islands of Singkep, Bangka and Billiton (Indonesia).

Exploration for submarine tin placers in submerged valley floors 50 m below mean sea level on the Sunda Shelf began in the 1930s. This shelf was exposed during Pleistocene low stands of sea level and eroded by coastal streams. Placers were concentrated along these former drainage courses as well as in the valleys on the present islands. Much of the alluvial material in the placers is cemented by iron hydroxides and forms a coherent conglomerate called kaksa, in which cassiterite is distributed between coarse fragments of quartz and sandstone (Adam, 1932, 1933, 1960; Krol, 1960; Sainsbury, 1969; Zaalberg, 1970).

According to Aleva (1973), the first seagoing bucket dredge started operations off Belitung in 1938, but now about ten such dredges are in operation around the tin islands in Indonesia. Shallow penetration continuous profiling techniques have been employed successfully to facilitate the exploration and exploitation of submarine tin placers off the coasts of Billiton and Singkep (Van Overeem, 1960a,b, 1970). Work started more than two decades ago in water depths between 2 m and 30 m in those areas where the thickness of overburden ranged from 0 to 50 m.

At Billiton the bedrock consists of weathered Permo-Carboniferous sediments intruded by igneous strata of Cretaceous age. An identical series of igneous rocks, intruded into a Cambrian or Precambrian series of schists, occur at Singkep.

Cassiterite, eroded from numerous tin lodes in Cornwall, England, has been transported to the sea since Tertiary times (Dunham and Sheppard, 1970; Dunlop and Meyer, 1973). Prospecting for tin in the sands of St Ives Bay has continued since 1962 by means of drilling, hydraulic probe and dredging (Loo, 1968; Dunham and Sheppard, 1970), while geochemical and geophysical mineral exploration experiments have been conducted in Mount's Bay, Cornwall (Tooms et al., 1965, 1969; Penhale and Hollick, 1968).

The Camborne–Redruth mining area, drained by the Red River, is the major source of the offshore tin placer deposits in St. Ives Bay. These have been worked on a small scale since 1967 and yield a tin concentration which averages between 0.19% and 0.22%. Prospecting licences for superficial deposits have been issued for the entire seabed around Cornwall as far as the 30-fathom line.

Tooms (1970) has applied a modified hypothesis of hydraulic equivalents in research on the distribution of tin in marine sediments and over lodes in Mount's Bay. He suggests that in exploration for tin lodes and for relict eluvial heavy-mineral deposits analytical data should be expressed as hydraulic equivalents.

From the results that have been obtained so far off the coast of Cornwall it appears that geochemical techniques combined with geophysical methods may be able to improve prospecting for detrital and authigenic minerals on the continental shelf.

For the past decade there has been growing speculation about the economic potential of the Australian continental shelf (Layton, 1966). Some of the information released by government and industry has been summarized by Noakes (1970). The results of shallow seismic profiling and reconnaissance drilling for tin off the coast of Tasmania and for alluvial gold off Bermagui on the New South Wales coast, where records of mining last century reported rich gold values in beach sands, have not been sufficiently encouraging to justify the continuation of three exploration programmes (Brown, 1971). Perhaps it is not surprising that the offshore exploration emphasis has changed recently from the search for phosphates, tin and gold to the viability of recovering zircon, monazite, chromite and the titanium ores, rutile and ilmenite, of which Australia is already a major producer.

A sparker sub-bottom profiling system with a variable energy level from 500 to 1000 joules has been used in the search for heavy minerals off the east coast of Australia. This system is capable of defining sedimentary strata 1.6 m thick with low-density control. This is coupled with a laser investigation system accurate to ±1.6 m in 8 km and a high-speed vibration drill capable of drilling to 12 m below the sea floor in 36.5 m or so of water (Brown and MacCulloch, 1970). Some low-grade rutile–zircon deposits have been discovered off the New South Wales coast but drilling has not realised the full economic potential of relict strandlines. No commercial mining operation has been announced at the time of writing and the possibility of mining offshore has not yet been confirmed, even if it is strongly indicated by the results of reconnaissance drilling by at least fourteen companies which have been in operation since 1966 (Brown, 1971; Hails, 1972). The writer has already expressed some reservation about an optimistic forecast in the light of the most recent estimates which indicate that a grade of 0.21– 0.22% total rutile plus zircon can be mined offshore (Hails, 1972). This grade is lower than that mined onshore at present. The important factor, however, is whether the available data are sufficiently reliable and conclusive to confirm that the grades are maximum or minimum percentage values.

The significant variations in the nature and occurrence of heavy mineral concentrates in modern beach, as well as in the Holocene and Pleistocene barrier deposits, cannot be ignored in this context either. If it is borne in mind that bores sunk across barrier beach ridges on some sectors of the New South Wales coast have been little more than 20 m or so apart, the problem of offshore exploration is readily apparent. The preliminary results obtained from offshore drilling programmes indicate that more closely spaced boreholes on fixed grids will be necessary before the grade and economic value of relict strandline deposits can be fully assessed, together with their total rutile–zircon content. Obviously, much will depend upon the quality control of new techniques, and whether or not such techniques can clearly delimit favourable prospecting areas, rather than widely distributed small pockets of unusually high heavy mineral concentrates (Hails, 1972).

It appears likely that some type of drag-dredge or hydraulic mining system capable of operating to the edge of the continental shelf and in a depth of water exceeding 200 m could be most usefully employed in the future. However, success in offshore mining will depend very much upon engineering expertise and how it can cope with any "overburden" that may have to be removed. The accuracy and reliability of offshore surveys in locating profitable heavy-mineral seams in relict strandline deposits must therefore be carefully appraised (see, also, Holmes, 1971).

Tanner et al. (1961) and Tanner (1962) have evaluated the use of equilibrium concepts in the search for heavy minerals on three shoals off the Chattahoochee—Apalachicola Rivers, Florida Panhandle. The thesis they have advanced is that heavy mineral content increases with depth in shoal sands. If their contention is correct, concentrations in excess of 4% should be encountered within about 5 m of the sand surface and within about 10 m of the water surface. They have concluded that one shoal (Cape St George) probably contains about $4 \cdot 10^7$ m^3 of heavy minerals, including perhaps $2 \cdot 10^6$ m^3 each of rutile and zircon and about $5 \cdot 10^5$ m^3 of monazite. However, it is debatable whether or not such shoals are really attractive areas for exploitation. Tanner argues that this particular area of the Florida Panhandle illustrates the operation of a hydrodynamic method of concentration.

In summary, application of the placer groups, proposed by Emery and Noakes (1968), to sea-floor exploration seems to indicate that heavy heavy minerals are likely to be located along submerged river channels and in submerged relict beach deposits if they are near primary sources, and light heavy minerals in submerged beaches of the nearshore zone. Of course, as Emery and Noakes correctly state, and as the writer has summarized here, best prospects for any group of placers on the sea floor are in areas adjacent to localities of existing terrestrial economic placers.

ALLUVIAL (STREAM) PLACERS

Rubey (1933) considered fall velocity to be the most important segregating mechanism accounting for the close association of small high-density grains with low-density but larger mineral grains in a sediment, together with such factors as density and hardness of the minerals, differences in original sizes of the grains in the source rock, degree of abrasion during transportation and degree of sorting at the site of deposition. On the other hand, Rittenhouse (1943) attributed the distribution of heavy minerals in streambeds to varying hydraulic conditions at the time and place of deposition, equivalent hydraulic size of each of the heavy minerals, availability of the minerals, and unknown factors.

Wertz (1949) concluded that there is a logarithmic sequence of rich erosion areas alternating with barren sedimentational areas in river placer deposits. More recently, the results of Brady and Jobson's (1973) experimental study of heavy-mineral segregation under alluvial-flow conditions shows that bed configuration and grain density are the most important factors affecting local segregation of heavy minerals in an open channel flow.

The Van Horn Sandstone, west Texas, U.S.A., is an ancient alluvial fan system, varying in thickness between 9 m and 212 m, in which gold has been reported. Recent analyses made by McGowan and Groat (1971) have revealed no significant quantities of this placer, although a prospecting model, applicable to similar but ore-bearing alluvial fan systems, might be developed from the data pertaining to the concentration of other

TABLE III

Characteristics of alluvial placers (after Kartashov, 1971)

	Autochthonous		Allochthonous
	bottom	above-bottom	
(1) Are represented by:	channel, valley, terrace and watershed placers	valley, terrace, and watershed placers	point-bar, delta, river-plain, valley, terrace, and watershed placers
(2) Occur:	adjacent to their ore sources		more or less far from ore sources, being separated from them and from autochthonous placers by zones of dispersion of "placer" minerals
(3) Are concentrated:	at the base of an instrative or substrative alluvium and in the crevices of a bedrock	at the base of a perstrative alluvium and within constrative strata, in the same parts of valleys as bottom placers	in surficial horizons of a perstrative alluvium and within constrative strata, downstream of autochthonous placers
(4) Consist of mineral grains:	received directly from ore sources or redeposited from older placers and not carried out by rivers from concentration zones		brought by rivers into concentration zones
(5) Accumulate during:	entire time of destruction of primary ore deposits, embracing, as a rule, many stages of river development		the last equilibrium and/or aggradation stages of river development
(6) An enclosing alluvium is formed during:	downcutting stages or transition from them to equilibrium stages		equilibrium and/or aggradation stages
(7) An enclosing alluvium being rewashed during a downcutting stage, is:	not destroyed but displaced to the level of a new bedrock bottom	displaced to the level of a new bedrock bottom and added to bottom placers	completely destroyed
(8) The mechanism of concentration of "placer" minerals:	does not essentially depend upon hydrodynamic properties of flowing water		depends to a great extent upon hydrodynamic properties of flowing water

heavy minerals in the respective facies, particularly that of the distal-fan.

The *Principles of Placer Geology* by Bilibin (1938) is a standard reference both within and outside the Soviet Union. More recently some of his views have been revised and some new concepts concerning alluvial placers have been introduced by Soviet geologists. According to Kartashov (1971), alluvial placer deposits can be divided into autochthonous and allochthonous subtypes (Table III). The former contain large heavy-mineral grains that are almost immovable by streams and, consequently, are located adjacent to primary ore deposits and concentrated at the base of alluvium, or in the crevices of its bedrock. Such placers accumulate during several stages of river development. These "bottom" autochthonous placers are distinguished by Kartashov from "above-bottom" autochthonous deposits which consist of smaller grains transported by rivers during the last stages of equilibrium and/or aggradation, and concentrated in the higher horizons of alluvium. "Above-bottom" placers may be reworked and incorporated with "bottom" placers during stages of downcutting. On the other hand, allochthonous placers, which contain finer heavy-mineral grains moved by a river as a part of its alluvial load, form some distance downstream from primary ore deposits or autochthonous placers and occur in surficial parts of channel alluvium. Allochthonous placers are formed on the surfaces of point-bars and channel floors where the rivers carrying loose material with finer grains of "placer" minerals regularly deposit their load but also regularly remove the lighter minerals and rock fragments. In aggrading rivers several point-bar concentrations may be superimposed to form thick allochthonous placers.

In contrast to the autochthonous placers, which form under practically any hydrodynamic condition providing the rivers receive sufficiently large mineral grains from sources of ore, the allochthonous placers require definite hydrodynamic conditions for their formation. Although the same placer minerals can form both autochthonous and allochthonous placers, every mineral prefers a certain subtype of placer (Kartashov, 1971). But some of them do not demonstrate this preference too clearly. For example, autochthonous gold placers predominate over allochthonous ones, but the latter, particularly the point-bar variety, are not uncommon. Alluvial placers of some other minerals, such as zircon, appear to belong to an allochthonous subtype alone.

The division into autochthonous and allochthonous subtypes applies not only to alluvial placers, but similar subtypes can also be clearly identified among coastal placers of lacustrine or marine origin, according to the findings of Kartashov. Most of these placers consist of mineral grains brought down by rivers and later displaced by wave action and other coastal processes (see also Harrison, 1954).

Within the last few years, it has become increasingly apparent that the interpretation and understanding of large fluvial piedmont deposits is critical to the future development of major gold resources (Mullens and Freeman, 1957; Peterson et al., 1968; Sestini, 1973; Galloway and Brown, 1973; Schumm, 1976).

Reference has already been made, in passing, to alluvial cassiterite placers within the Southeast Asian tin belt (see, also, Palmer, 1942). These mainly occur on the floor of wide and shallow valleys in association with a layer of concentrated heavy minerals on chemically weathered bedrock as a sandy to gravelly bed, varying in thickness from a few centimetres to several metres. A decade ago about 60% of the world's tin production came from such placers (Shelton, 1965).

Broadhurst and Batzer (1965) have summarized the situation regarding valuation and tin recovery in Malaysia in the mid-1960s and have concluded that the principal uncertainty in valuation is core measurement, with respect to alluvial deposits. Large-scale hydraulic mining practice, for the exploitation of alluvial tin, has undergone a number of major changes over the years culminating in the almost universal adoption of water jets to break ground and gravel pumps to remove the spoil so produced (Pakianathan and Simpson, 1965). According to Pakianathan and Simpson, hydraulic gravel-pump mining has a number of advantages over other alluvial mining methods. These can be summarized as follows. Firstly, undulating ground can be worked provided there is sufficient land for tailings disposal, although the presence of swamps could preclude such operations. Secondly, deep pockets in bedrock and local enrichments of limited area but considerable depth, as located in contact zones, can be exploited by this means. Thirdly, working faces are visible at all times. Fourthly, selective mining can be practised. Lastly, capital cost of equipment is lower than for other types of mining.

The problem of evaluating potentially economic auriferous stream valley (alluvial) deposits has been studied by placer mining engineers since the turn of the century and the introduction of bucket-line dredges. Daily (1962) has described a reliable procedure for estimating the concentration of gold in a soil and according to his results recoveries have been within 5% of those predicted. The procedures he described — establishment from prospect drill log data of depth of mining section and unit gold content, areal valuation and application of management and experience factors — are equally applicable to valuation of alluvial platinum-bearing deposits, with some modification to tin-bearing deposits, and under certain conditions to heavy-mineral concentrates.

McFarland (1965) and Hester (1970) have investigated the placer gold deposits in the Klondike area, Yukon Territory, when churn drilling on a widely spaced but regular pattern has been the principal technique employed for the exploration and valuation of alluvial deposits, despite the prevailing permafrost. As a result of this environmental problem, in sub-arctic climatic conditions, operations are restricted to six months of the year between April and November.

Valleys in the non-glaciated areas of the Yukon are buried by loess, commonly referred to as "muck", which varies between 1.5 m and 9 m in thickness, and derives from the outwash fans of the glaciers in the Ogilvie Mountains to the north. Gold particles occur in economic quantities near the base of valley gravels and on some of the remnant valley benches. No gold has been found in the loess but the cost of thawing and removing the material by hydraulic stripping has to be carefully evaluated. Gravel and bedrock are thawed by circulated cold water which is introduced into the ground under pressure. In the mid-1960s production averaged 40,000–50,000 oz. gold per annum from 4,000,000 yd^3 of material. Heavy minerals found with the gold are cassiterite, in the form of "wood tin", ilmenite and magnetite. According to Hester (1970), precise comparison of evaluations with actual values recovered, when the dredge is used for mining, shows a consistent undervaluation of low-grade deposits and an overvaluation of high-grade deposits. Variations in the size of the predominantly coarse gold particles, and the extremely low overall content of gold in samples, are considered to be the main cause of this bias. Hester also claims that the maps of the distribution of gold show that the drill-holes have been too widely spaced to disclose the many variations in gold distribution. Generally, drilling has served to differentiate areas of economic value from waste, but it has contributed little to precise knowledge of gold content.

The possibility of recovering economic placers from tills, and most other glacial deposits, is seemingly remote according to the results of studies completed hitherto. Bayrock (1972), for example, has examined the tills of central Alberta and has concluded from statistical computations that on average about 95% of the heavy minerals have been derived from the Canadian Shield and that it is therefore virtually impossible to use such minerals in studies of correlation, and for differentiating surface tills. It also follows, from a comparison with marine (offshore and beach) and alluvial (also in formerly glaciated, but now permafrost areas), and to a lesser degree eluvial, deposits, that those placers in tills and most other glacial sediments are relatively unsorted regardless of geologic age and the number of occasions reworking may have ensued.

Peterson et al. (1968) have tested and used shallow seismic refraction in conjunction with gravity methods in an attempt to determine the configuration of a large Tertiary gravel-filled channel, and the size and extent of possible gold deposits remaining in situ, in the area between the South and Middle Yuba Rivers in northern Nevada County, California. The deposits in this particular channel were once the site of some of the most productive hydraulic gold mines in California between the 1850s and 1884. The gravel is partly covered by the remnants of an extensive sheet of volcanic rocks. The lower horizons of the gravel contain the highest values of placer gold. Historically, a court decision halted the dumping of debris into streams tributary to the Sacramento and San Joaquin Rivers, after which small hydraulic mines operated sporadically and the high-grade deposits immediately above bedrock were mined locally by underground methods.

Alluvial mining has been undertaken in various ways in almost every country in South America, but the largest output has come from Colombia with gold as the principal product, followed by platinum (O'Neill, 1965). The three main areas of present-day alluvial mining in Colombia are the Nechi River in the vicinity of Zaragoza, a tributary of the San Juan River near Istmina Condoto and Novita, and the Telembi River near Barbacoas very near the Ecuador border (Fig. 5).

Almost a decade ago inflation, accompanied by an unfavourable balance of payments, seriously restricted the importation of essential equipment and spare parts into Colombia for alluvial mining. This economic situation, nevertheless, resulted in steadily

Fig. 5. Map to show distribution of tin and platinum in Colombia and Bolivia. (After Cruickshank, 1965.)

improved operating efficiencies and recoveries with the result that mining companies were able to continue on a profitable basis (O'Neill, 1965). Platinum was sold to the U.S.A. at that time and gold in the London market. The reader is referred to O'Neill's paper for general information on exploration by drilling, costs and difficulties, and recovery systems, including dredging. O'Neill also examines the cost of equipping mines in remote locations in Bolivia where it is necessary to construct airfields and to fly everything into the area to be developed.

Considerable discussion has arisen over Reid's (1974) model of Early Precambrian transport of diamonds from West African kimberlites (2.1–2.3 billion years old) to Surinam, South America (Hastings, 1974; Leo and Coonrad, 1974; Schönberger, 1974; Schönberger and De Roever, 1974). In particular, the assumptions have been challenged that: (1) kimberlite sources in West Africa are both extensive and old enough to have supplied the Rosebel and Roraima diamonds in Surinam; (2) diamonds have been transported alluvially over distances up to 3500 km; and (3) widespread diamonds in the

Roraima Group have been derived from comparatively small outcrop areas of the Rosebel Formation in Surinam. In fact, Leo and Coonrad (1974) argue from geological records and fieldwork in Liberia, and from Hall's (1972) work in Sierra Leone, that diamond concentrations diminish rapidly downstream from kimberlites and that significant deposits are rarely found more than about 100 km from the source.

According to Hall, the alluvial diamondiferous deposits along the Sewa River, Sierra Leone, extend approximately 200 km downstream from the Koidu kimberlite, but other local sources of diamonds are contributing to the apparent dispersion train. Recently, kimberlite sources in Liberia have been identified near alluvial deposits that formerly were assumed to have been transported from afar. Leo and Coonrad (1974) point out that these observations refer to dispersal periods that may be short compared with the potential dispersal time of several hundred million years inherent in Reid's hypothesis, but conclude it appears unlikely that the same diamonds should appear in quantity 2000—3500 km from their source, irrespective of the time available for their dispersion, because concentrations of diamonds in Liberia and Sierra Leone dwindle so markedly downstream. Nevertheless, the fact remains that diamonds can be located considerable distances from their primary source(s) because of their resistance to mechanical abrasion. Diamonds were first discovered along the Southwest African coast around 1908 and in Pleistocene marine terraces and stranded beach deposits some twenty years later. According to Cruickshank (1965), diamondiferous gravels, trapped in old marine terraces and submerged river beds which are invisible on land but exist offshore, occur along a 1600-km sector of the coast in the region of the Orange River. However, it was not until 1961 that the Marine Diamond Corporation launched the first sea-going mining unit a year later.

Hoyt et al. (1968) and Oostdam (1969) have undertaken detailed "sparker" surveys of the inner continental shelf off the Orange River as part of an exploration and evaluation programme for offshore diamonds and have also described in some detail the offshore sediments and valleys of the Orange River (but see also Stocken, 1962; Webb, 1965). Speculation, nevertheless, still surrounds the original source of the diamonds along this coast. All available evidence suggests that the Orange River has been the major agent for the transport of the diamonds to the sea at a time when base level was substantially lowered during the Pleistocene glaciation (Murray, 1970). So far diamonds have been recovered from the sea floor in water depths of 11 m—31 m by airlift pumps using water jets to loosen the host material. The dredged material is forced up large-diameter rubber pipes and the oversize is screened off and returned to the sea. Sophisticated survey techniques have been employed which include the use of Tellurometer and Hydrodist electronic distance and position fixers. Marine prospecting units have plotted to a margin of less than 2 m and accurate mining maps of the ocean floor, as well as the exact location of the marine placer deposits of diamonds, have been compiled from aerial photographic surveys. One group of companies has a 614-km concession which covers the area between the mouth of the Olifants River on the Cape Coast

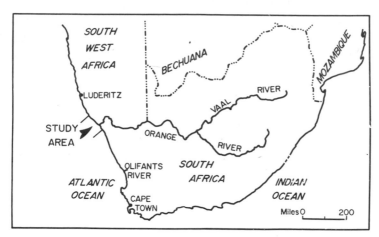

Fig. 6. Detailed map to show diamond producing areas of South West Africa and South Africa. (After Hoyt et al., 1968.)

to Luderitz in South West Africa, and which extends from low-water mark (LWM) to about 6 km offshore (Fig. 6).

Alluvial deposits offshore are mined at present by dredging using bucket ladders, draglines, clamshells, hydraulic dredges and airlifts. A major problem, however, as with most marine exploration, is that of position fixing, not to mention waste disposal and legislation — important to environmentalists. Yet, despite the capital investment involved in a supposedly lucrative operation, marine diamond operations currently account for less than 3% of the total mineral values annually produced from offshore and beach mining (Cruickshank et al., 1968). This fact reflects that the numerous diamonds on the modern, stranded (marine terraces) and submerged beaches are smaller than those of the streams, and therefore have less unit and total value (Nesbitt, 1967).

BEACH PLACERS

Most marine placers are essentially beach or submerged (relict) beach placers. The winnowing action of waves on the foreshore, with the removal of light grains, invariably results in the concentration of "lag" deposits of which gold, magnetite, garnet and diamonds, for example, are typical examples (see also Tourtelot, 1968). May (1973) has proposed, as a result of wavetank experiments, a model for the selective transport of heavy minerals to the shoreline by shoaling waves, but field data are now required to verify his preliminary results. Some care should be exercised with this and similar experiments related to hydraulic fractionation of heavy mineral suites, because incorrect conclusions can be made about the true value of beach placers as indicators of differential movement over a long time scale unless diagnostic constituents are derived from known source areas (Hails, 1974). Dubois (1972) suggests that as the per cent of mid-foreshore sediment increases from light to heavy density, the foreshore slope likewise increases. Accordingly, the heavy-mineral content of foreshore deposits is considered as an active variable influencing the foreshore slope along with grain size, wave steepness and wave length.

Although less common than in the young placer deposits mentioned elsewhere in this paper, gold occurs in beach deposits distributed from Polar through temperate to tropical latitudes. In this respect the beach at Nome, Alaska, is a classical example, where a narrow strip 60 m or so wide produced more than U.S. $ 2 million in fine gold up to the mid-1960s (Cruickshank, 1965). Other placer deposits in Alaska, but now more or less exhausted, have been located in stranded (raised) beaches (Cobb, 1973). Reference has been made to other abandoned beach deposits in other sections of the paper.

Australia is probably the world's largest producer of "black sand" natural heavy-mineral concentrates. Commercial production of rutile, zircon, ilmenite and monazite commenced on the east coast (Queensland and New South Wales) in 1934 and on the southwest coast (Western Australia) in 1965. Australia produced 95% of the world's rutile concentrates, estimated as 386,000 tonnes in 1971. Sierra Leone, India, Ceylon (Sri Lanka), and Brazil supplied the balance, but now mining operations are suspended in Sierra Leone. In the same period, 77% of the world's zircon and 25% of the world's monazite together with 45,000 tonnes of ilmenite were produced annually in Australia (Ward, 1972).

Following the depletion of high-grade heavy mineral reserves on the east coast in the vicinity of the Queensland-New South Wales state border in the 1950s, attention focussed on lower-grade, less accessible deposits requiring larger mining units with the inevitable result that production costs have risen. In southeastern Queensland, the source of mining activity for almost a decade has moved from Southport-Coolangatta to off-shore, wind-blown disseminated deposits of the higher dunes of Curtis, Fraser, Moreton and Stradbroke Islands. The high-dunal deposits of North Stradbroke Island have been a viable mining operation since 1966 with the use of light-weight, portable, hydraulically operated drilling equipment capable of testing to depths of 100 m or more, and the introduction of selective earth-moving techniques (see also Macdonald, 1966).

According to Ward (1972), "production from the east coast of Australia might be expected to taper off by the early 1980s and, despite development of the recently discovered deposits at Eneabba, Western Australia, the domestic rutile–zircon industry is expected to move from a position of pre-eminence to one, which, while still important, will find increasing difficulty in coping with increased world demand". On the other hand, Ward's forecast is more optimistic for the domestic ilmenite industry as there is potential for increased production. However, this will be mainly dependent upon the ability of Western Australia producers to not only maintain but also expand their share of the growing world market. To this extent, the success and expertise of east-coast producers in marketing a saleable ilmenite product is essential.

Heavy-mineral concentrates of beach placers as well as the processes leading to their concentration, have been described by several workers including Baker, 1945; Beasly, 1948, 1950; Connah, 1948, 1962; Fisher, 1949; Gardner, 1955; Whitworth, 1956; Mackay, 1956; Ludwig and Vollbrecht, 1957; Rao, 1957; Ward, 1957, 1965; Hails, 1964, 1969, 1972; Welch, 1964; Johns, 1966. Detailed examination of barrier beach placers by some of these researchers has shown that the highest concentrations of rutile and zircon are at the base of frontal dunes of an outer barrier, immediately behind the beach, and also towards the northern ends of bays (see Beasly, Gardner, Whitworth, Connah, and Hails, and Fig. 7 for a schematic cross-section of a typical placer deposit). This distribution is directly related to marked variations in wave energy along the coast caused by the refraction of a predominantly southeasterly swell around the headlands at each end of arcuate bays, and also to the incidence of cyclonic storms which are usually accompanied by southerly gale-force winds. Wave energy is higher, and barrier erosion therefore more pronounced, at the northern end of the bays in marked contrast to the southern end, which is more sheltered by the headlands from southeasterly wave attack.

Thick lenses or seams of wave-concentrated minerals, which are particularly evident in open cuts and along frontal dunes, are invariably buried beneath notably thinner seams of wind-deposited minerals.

The occurrence of similar "black sand" deposits, containing zircon, rutile, ilmenite and monazite in particular, in the British Isles is extremely limited compared with that of Australia and Southwest Africa.

Apart from one or two seemingly promising areas, such as the Northumberland coast around Budle Bay which is being exploited by the Institute of Geological Sciences, it seems unlikely that there will be commercial quantities available (Dunham and Sheppard, 1969; see also Dunham, 1967). This is also true for several other areas where the occurrence of heavy-mineral concentrates formerly supported beach placer mining. Such deposits were productive for several years along the southwest Oregon coast where black sand deposits have been mined for nearly a century, after gold and platinum were discovered initially in 1852 and again in the 1870s in ancient abandoned (raised) beaches (Pardee, 1934; Twenhofel, 1943). Renewed interest in them was caused by the industrial depression after 1929 (Oregon State Publ., 1957). The presence of chromite, a strate-

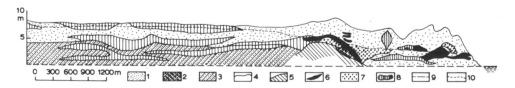

Fig. 7. Schematic cross-section of an Australian placer. 1 = sands; 2 = gravel and pebbles; 3 = consolidated sands; 4 = boundary of consolidated sands; 5 = compact rocks of the Upper Paleozoic; 6 = sands with a heavy-minerals content >180 kg/ton; 7 = the same ranging from 180 to 72 kg/ton; 8 = the same <72 kg/ton; 9 = sea level; 10 = groundwater table. (After Gardner, 1955.)

gic mineral, has also stimulated interests lately since the decline of placer mining for gold and platinum (Twenhofel, 1943; Griggs, 1945). Evidence from preliminary geological and geophysical data indicate that magnetite-bearing placer deposits may exist on the southern Oregon continental shelf in water depths ranging between 18 m and 183 m (Kulm et al., 1968a,b). These deposits are thought to be relict, and are located off the mouths of the Rogue and Sixes Rivers, which have sources of heavy metals in their respective catchment areas.

Although monazite occurs in beach deposits along the Pacific coast of California, these placers are regarded to be of no commercial importance particularly in the light of the low tenor in ThO_2 (i.e., 3.5—4.4%) and the large readily exploitable deposits in India.

On the other hand, potentially valuable beach sands occur along some 402 km of the westcoast (Westport district) of South Island, New Zealand, from Karamea to Jacksons Bay (Nicholson et al., 1958; Hudson, 1960; Nicholson, 1967; Fig. 1). Nicholson et al. (1958, 1966), following the previous work of Hutton (1950) and Marshall et al. (1958), estimated that upwards of 1000 million tonnes of dredgeable sand, containing significant amounts of ilmenite and minor accessory zircon and monazite, exist on the modern beaches and along relict Pleistocene and Recent shorelines at various levels up to a maximum elevation of 152 m above present mean sea level. It is also possible that gold, scheelite, rutile and cassiterite can be extracted economically in some localities.

The northern beaches of the Nile delta, covering a distance of 250 km, contain 70—90% heavy minerals, mainly ilmenite, magnetite, zircon, rutile, monazite and garnet (Hilmy, 1951; Higazy and Naguib, 1958; Khairy et al., 1964, for example). Estimation of the ore reserves has been made by churn water jet drilling along parts of the coast to depths of 15 m at a frequency of about 1 km. The geological and radiometric surveys conducted along the whole Mediterranean beach of the Nile delta have proved the occurrence of differential deposition of the various minerals, as elsewhere.

The persistence of certain heavy minerals in some sediments is determined by their chemical and mechanical stability over geologic time. The writer (Hails, 1969) has argued, for example, that although staurolite has an extremely low stability to weathering index compared with zircon and tourmaline, its occurrence in beach and dune sands on the mid-north New South Wales coast, Australia, is attributed to its resistance to abrasion. Despite the relatively high position of biotite in Pettijohn's (1941) "Order of Persistence" table, it appears to be an unstable mineral in these particular deposits because it is entirely absent from all the samples analysed by the writer, including Holocene barrier sands and fluviatile sediments.

According to Baker (1962), though, differences in stability of any mineral undoubtedly exist in different areas according to variations in climate, topography and vegetation. On the other hand, Allen (1948) has stated that the resistance of minerals to weathering depends upon the particular variety of the mineral in question. Dryden and Dryden

(1946) in their study of the comparative rates of weathering of common heavy minerals in situ found that garnet is the least resistant to chemical alteration, even though most other workers list this mineral as highly stable. It is generally accepted that the mechanical stability of minerals like tourmaline, rutile and zircon, for example, is controlled in the main by their physical properties, particularly their hardness, cleavage and tenacity.

ELUVIAL, RESIDUAL AND FOSSIL PLACERS

Eluvial placers are mainly located on hill-slopes where the material of the parent lode is disintegrated by weathering and moved downhill by the process of *creep,* with the result that the heavier placers are concentrated more than the lighter ones.

In the case of tin, eluvial placers are formed by the chemical decay of tin-bearing strata and the gravity separation of cassiterite and other heavy resistates as the decayed mantle moves downslope under the influence of sheetwash, gravity, and, locally, frost action. This type of placer grades imperceptibly into residual placers upslope, and often into alluvial (stream) placers downslope. Some eluvial placers are richer than residual placers because of active gravity sorting but, even so, many of them contain coarse rubble, and, in cold regions (Alaska and Siberia), fragments of unoxidized deleterious constituents such as arsenopyrite (Brobst and Pratt, 1973).

Eluvial deposits overlying, and subjacent to, bedrock mineralization undoubtedly will have been reworked to some extent during eustatic changes of sea level. Fine light material will have been elutriated and the heavy minerals will have been incorporated into the marine sediments to at least a limited degree, and their concentration can be expected to decrease with distance from the original (terrestrial) source (Tooms, 1970).

Residual placers accumulate in situ immediately above a bedrock source of cassiterite by the chemical decay and removal of the rock minerals. According to Brobst and Pratt (1973), in some areas such as Kitololo in the Democratic Republic of the Congo, residual tin placers grade downward into weathered lodes and are mined either as placers or as open-pit lodes. In a genuine residual placer, the enriched zone contains not only heavy minerals like cassiterite and columbite—tantalite, but also chemically resistant light minerals such as beryl. On the other hand, in Indonesia, for example, residual placers commonly are cemented tightly by hydrous iron oxides, forming the "kaksa" of industry, which must be crushed to free the cassiterite. Residual placers occur widely in the deeply weathered strata of the southeastern United States. Any type of placer referred to in this chapter may become fossil by burial beneath younger sediments or lava. Second-cycle alluvial placers may occur as a result of uplift and renewed erosion of fossil placers along disrupted drainages.

In contrast, the gravels of some fossil placers may be strongly lithified so that they must be mined by lode-mining methods. Sometimes, post-depositional changes may affect the value of tin, as in the case of partial solution of cassiterite.

As mentioned earlier, large resources of monazite have been discovered in fossil placers in the upper sections of littoral marine sandstone of Late Cretaceous age in the San Juan basin, Colorado and New Mexico (Chenoweth, 1956, 1957) and in sedimentary strata of comparable age in Wyoming and Montana (Murphy and Houston, 1955). Outcrops of the fossil placers are discontinuously exposed for at least 1150 km subparallel to the Rocky Mountains. These fossil placers may contain Colorado's largest known resources in monazite. In addition to monazite, fossil placers contain ilmenite, garnet, zircon, tourmaline, magnetite and rutile.

REFERENCES

Adam, J.W.H., 1932. Kaksa-genese. *Mijning. Ned. Indie*, 1932: 2.
Adam, J.W.H., 1933. Kaksa-genese. *Mijning. Ned. Indie*, 1933: 1, 2, 3.
Adam, J.W.H., 1960. On the geology of the primary tin-ore deposits in the sedimentary formation of Billiton. *Geol. Mijnbouw*, 39: 405–426.
Aksenov, A.A., Nevessky, E.N., Pavlidis, Yu.A. and Shcherbakov, F.A., 1965. Questions of the formation of near-shore marine placers. *Tr. Inst. Okeanol., Akad. Nauk, S.S.S.R.*, 76: 5–53.
Aleva, G.J.J., 1973. Aspects of the historical and physical geology of the Sunda shelf essential to the exploration of submarine tin placers. *Geol. Mijnbouw*, 52: 79–91.
Aleva, G.J.J., Fick. L.J. and Krol, G.L., 1973a. Some remarks on the environmental influence on secondary tin deposits. *Bur. Miner. Res., Geol. Geophys. Bull.*, 141: 163–172.
Aleva, G.J.J., Bon, E.H., Nossin, J.J. and Sluiter, W.J., 1973b. A contribution to the geology of part of the Indonesian tinbelt: the sea areas between Singkep and Bangka Islands and around the Karimata Islands. *Geol. Soc. Malaysia, Bull.*, 6: 257 271.
Allen, V.T., 1948. Weathering and heavy minerals. *J. Sediment. Petrol.*, 18: 38–42.
Baker, G., 1945. Heavy black sands on some Victorian beaches, *J. Sediment. Petrol.*, 15: 11–19.
Baker, G., 1962. Detrital heavy minerals in natural accumulates. *Aust. Inst. Min. Metall., Monogr. Ser.*, 1: 146 pp.
Bateman, J.D., 1958. Uranium-bearing auriferous reefs at Jacobina, Brazil. *Econ. Geol.*, 53: 417–425.
Bayrock, L.A., 1962. Heavy minerals in till of central Alberta. *J. Alberta Soc. Pet. Geol.*, 10: 171–184.
Beasley, A.W., 1948. Heavy-mineral beach sands of southern Queensland, 1. *Proc. R. Soc. Qld.*, 59: 109–140.
Beasley, A.W., 1950. Heavy-mineral beach sands of southern Queensland, 2. *Proc. R. Soc. Qld.*, 61: 59–104.
Bilibin, Y.A., 1938. *Principles of Placer Geology*. G.O.N.T.I., Moscow-Leningrad, 505 pp.
Brady, L.L. and Jobson, H.E., 1973. An experimental study of heavy-mineral segregation under alluvial-flow conditions. *U.S. Geol. Surv. Prof. Pap.*, 562-K: 38 pp.
Broadhurst, J.K. and Batzer, D.J., 1965. Valuation of alluvial tin deposits in Malaya, with special reference to exploitation by dredging. In: *Opencast Mining, Quarrying and Alluvial Mining*. Inst. Min. Metall., London, pp. 97–113.
Brobst, D.A. and Pratt, W.P., 1973. United States mineral resources. *U.S. Geol. Surv. Prof. Pap.*, 820: 722 pp.
Brown, G.A., 1971. Offshore mineral exploration in Australia. *Underwater J.*, 3: 166–176.
Brown, G.A. and MacCullough, C., 1970. Investigations for heavy minerals off the east coast of Australia. *6th Annual Conf. Mar. Technol. Soc.*
Brown, G.F., Buravas, S., Charaljavanaphet, J., Jolichandra, N., Johnston, W.D., Jr., Stresthaputra, V. and Taylor, G.C., Jr., 1951. Geologic reconnaissance of the mineral deposits of Thailand. *U.S. Geol. Surv. Bull.*, 984: 183 pp.

Chenoweth, W.L., 1956. Radioactive titaniferous heavy mineral deposits in the San Juan Basin, New Mexico and Colorado. *Geol. Soc. Am. Bull.,* 67: 1792 (abstr.).

Chenoweth, W.L., 1957. Radioactive titaniferous heavy mineral deposits in the San Juan Basin, New Mexico and Colorado. In: *New Mex. Geol. Soc. Guideb., 8th Field Conf.,* pp. 212–217.

Clifton, H.E., Hubert, A. and Phillips, R.L., 1967. Marine sediment sample preparation for analysis for low concentrations of fine detrital gold. *U.S. Geol. Surv. Circ.,* 545: 11 pp.

Cobb, E.H., 1973. Placer deposits of Alaska. *U.S. Geol. Surv. Bull.,* 1374: 200 pp.

Connah, T.H., 1948. Reconnaissance survey of black sand deposits, southeast Queensland. *Qld. Gov. Min. J.,* 49: 561 pp.

Connah, T.H., 1962. Beach sand heavy-mineral deposits of Queensland. *Qld. Dep. Mines Publ.,* 302: 31 pp.

Cruickshank, M.J., 1965. Mining offshore alluvials. In: *Opencast Mining. Quarrying and Alluvial Mining.* Inst. Min. Metall., London, pp. 125–155.

Cruickshank, M.J., 1969. Mining and mineral recovery. *Undersea Technol. Handb.,* Repr. E-7.

Cruickshank, M.J., 1974. Mineral resources potential of continental margins. In: C.A. Burk and C.L. Drake (Editors), *The Geology of Continental Margins.* Springer, Heidelberg, pp. 965–1000.

Cruickshank, M.J., Romanowitz, C.M. and Overall, M.P., 1968. Offshore mining – present and future. *Eng. Min. J.,* 169: 84–91.

Cruickshank, M.J., Corp, E.L., Terichow, O. and Stephenson, D.E., 1969. Environment and technology in marine mining. *J. Environ. Sci.,* April: 3–11.

Daily, A., 1962. Valuation of large, gold-bearing placers. *Eng. Min. J.,* 163: 80–88.

Davis, E.G. and Sullivan, G.V., 1971. Recovery of heavy minerals from sand and gravel operations in the southeastern United States. *U.S. Bur. Mines Rep. Invest.,* 7517: 25 pp.

Dirkle, E.C. and Yoho, W.H., 1970. The heavy-mineral orebody of Trail Ridge, Florida. *Econ. Geol.,* 65: 17–30.

Dubois, R.N., 1972. Inverse relation between foreshore slope and mean grain size as a function of the heavy-mineral content. *Geol. Soc. Am. Bull.,* 83: 871–876.

Dunham, K.C., 1967. Economic geology of the continental shelf around Britain. In: *The Technology of the Sea and the Sea Bed, 2.* H.M.S.O., London, pp. 328–341.

Dunham, K.C., 1969. Sedimentary ore deposits – a conspectus. In: *Sedimentary Ores: Ancient and Modern – Univ. Leicester, Dept. Geol., Spec. Publ.,* 1: 5–12.

Dunham, K.C. and Sheppard, J.S., 1970. Superficial and solid mineral deposits of the continental shelf around Britain. In: M.J. Jones (Editor), *Mining and Petroleum Geology – Proc. 9th Commonw. Congr.* Inst. Min. Metall., London, pp. 3–25.

Dunlop, A.C. and Meyer, W.T., 1973. Influence of Late Miocene–Pliocene submergence on regional distribution of tin in stream sediments, southwest England. *Trans. Inst. Min. Metall.,* 82: B62–B64.

Dryden, A.L. and Dryden, C., 1946. Comparative rates of weathering of some common heavy minerals. *J. Sediment. Petrol.,* 16: 91–96.

Emery, K.O. and Noakes, L.C., 1968. Economic placer deposits of the continental shelf. *Tech. Bull. ECAFE,* 1: 95–111.

Fisher, N.H., 1949. Cape Everard heavy mineral deposits. *Bur. Miner. Resour. Aust. Rec.,* 1949/30 (unpubl.)

Galloway, W.E. and Brown, L.F., Jr., 1973. Depositional systems and shelf-slope relations on creatonic basin margin, uppermost Pennsylvanian of north-central Texas. *Am. Assoc. Pet. Geol. Bull.,* 57: 1185–1218.

Gardner, D.E., 1955. Beach sand heavy mineral deposits of Eastern Australia. *Bur. Miner. Resour. Aust. Bull.,* 28: 1–103.

Gold Producers Committee, 1967. The outlook for gold mining – the role of the gold mining industry in the future economy of South Africa. (Gold Producer's Committee, Chamber of Mines, Transvaal and Orange Free State, 63 pp.)

Griffith, S.V., 1960. *Alluvial Prospecting and Mining,* 245 pp.

Griggs, A.B., 1945. Chromite-bearing sands of the southern part of the coast of Oregon. *U.S. Geol. Surv. Bull.,* 945-E, pp. 113–150.

Hails, J.R., 1964. A reappraisal of the nature and occurrence of heavy mineral deposits along parts of the east Australian coast. *Aust. J. Sci.,* 27: 22–23.

Hails, J.R., 1969. The nature and occurrence of heavy minerals in three coastal areas of New South Wales. *Proc. R. Soc. N.S.W.,* 102: 21–39.

Hails, J.R., 1972. The problem of recovering heavy minerals from the sea floor – an appraisal of depositional processes. *24th Int. Geol. Congr., Sect. 8:* 157–164.

Hails, J.R., 1974. A review of some current trends in nearshore research. *Earth-Sci. Rev.,* 10: 171–202.

Hall, P.K., 1972. The diamond fields of Sierra Leone. *Sierra Leone Geol. Surv. Bull.,* 5: 133 pp.

Harrison, H.L.H., 1954. *Valuation of Alluvial Deposits.* Mining Publications Ltd., London, 308 pp.

Hastings, D.A., 1974. Proposed origin for Guianian diamonds: comment. *Geology,* 2: 475–476.

Hester, B.W., 1970. Geology and evaluation of placer gold deposits in the Klondike area, Yukon Territory. *Trans. Inst. Min. Metall., Sect. B,* 79: 60–67.

Higazy, R.A. and Naguib, A.G., 1958. A study of the Egyptian monazite-bearing black sands. *Proc. 2nd U.N. Conf. Peaceful Use At. Energy,* 21: 658–662.

Hill, P.A. and Parker, A., 1970. Tin and zirconium in the sediments around the British Isles. A preliminary reconnaissance. *Econ. Geol.,* 65: 409–416.

Hill, P.A. and Parker, A., 1971. Detrital zircon in the Thames estuary: the economic potential. *Econ. Geol.,* 66: 1072–1075.

Hilmy, M.E., 1951. Beach sands of the Mediterranean coast of Egypt. *J. Sediment. Petrol.,* 21: 109–120.

Holmes, C.W., 1971. Zirconium on the continental shelf – possible indicator of ancient shoreline deposition. *U.S. Geol. Surv. Res. Circ.,* 7–12.

Hoyt, J.H., Oostdam, B.L. and Smith, D.D., 1968. Offshore sediments and valleys of the Orange River (South and South West Africa) *Mar. Geol.,* 7: 69–84.

Hudson, S.B., 1960. Concentration of ilmenite in beach sands from Greymouth, New Zealand. *C.S.I.R.O. Ore-Drilling Invest. Rep.,* 592.

Hutton, C.O., 1950. Studies of heavy detrital minerals. *Geol. Soc. Am. Bull.,* 61: 635–710.

Hutton, G.H., 1921. Valuation of placer deposits. *Min. Sci. Press, S. Francisco,* 123: 365–368.

Ingham, F.T. and Bradford, E.F., 1960. Geology and mineral resources of the Kinta Valley, Perak. *Malaya Geol. Surv. Distr., Mem.,* 9: 347 pp.

Inman, D.L., 1949. Sorting of sediments in the light of fluid mechanics. *J. Sediment. Petrol.,* 19: 51–70.

Johns, R.K., 1966. Heavy-mineral sands – Kangaroo Is. *Dep. Mines, S.Aust., Min. Rev.,* (for half year ended 30 June 1966).

Kartashov, I.P., 1971. Geological features of alluvial placers. *Econ. Geol.,* 66: 879–885.

Kavanagh, P.M., 1968. Have 6,000 years of gold mining exhausted the world's gold reserves? *Can. Min. Metall. Bull.,* 61: 553–558.

Khairy, E.M., Hussein, A.M.K., Nakhla, F.M. and El Tawil, S.Z., 1964. Analysis and composition of Egyptian ilmenite ores from Abu Ghalaga and Rosetta. *J. Geol. U.A.R.,* 8: 1–9.

Koppel, V.H. and Saager, R., 1973. Lead-isotope evidence on the detrital origin of Witwatersrand pyrites and its bearing on the provenance of the Witwatersrand gold. *Econ. Geol.,* 69: 318–331.

Krol, G.L., 1960. Theories on the genesis of the kaksa. *Geol. Mijnbouw,* 10: 437–443.

Kulm, L.D., Heinrichs, D.F., Buehrig, R.M. and Chambers, D.M., 1968a. Evidence for possible placer accumulations on the southern Oregon continental shelf. *Ore Bin,* 30: 81–104.

Kulm, L.D., Scheidegger, K.F., Byrne, J.V. and Spigai, J.J., 1968b. A preliminary investigation of the heavy-mineral suites of the coastal rivers and beaches of Oregon and Northern California. *Ore Bin,* 30: 165–184.

Layton, W., 1966. Prospects of offshore mineral deposits on the eastern seaboard of Australia. *Mineral. Mag.,* 115: 344–351.

Lee, G.S., 1968. Prospecting for tin in the sands of St. Ives Bay, Cornwall. *Trans. Inst. Min. Metall. Sect. A,* 77: 49–64.

Leo, G.W. and Coonrad, W.L., 1974. Proposed origin for Guianian diamonds: comment. *Geology,* 2: 336.

Ludwig, G. von and Vollbrecht, K., 1957. Die allgemeinen Bildungsbedingungen littoraler Schwer-
mineralkonzentrate und ihre Bedeutung für die Auffindung sedimentärer Lagerstätten. *Geol.
Jahrb.*, 6: 233–277.
Lukashev, K.I., 1970. *Lithology and Geochemistry of the Weathering Crust*. Israel Prog. Sci. Transl.,
Jerusalem, 368 pp.
Macdonald, E.H., 1966. The testing and evaluation of Australian placer deposits. *Aust. Inst. Min.
Metall. Proc.*, 218: 25–45.
Mackay, N.J., 1956. Occurrence of black sands on the north coast of Melville Island, Northern Ter-
ritory. *Bur. Miner. Resour. Aust. Rec.*, 1956/21 (unpubl.)
Marshall, T., Suggate, R.P. and Nicholson, D.S., 1958. Borehole survey of ilmenite-bearing beach
sands at Cape Foulwind, Westport, N.Z. *N.Z. J. Geol. Geophys.*, 1: 318–324.
May, J.P., 1973. Selective transport of heavy minerals by shoaling waves. *Sedimentology*, 20: 203–
211.
McFarland, W.H.S., 1965. Operations of the Yukon Consolidated Gold Corporation, Canada. In:
Opencast Mining, Quarrying and Alluvial Mining. Inst. Min. Metall., London, pp. 180–195.
McGowan, J.H. and Groat, C.G., 1971. Van Horn Sandstone, west Texas: an alluvial fan model for
mineral exploration. *Bur. Econ. Geol., Univ. Texas, Austin*, 72, 57 pp.
McKelvey, V.E., Wang, F.H., Schweinfurth, S.P. and Overstreet, W.C., 1968. Potential mineral re-
sources of the U.S. outer continental shelf. (Public Land Law Review Commission; Study of
OCS lands of U.S., Dept. Comm., P.B. 188717; 117 pp.)
Mertie, J.B., Jr., 1953. Monazite deposits of the southeastern Atlantic States. *U.S. Geol. Surv. Circ.*,
237: 31 pp.
Mullens, T.E. and Freeman, V.L., 1957. Lithofacies of the Salt Wash Member of the Morrison For-
mation, Colorado Plateau. *Geol. Soc. Am. Bull.*, 68: 505–526.
Murphy, J.F. and Houston, R.S., 1955. Titanium-bearing black sand deposits of Wyoming. In: *Wyo.
Geol. Assoc. Guideb., 10th Annu. Field Conf. Green River Basin*, pp. 190–196.
Murray, L.G., 1970. Exploration and sampling methods employed in the offshore diamond industry.
In: M.J. Jones (Editor), *Mining and Petroleum Geology – Proc. 9th Commonw. Congr.* Inst. Min.
Metall., London, pp. 71–94.
Nesbitt, A.C., 1967. Diamond mining at sea. *Proc. 1st World Dredging Conf., N.Y.*, pp. 697–725.
Nicholson, D.S., 1967. Distribution of economic minerals in South Is. west coast beach sands. *N.Z.
J. Sci.*, 10: 447–456.
Nicholson, D.S., Martin, W.R.B. and Cornes, J.J.S., 1958. Ilmenite deposits in New Zealand. *N.Z.
J. Geol. Geophys.*, 1: 611–616.
Nicholson, D.S., Shannon, W.T. and Marshall, T., 1966. Separation of ilmenite, zircon and monazite
from Westport Beach sands. *N.Z. J. Sci.*, 9: 586–598.
Noakes, L.C., 1970. Mineral resources offshore with special reference to Australia. *Aust. Miner.
Ind. Q. Rev.*, 23: 38–57.
Noakes, J.E. and Harding, J.L., 1971. New techniques in sea-floor mineral exploration. *J. Mar. Tech-
nol.*, 5 (6): 41–44.
Noakes, J.E., Harding, J.L. and Spaulding, J.D., 1974. Locating offshore mineral deposits by natural
radioactive measurements. *Mar. Technol. Soc. J.*, 8 (5): 36–39.
O'Neill, P.H., 1965. Gold and platinum dredging in Colombia and Bolivia, South America. In: *Open-
cast Mining, Quarrying and Alluvial Mining*. Inst. Min. Metall., London, pp. 156–195.
Oostdam, B.L., 1969. Geophysical exploration for marine placer deposits of diamonds. *Proc. Cont.
Centennial Congr. Min. Groundwater Geophys.*
Oregon State Publications, 1957. Oregon gold placers. *Dep. Geol Miner. Ind., Misc. Pap.*, 5: 14 pp.
Osberger, R., 1967. Prospecting tin placers in Indonesia. *Mineral. Mag.*, 117: 97–105.
Osberger, R., 1968. Billiton tin placers: types, occurrences and how they were formed. *World Miner.*,
June: 34–40.
Overeem, A.J.A. van, 1960a. The geology of the cassiterite placers of Billiton, Indonesia. *Geol. Mijn-
bouw*, 39: 444–457.
Overeem, A.J.A. van, 1960b. Sonic underwater surveys to locate bedrock off the coasts of Billiton
and Singkep, Indonesia. *Geol. Mijnbouw*, 39: 464–471.

Overeem, A.J.A. van, 1970. Offshore tin exploration in Indonesia. *Trans. Inst. Min. Metall., Sect. A*, 79: 81–85.

Overstreet, W.C., 1967. The geologic occurrence of monazite. *U.S. Geol. Surv., Prof. Pap.*, 530: 327 pp.

Overstreet, W.C., Theobald, P.K., Jr. and Whitlow, J.W., 1959. Thorium and uranium resources in monazite placers of the western Piedmont, North and South Carolina. *Min. Eng.*, 11: 709–714.

Overstreet, W.C., Yates, R.G. and Griffiths, W.R., 1963. Heavy minerals in the saprolite of the crystalline rocks in the Shelby Quadrangle, North Carolina. *U.S. Geol. Surv. Bull.*, 1162-F: 31 pp.

Overstreet, W.C., White, A.M., Whitlow, J.W., Theobald, P.K., Jr., Coldwell, D.W. and Cuppels, N.P., 1968. Fluvial monazite deposits in the southeastern United States. *U.S. Geol. Surv., Prof. Pap.*, 568: 85 pp.

Overstreet, W.C., Warr, J.J., Jr. and White, A.M., 1969. Thorium and uranium in detrital monazite from the Georgia Piedmont. *Southeast. Geol.*, 10: 63–76.

Overstreet, W.C., White, A.M. and Warr, J.J., Jr., 1970a. Uranium-rich monazites in the United States. *U.S. Geol. Surv., Prof. Pap.*, 700-D: 169–175.

Overstreet, W.C., Warr, J.J., Jr. and White, A.M., 1970b. Influence of grain size on percentages of ThO_2 and U_3O_8 in detrital monazite from North Carolina and South Carolina. *U.S. Geol. Surv., Prof. Pap.*, 700-D: 207–216.

Overstreet, W.C., Warr, J.J., Jr. and White, A.M., 1971. Possible petrogenic relations of thorium, uranium, and Ce/(Nd + Y) in detrital monazite from Surry and Stokes Counties, North Carolina. *Southeast. Geol.*, 13: 99–125.

Pakianathan, S. and Simpson, P., 1965. General practices on large-scale hydraulic mines in Malaysia. In: *Opencast Mining, Quarrying and Alluvial Mining*. Inst. Min. Metall., London, pp. 114–124.

Palmer, A.G., 1942. The estimation of gold and tin alluvials in Malaya. *Proc. Aust. Inst. Min. Metall.*, 128: 201–220.

Pardee, J.T., 1934. Beach placers of the Oregon coast. *U.S. Geol. Surv. Circ.*, 8: 41 pp.

Penhale, J. and Hollick, C.T., 1968. Beneficiation testing of the St. Ives Bay, Cornwall, tin sands. *Trans. Inst. Min. Metall., Sect. A*, 77: 65–73.

Peterson, D.W., Yeend, W.E., Oliver, H.W. and Mattick, R.E., 1968. Tertiary gold channel gravel in northern Nevada County, California. *U.S. Geol. Surv., Circ.*, 566: 22 pp.

Pettijohn, F.J., 1941. Persistence of heavy minerals and geologic age. *J. Geol.*, 49: 610–625.

Plant, J. and Coleman, R.F., 1972. Application of neutron activation analysis to the evaluation of placer gold concentrations. In: M.J. Jones (Editor), *Geochemical Exploration 1972 – 4th Int. Geochem. Explor. Symp.* Inst. Min. Metall., London, pp. 373–381.

Ras, C.B., 1957. Beach erosion and concentration of heavy-mineral sands. *J. Sediment. Petrol.*, 27: 143–147.

Reid, A.R., 1974. Proposed origin for Guianian diamonds. *Geology*, 2: 67–68.

Rittenhouse, G., 1943. Transportation and deposition of heavy minerals. *Geol. Soc. Am., Bull.*, 54: 1725–1780.

Roe, F.W., 1953. The geology and mineral resources of the neighbourhood of Kuale Selangor and Rasa, Selangor, Federation of Malaya, with an account of the geology of Batu Arang coalfield. *Malaya Geol. Surv. Mem.*, 7: 163 pp.

Roscoe, S.M., 1957. Geology and uranium deposits, Quirke Lake – Elliott Lake, Blind River area, Ontario. *Can. Geol. Surv. Pap.*, 56-7: 21 pp.

Rubey, W.W., 1933. The size distribution of heavy minerals within a waterlain sandstone. *J. Sediment. Petrol.*, 3: 3–29.

Sainsbury, C.L., 1969. Tin resources of the world. *U.S. Geol. Surv., Bull.*, 1301: 55 pp.

Schönberger, H., 1974. Diamond exploration in central-eastern Suriname. *Geol. Mijnbouwk. Dienst Suriname, Med.*, 23.

Schönberger, H. and De Roever, E.W.F., 1974. Possible origin of diamonds in the Guiana Shield. *Geology*, 2: 474–475.

Schuiling, R.D., 1967. Tin belts on the continents around the Atlantic Ocean. *Econ. Geol.*, 62: 540–550.

Schumm, S.A., 1976. Fluvial geomorphology: principles and applications. In: J.R. Hails (Editor), *Applied Geomorphology*. Elsevier, Amsterdam, (In Press).

Sestini, G., 1973. Sedimentology of a palaeoplacer: the gold-bearing Tarkwaian of Ghana. In: G.C. Amstutz and A.J. Bernard (Editors), *Ores in Sediments*. Springer, Berlin, pp. 275–306.

Shelton, J.E., 1965. Tin. *U.S. Bur. Mines Miner. Yearb., 1964*, 1: 1057–1073.

Stocken, C.G., 1962. The diamond deposits of the Sperrgebiet, South West Africa. *Geol. Soc. S. Afr., Field Trip 5th Annu. Congr.*, 15 pp.

Tanner, W.F., Mullins, A. and Bates, J.D., 1961. Possible masked heavy-mineral deposit, Florida Panhandle. *Econ. Geol.*, 56: 1079–1087.

Tanner, W.F., 1962. The use of equilibrium concepts in the search for heavy minerals. *Am. Inst. Min. Metall. Pet. Eng. Soc., Min. Eng., Trans.*, 223: 395–399.

Tooms, J.S., 1970. Some aspects of exploration for marine mineral deposits. In: M.J. Jones (Editor), *Mining and Petroleum Geology – Proc. 9th Commonw. Congr.* Inst. Min. Metall., London, pp. 285–296.

Tooms, J.S. et al., 1965. Geochemical and geophysical mineral exploration experiments in Mounts Bay, Cornwall. *Colston Pap.*, 17: 363–391.

Tooms, J.S., Summerhayes, C.P. and Cronan, D.S., 1969. Geochemistry of marine phosphates and manganese deposits. *Oceanogr. Mar. Biol.*, 7: 49–100.

Tourtelot, H.A., 1968. Hydraulic equivalence of grains of quartz and heavier minerals and implications for the study of placers. *U.S. Geol. Surv., Prof. Pap.*, 594-F: 13 pp.

Twenhofel, W.H., 1943. Origin of the black sands of the coast of southwest Oregon. *State Ore., Dep. Geol. Miner. Ind., Bull.*, 24: 25 pp.

U.S. Bureau of Mines, 1972. Commodity data summaries. (Published January, 1973.)

Vlasov, K.A., 1968. *Geochemistry and Mineralogy of Rare Elements and Genetic Types of their Deposits*, 3: 614–633.

Ward, J., 1957. Occurrence of heavy mineral beach sands in the vicinity of Point Blaze, Northern Territory. (Bur. Miner. Resour. Aust. Rec., 1957/88, unpublished).

Ward, J., 1965. Heavy-mineral beach sands of Australia. In: J. McAndrew (Editor), *Geology of Australia Ore Deposits – 8th Commonw. Min. Metall. Congr., Melbourne*, pp. 53–54.

Ward, J., 1972. Australian mineral industry. *Bur. Miner. Resour. Geol. Geophys.*, 25: 23 pp.

Webb, B., 1965. Technology of sea diamond mining. *Mar. Technol. Soc., Proc. 1st Annu. Conf.*, June, 8–23.

Welch, B.K., 1964. The ilmenite deposits of Geographe Bay. *Proc. Aust. Inst. Min. Metall.*, 211.

Wertz, J.B., 1949. Logarithmic pattern in river placer deposits. *Econ. Geol.*, 44: 193–209.

Whitworth, H.F., 1956. The zircon–rutile deposits on the beaches of the east coast of New South Wales, with special reference to their mode of occurrence and the origin of the minerals. *Tech. Rep. Dep. Mines, N.S.W.*, 4: 60 pp.

Wilcox, S.M., Mead, W.J. and Sorensen, P.E., 1972. Economic potential of marine placer mining. *Ocean Ind.*, August, 27–28.

Zaalberg, P.H.A., 1970. Offshore tin dredging in Indonesia. *Trans. Inst. Min. Metall.*, 79: A86–A95.

Zeschke, G., 1961. Prospecting for ore deposits by panning heavy minerals from river sands. *Econ. Geol.*, 56: 1250–1257.

Zimmerle, W., 1973. Fossil heavy-mineral concentrations. *Geol. Rundsch.*, 62: 536–548.

Chapter 6

MINERAL (INORGANIC) RESOURCES OF THE OCEANS AND OCEAN FLOORS: A GENERAL REVIEW

WOLFGANG SCHOTT

INTRODUCTION

Many occurrences of mineral resources in ocean water, on the ocean floor and in its deeper parts have long been well-known. This is readily understandable, because numerous ancient mineral products which are exploited on land were originally formed in marine milieus as far back as the Precambrian, e.g., marine sedimentary iron ores, rock salt, potassium, phosphate and manganese, as well as petroleum and natural gas.

This book deals with ores in sediments, and in sedimentary and volcanic rocks. According to the Concise Oxford Dictionary (Fowler and Fowler, 1964), ore is defined as "solid native mineral aggregate from which valuable constituents not necessarily metal may be usefully extracted". Therefore, in this chapter only mineral resources of *inorganic* origin in ocean water and on the ocean floor are discussed and not those of organic derivation, although the existence of petroleum, natural gas and, to some extent, of coal in the subsurface of the ocean shelves is of greater economic importance. The amount of oil and gas produced in 1969 represented more than 90% by value of all mineral resources obtained from the oceans and ocean floors.

Ore deposits in bed-rock formations of the shelf region, such as those which are exploited near Cornwall (England) and Newfoundland for example, are also irrelevant, as these are merely extensions of discoveries on the nearby mainland.

The purpose of this chapter is to give a general review of the (inorganic) mineral resources of the oceans and the ocean floors. In this respect, not only are the economically important products which are already in use discussed, but also those materials which, in their oceanic environment, can become of economic value in the near or more remote future. For a better understanding of the existence of such materials, some deposits have also had to be reviewed which will not become of economic value.

Since some contributions in this book are specifically dedicated to Recent marine ferromanganese deposits and Recent phosphorite deposits, these mineral resources are only briefly discussed in this chapter. (For details on Recent marine and lacustrine manganese deposits see Chapters 7 and 8 by Glasby/Read and by Callender/Bowser, respectively, in Volume 7.)

The litarature on the mineral (inorganic) resources of the oceans and ocean floors is extremely extensive, so that a choice had to be made and, therefore, the bibliography added to this chapter consists mainly of recent publications.

RAW MATERIALS FROM OCEAN WATER

In this section raw materials form *normal* ocean water with a salinity of about 35‰ will be discussed, whereas brines, which e.g. can appear as hydrothermal exhalations on the sea floor, are reviewed in the section on "Hydrothermal sedimentary metallic deposits" (p. 270; see also Chapter 4 by Degens and Ross, Vol. 4).

In addition to hydrogen and oxygen, ocean water contains an average of 3.5% of various other elements in solution, this value of 3.5% being the salinity of seawater (35 g/kg = 35‰ in oceanography; Sverdrup et al., 1946; Dietrich, 1963). Its main components are common salt (NaCl), magnesium chloride ($MgCl_2$), magnesium sulphate ($MgSO_4$), and calcium sulphate ($CaSO_4$). The salinity has a nearly constant composition of elements in solution. The 92 stable elements which occur in nature must exist in ocean water; up to now over 75 elements have been quantitatively recorded. Except for hydrogen and oxy-gen, the 11 elements reported in Table I determine the composition of the salinity and they account for 99.997% of all dissolved salts. Thus, including hydrogen and oxygen, seawater consists of 13 main components. The remaining elements occur only as traces (Table II). In 1 kg of seawater with a salinity of 35‰, these trace elements together form less than 3 mg, which is less than 1‰ of the salinity. Furthermore, the industrially important metals in ocean water nearly always occur in very small quantities.

At the present time only six of the elements dissolved in seawater are of greater economic interest as raw materials; namely hydrogen and oxygen in the form of fresh-water, sodium and chloride together as salt, and magnesium and bromine.

The continuously increasing need for drinking and industrial water has strongly stimu-lated the construction of seawater desalination plants, not only in arid and semi-arid areas, but also in temperate regions. Only after the Second World War has the technology

TABLE I

Main components of ocean water (35‰ salinity) (after Riley and Chester, 1971)

Element	Concentration (mg/l)	Element	Concentration (mg/l)
Oxygen (O)	856,000	calcium (Ca)	422
Hydrogen (H)	107,800	potassium (K)	416
Chlorine (Cl)	19,870	bromine (Br)	68
Sodium (Na)	11,050	carbon (C)	28
Magnesium (Mg)	1,326	strontium (Sr)	8.5
Sulphur (S)	928	boron (B)	4.5
		fluorine (F)	1.4

TABLE II

The average concentration of some trace elements in seawater (after Goldberg, 1963)

Element	Concentration (mg/l)	Element	Concentration (mg/l)
Zinc (Zn)	0.01	titanium (Ti)	0.001
Iron (Fe)	0.01	antimony (Sb)	0.0005
Aluminium (Al)	0.01	cobalt (Co)	0.0005
Molybdenum (Mo)	0.01	silver (Ag)	0.0003
Tin (Sn)	0.003	cadmium (Cd)	0.00011
Copper (Cu)	0.003	chromium (Cr)	0.00005
Uranium (U)	0.003	lead (Pb)	0.00003
Nickel (Ni)	0.002	mercury (Hg)	0.00003
Vanadium (V)	0.002	gold (Au)	0.000005
Manganese (Mn)	0.002		

to obtain freshwater from seawater been developed, and nowadays several methods exist, although, to some considerable extent, they still need further technical improvement in order to render their operation more economic. At present, distillation methods are of the greatest practical importance and in many places they are already being applied in very large installations, some of which produce about 20,000 tonnes[1] of freshwater per day; however, their energy requirements are very great. Freshwater plants using electro-dialysis (electro-osmosis) are also in operation in various parts of the world. In areas where the daily amount of sunshine is greater, solar energy can be used for the evaporation and condensation of water, if only small amounts are required (solar seawater desalination). Plants of this type are in use, e.g., in Australia and Greece. Applications of hyperfiltration (reversed osmosis) and freezing methods are still in the development and experimental stages, respectively (Ullmann, 1967, vol. 18, pp. 458–474). As to 1974 the number of operative desalination plants is estimated at 800. In 1968 about $142 \cdot 10^6$ tonnes of freshwater were thus produced (Table III; Shigley, 1968). (Further literature: Girelli, 1965; 1st–4th Int. (European) Symp. on freshwater from the sea (last symposium Heidelberg, 1973); U.S. Dep. Int. Saline Water Conversion Report, 1965.)

Eventually, in the near or more remote future, seawater desalination plants could also be utilized for the production of several economically important marine trace elements which can be extracted from salt brines (see Mero, 1965, p. 43; Ullmann, 1967, vol. 18, p. 474).

The production of common salt (NaCl) from seawater is widespread in arid and semi-arid areas as well as in temperate regions with dry, warm summers. Already in ancient times, the human needs for salt were to a great extent covered by this method; in China apparently already at some time prior to 2200 B.C. (Andrée, 1920, pp. 573–576; Mero,

---

[1] 1 tonne = 1 metric ton = 1,000 kg.

Fig. 1. Salt garden in the Lagõa de Obidos near the village of Arelho, Portugal. (Photo W. Schott, 1951.)

1965, p. 25). Up to present, these production methods have hardly changed in their principle. The seawater evaporates to a brine under strong solar radiation in artificially constructed ponds within bays (Fig. 1). The crystallization of the various salts occurs on the basis of their different solubilities in seawater. To separate the precipitations of

TABLE III

World production from ocean water (W.F. McIlhenny, 1968: see Shigley, 1968, fig. 6)

	Production (tonne/year)	% of world production
Sodium chloride	35,000,000	29
Magnesium		
metal	106,000	61
compounds	690,000	6
Bromine	102,000	70
Fresh water	142,000,000**	59*

* Of the "manufactured" water.
** According to a very rough estimate in "Water Desalination Report", January 3rd, 1974, a total of about $1.6 \cdot 10^6$ m^3 per day was produced at the beginning of 1970 (the annual value of these commodities from seawater approaches $ 400 \cdot 10^6$). (From Shigley, 1968.)

different salts (e.g. calcium sulphate from sodium chloride), the brine is channelled to another pond after some time when the concentration of the brine has increased further. In this way, calcium sulphate, sodium chloride (over 99% NaCl), magnesium compounds, bromine and other salts can be separated. The brine after precipitation of sodium chloride is called "bittern". In some countries, e.g. in Japan, common salt is obtained to some extent in vacuum evaporation plants using brine produced from seawater in salt ponds. In polar regions the salt is separated from seawater by freezing techniques; the ice is practically salt-free. The remaining solution is salt-enriched and sodium chloride can be extracted form it. This method is applied, e.g., in the White Sea area (Ullmann, 1960, vol. 12, p. 669).

In 1968 the world production of common salt from seawater amounted to about $35 \cdot 10^6$ tonnes, which formed 29% of the total world salt production (Table III). In the United States, about 5% of the annual salt consumption is produced from ocean water on the West Coast, principally in the San Francisco Bay area, while in India more than 70% and in Japan nearly 100% of the salt demand is covered from seawater. (For details, see Ullmann, 1960, vol. 12, pp. 666–671; Mero, 1965, pp. 25–31.)

Natural precipitation of sodium chloride from ocean water in so-called salt pans (salinas) by solar radiation is a well-known fact in several coastal areas. An interesting example is the Rann of Kutch, north of peninsular Kutch in India, approximately 170 km ESE of the mouth of the river Indus. The Rann of Kutch, covering an area of about

Fig. 2. Rann of Kutch, 12 km east of the village of Lakhpat, India. (Photo W. Schott, 1957.)

23,300 km^2, was a shallow bay of the Arabian Sea during the Pleistocene. In the fourth century B.C., when Alexander the Great invaded India, it was still deep enough for sailing ships. At present it is a large flat saline desert for a great part of the year during the dry NE monsoon, the surface being covered by a hard salt layer (Fig. 2). Salt beds up to 20 cm thick have been observed in the Recent mud. During the rainy season of the SW monsoon the Rann is covered with a few centimeters of water preponderantly caused by periodic flooding of seawater through the Cori Creek and the Gulf of Kutch (Krishnan, 1956, p. 45; Lotze, 1957, p. 134).

Even small natural-salt precipitations from seawater are locally used by man. In rocky coastal areas, salt can form thin layers in fissures, when the seawater spray evaporates; it is produced in this way for example in Norway and on the Sinai Peninsula (Von Buschman, 1906, 1909). Major quantities of natural precipitations of sodium chloride from seawater, which originated in many places during the geologic past, have been recorded. They appear as salt diapirs in some regions in the subsurface of the continents, and beneath Recent shelf floors and deep-sea floors. The most famous examples of deep-sea salt diapirs are those at a water depth of about 3,500 m in the abyssal plain of the Gulf of Mexico which form small topographic prominences on the sea floor, the so-called "Sigsbee-Knolls". The age of this salt is estimated at Middle to Late Jurassic to possibly Late Cretaceous (Ewing, J.I. et al., 1962; Ewing, M. et al., 1969).

The production of magnesium and magnesium compounds (MgO, Mg(OH)$_2$, MgCl$_2$) from seawater has gained economic importance in several countries. This metal was extracted from seawater for the first time in England. Today the total needs for magnesium in the United States are covered by this technique: e.g. in the important plant at Freeport, Texas. In many parts of the world, notably in England and the United States, magnesium compounds are also manufactured from seawater (see Table III). Initially, magnesium compounds were mainly produced as a by-product from the bittern in salt gardens (Ullmann, 1960, vol. 12, pp. 85–90; Mero, 1965, pp. 34–39).

As the demand for bromine is ever-increasing, this metal is nowadays mainly extracted from ocean water and not only from salt seas and fossil salt deposits (Table III). Huge plants, such as those in Freeport, Texas, where approximately 80% of the annual bromine consumption of the U.S.A. are produced, make the United States the greatest supplier in the world of ocean-extracted bromine. Other sizeable plants where ocean water is utilized are in Great Britain, South Africa and other countries (Ullmann, 1953, vol. 4, pp. 730–742; Mero, 1965, pp. 31–54).

Metals of economic importance, but which are only dissolved in trace quantities in seawater (see Table II), have continuously been subjects of research. For example, during the nineteen twenties, the famous chemist and Nobel prize winner Fritz Haber thoroughly investigated the possibility of producing gold from seawater (Haber, 1927; see also Ryabinin et al., 1974). Today it is uranium that attracts attention, and in various countries, such as England and Japan, attempts are being made using various techniques to extract uranium dissolved as carbonate, but this is still in the experimental stage. Today marine-produced uranium could not compete with normally mined uranium because the cost is several times as great.

The suspended matter in ocean water also contains very small quantities of some metals, such as e.g., manganese, lead and iron. However, the filtering of these very fine particles from seawater is still a very difficult problem (Mero, 1965, p. 50).

MINERAL (INORGANIC) RESOURCES OF THE OCEAN FLOORS

Sedimentary mineral deposits on the ocean floor consist of reworked and locally exceptionally enriched minerals, or of minerals originating from chemical concentration or precipitation. Reworked and locally enriched minerals are the clastic (= detrital) placer deposits. They are formed by mechanical processes, which are caused by ocean currents and waves. The materials of these mineral resources are derived primarily from massive detritus from the mainland. This detrital material is carried into the oceans either as a result of coastal erosion or − chiefly − by river transportation of erosional products after atmospheric weathering of rock masses on continents and islands. Ocean currents and chemical modifications and differentiations in ocean water cause this material to be very extensively sorted before deposition as sediment on the ocean floor, nearshore, on the shelf or in the deep sea. Besides the normal marine-sedimentary deposits, these mechanical and chemical processes, which in addition are to some extent influenced to varying degrees by biological, volcanic and hydrothermal processes, produce the mineral resources of the ocean floor (Ludwig and Vollbrecht, 1957; Mero, 1965, 1971; Keiffer, 1968; Love, 1969; McKelvey and Wang, 1969; Wenk, 1969; W. Schott, 1970a, b, 1974; Seibold, 1970, 1973; Bäcker and Schoell, 1974; Cruickshank, 1974; Lüttig, 1974).

Locally enriched minerals

Sand and gravel. Especially in the shallow-water areas of the shelf, the detritus deriving from the mainland is continuously reworked. As a result of the mechanical sorting due to wave action and ocean currents, large deposits of pure gravel and sand of economic value can develop.

In some countries, the increasing need for sand and gravel as building materials has already led to exploitation of such submarine materials, particularly in British and North American waters (e.g., Pepper, 1958; Mero, 1965; Dunham, 1967, 1970; Schlee, 1968; Wenk, 1969; Dunham and Sheppard, 1969; Veenstra, 1970). Sand and gravel of the British shelf originate from rivers or from partly to totally reworked glacial and periglacial deposits of Pleistocene age, and as these materials occur abundantly around the British coast, they have been dredged in large quantities for many years. In 1969 some 60 vessels were operative in this industry and their yearly production was about $8 \cdot 10^6$ tonnes, 10% of the needs of the United Kingdom.

Fig. 3. Heavy-mineral deposits, beach at Pebane, Mozambique. Dark grey: ilmenite, rutile, zircon, monazite. Light grey and white: preponderantly quartz. (Photo W. Schott, 1970.)

Furthermore, in the United States sand and gravel, mostly of Pleistocene age, are taken in great quantities along the seashore and in bays, in particular, for example, off Long Island Sound, New Jersey and Florida. In 1968, the annual production was about $27 \cdot 10^6$ m^3. The estimated resources of sand and gravel on the U.S. continental terrace amounted to $14 \cdot 10^5$ tonnes (Cruickshank, 1974).

Heavy-mineral deposits (placer deposits).[1] Rocks eroded from the mainland frequently contain heavy minerals with a density of more than 2.89 g/cm^3, which distinctly exceeds that of quartz. Consequently, these heavy grains are separated from the quartz grains during the mechanical sorting by the ocean currents which usually run parallel to the coast, and locally they are concentrated into marine placer deposits of economic value. Normally heavy-mineral sands are well bedded but mostly lenticular and they vary in thickness from a few centimeters to several decimeters (Fig. 3).

Occurrences of heavy-mineral sands in nearshore and offshore regions may commonly be subdivided into: (a) dune placer; (b) beach placer alongshore (partially submerged, partially raised in beach terraces); (c) placer on sand banks; (d) submarine deep leads; and (e) placer caused by submarine erosion.

---

[1] For a detailed summary see Chapter 5 by Hails, this Vol., and for an example of ancient placer deposits see Chapters 1 and 2 by Pretorius, Vol. 7.

Beach placers and placers on sand banks are formed directly or indirectly by the sorting effect of the waves that break against the beach or on the sand banks (Seibold, 1974, p. 139). Strong tides or large tidal ranges may greatly influence this process and the wind may reinforce the sorting, whereby dune placers are formed at the beach predominantly by aeolian processes. As the result of positive shifting of the beach by regression of the ocean, placer deposits may occur in raised beach terraces. These marine heavy-mineral sands are of Recent, Subrecent or Pleistocene age.

Submarine placer deposits in the shelf region are predominantly of Subrecent or Glacial age. During the Pleistocene glacial ages, the sea level was considerably lowered as the result of increased ice formation in the polar regions. On the average it was 150 m below the present sea level, thus usually approaching the present shelf edge. This is demonstrated by the widespread occurrence of relict sediments of Postglacial and young Pleistocene age on the outer shelf (W. Schott and Von Stackelberg, 1965; Emery, 1966; Von Stackelberg, 1972). Thus, during the Pleistocene beach placers were able to form on the shelf. When the sea level rose again at the end of the Pleistocene, these placers were covered by the rising sea, and thereby were partly reworked, destroyed and partly covered by Recent sediments. The deep leads of the shelf region also belong to these Subrecent to Glacial mineral sands. These are Subrecent or Glacial buried river channels containing fluviatile mineral sands, which have been preserved under younger sediments. These river channels formed when the coasts were close by the present-day shelf edges. Occurrences of placer deposits due to submarine erosion are possible when weathered source rocks containing economic heavy minerals outcrop on the floor of the shelf.

The commonest minerals of economic value in placer deposits are zircon, ilmenite, rutile, monazite, magnetite, garnet, epidote, topaz, tourmaline, corundum, gold, platinum, cassiterite and clinozoisite. The first three minerals and cassiterite are the most common in nearshore and shelf regions. If the colour of the predominant minerals in mineral sands is dark, the placer is termed "black sands".

The economic detrital heavy minerals may be classified into three groups (Emery and Noakes, 1968, p. 96): (a) heavy heavy minerals (gold, tin, platinum; specific gravity 6.8–21); (b) light heavy minerals (zircon, ilmenite, rutile, monazite; specific gravity 4.2–5.3); and (c) gems (diamonds, rubies, sapphires; relatively low specific gravity of 2.9–4.1, but with extreme hardness).

Of the heavy heavy minerals in offshore regions, tin occurs predominantly in ancient river-deposits which are submerged below sea-level, and gold in nearshore areas. However, the occurrences of gold and also of platinum are not of economic value. The light heavy minerals can be transported far away from their source rocks, over distances of hundreds of kilometers. Therefore, Recent and Subrecent ocean beaches probably represent the optimum conditions under which the minerals of economic value can concentrate. As regards the gems, diamonds have been found in Recent beach sands and in submerged subrecent beach sands. (For further information about the possibilities of economic placer deposits occurring on ocean shelves see Emery and Noakes, 1968).

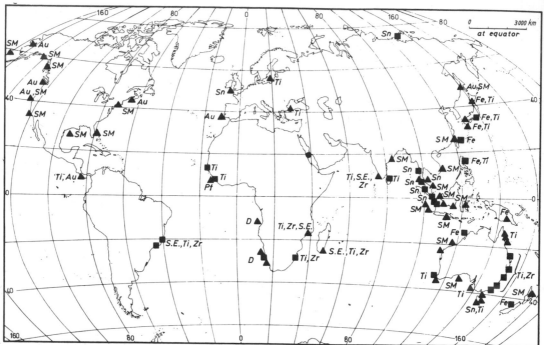

Fig. 4. Offshore placer deposits and hydrothermal metallic deposits of the oceans. *Au*: gold; *Pt*: plati-num; *Fe*: iron ore; *Sn*: tin; *Ti*: titanium minerals; *Zr*: zircon; *SE*: rare-earth minerals among which monazite; *SM*: heavy minerals, not further distinguished. (From W. Scott, 1970b, with supplements.) ■ = exploitation on beach or nearshore; ▲ = important exploration activity; ● = exploration of hydro-thermal metallic deposits.

In the three oceans, offshore placer deposits are widespread in shelf regions (Fig. 4). The most economically important occurrences are in southeast Asia and off the coasts of Australia. Southeast Asia is the most important tin-producing area of the world (Cissarz and Baum, 1960; Van Overeem, 1960, 1970; CCOP, 1967, 1968a, b; Osberger and Roma-nowitz, 1967; Osberger, 1968; Zaalberg, 1970; Hosking, 1971; MacDonald, 1971). The large tin province of this region stretches from northern Burma over peninsular Thailand and western Malaysia to the well-known Tin Islands of Indonesia (Singkep, Bangka and

Fig. 5. The offshore tin areas of southeast Asia (general distribution of the granites and tin belts). (From Hosking, 1971, fig. X-1.) 1 = Belugyun Island; 2 = Heinze Basin: cassiterite dredged from tide-ways; 3 = Spider Island (at mouth of Palauk River); 4 = Tenasserim delta, Lamp and neighbouring islands; 5 = Ranong and coast to south; 6 = Takuopa: suction dredged working offshore; 7 = Thai Muang; 8 = Phuket: dredges operating off east coast; 9 = Ko Phangan and Ko Samui; 9a = Rayong: cassiterite in beach sands and offshore; 10 = Langkawi islands; 10a = Ko Ra Wi and Ko La Dang (WNW of Langkawi); 10b = islands west of Kedah Peak; 11 = Lumut–Dindings: cassiterite in beach sands and offshore; 12 = Malacca: cassiterite in mainland beach sands and offshore; 13 = Bintan; 14 = The Tin Islands: substantial offshore mining and exploration activity; 15a = Submarine tin granites between Billiton and Borneo; 16 = Anambas and Natuna islands.

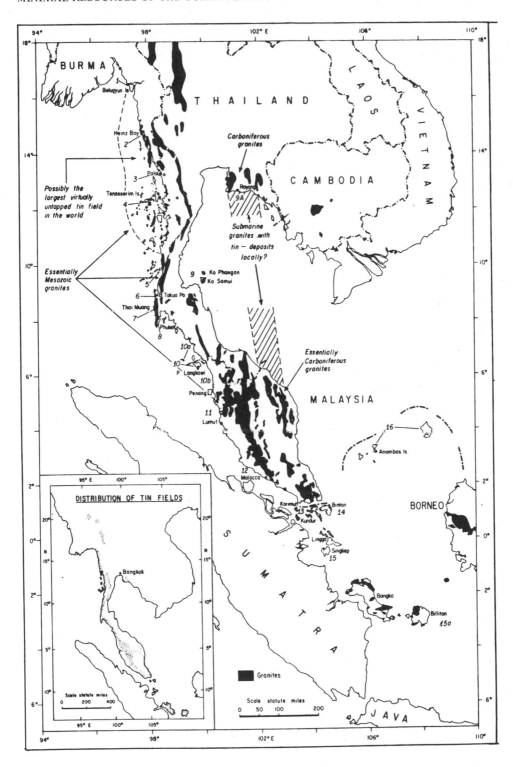

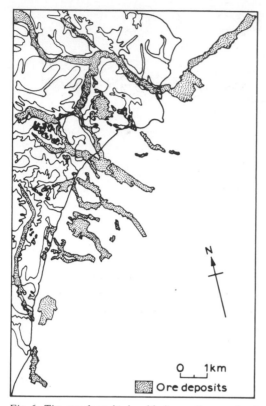

Fig. 6. Tin ore deposits in old river valleys on land and nearshore. East coast of Singkep, Indonesia. (From Zaalberg, 1970.)

Billiton: Fig. 5), a distance of more than 2,900 km. As far as is known, the source rocks of this tin placer are the granite intrusives of Mesozoic, mostly Late Jurassic–Cretaceous age which are widely distributed in the above-mentioned region.

The exploitation of this tin placer commenced on land in rivers. The tin accumulations in the sediments of the Kinta Valley, western Malaysia, form the most important tin-producing area (Ingham and Bradford, 1960; Newell, 1971). The offshore dredging of submarine tin first began in 1908 in the shoreline area of Phuket Island, Thailand (Hosking, 1971); it is mostly composed of the dense mineral cassiterite, SnO_2. A very considerable amount of cassiterite has been exploited from submarine placer deposits for many years. The main producing areas are those surrounding the Indonesian Tin Islands and Phuket Island. In various parts of the Indonesian Tin Islands outcrops of partially deeply weathered granitic intrusives occur. Large bucket, grab and suction dredges gather the submarine placer material. The world's largest tin dredger, "Bangka 1", flagship of the Indonesian bucket-dredge fleet, can dig to a depth of 40 m below sea level; its monthly capacity is about 305,000 m^3. Suction dredges can recover material from depths of about 65 m below sea level.

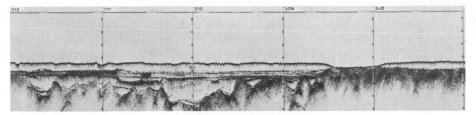

Fig. 7. Sonograms of the ocean floor. Pleistocene valley incised in granite bedrock (Indonesia) (time intervals indicate approx. 10 m depth, horizontal distance between position marks is about 600 m). (With permission of Fugro-Cesco, Leidschendam, The Netherlands.)

The most important offshore tin placers near the Indonesian Tin Islands are found in submerged river channels of Glacial or Postglacial age, covered by younger sediments (Fig. 6). Today the most strongly furcated river channel systems with deep leads are located by systematic surveying by means of seismic reflection profiling (e.g., sonic survey, Untung, 1968). Besides the water depth, the position of the river valleys and the thickness of the sediments covering the river deposits can be determined (Fig. 7).

In the area of this southeast Asian tin province, finds of other heavy minerals, such as zircon, ilmenite, rutile, tourmaline and monazite, have also been made, partly as by-products of tin mining. Zircon and titanium are chiefly mined in the beach and coastal areas of Australia (Fig. 4) (Gardner, 1955; Layton, 1966; Ward, 1972; etc.).

In 1934, the production of heavy minerals began on the east coast of Australia and on the southwest coast of Western Australia in 1956. The black sands occur along the coasts, principally as beach sands. Due to intensive mining, the high-grade deposits of black sands have been exhausted at very many locations on the east coast of Queensland and New South Wales and, as a result, the exploitation activities are now concentrated on extensive low-grade occurrences. Hence, the mineral sands industry in Queensland is at present concentrated on the offshore islands, i.e. Fraser, Stradbroke and Moreton Islands. The high dunes of North Stradbroke Island contain large quantities of rutile and zircon; today mining of this low-grade placer is economically justified.

Along the coast of northern New South Wales and southern Queensland, the shelf region (offshore of productive beaches) has been investigated in recent years. Fossil beach lines have been found at water depths of 40 m or more containing interesting quantities of zircon, rutile and ilmenite. One mining company has discovered an area of about 200,000 m^2 of placer deposits with overall grades of 1.5% of heavy minerals, up to 10 m thick and under a water depth of about 33 m (Emery and Noakes, 1968). Thanks to the numerous black-sand occurrences, Australia is the main producer of rutile, zircon, ilmenite and monazite in the world. Of the 1971 world production, Australia accounted for approximately 95% of the rutile, 87% of the zircon, about 50% of the monazite and 25% of the ilmenite. At the end of 1971, total production of saleable-grade concentrates amounted to 3,487,000 tonnes of rutile, 3,641,000 tonnes of zircon, 5,637,000 tonnes of ilmenite and 28,000 tonnes of monazite. In the light of the reserves, the black sands on

TABLE IV

Recoverable mineral sand resources of Australia (in thousands of tonnes, June 1972) (adapted from Ward, 1972)

	Rutile	Zircon	Ilmenite	Monazite
East Coast	6,183	5,693	13,396	42
Western Australia	1,667	5,830	34,870	195
Total Australia	7,924	11,579	49,489	238

the coast of Australia are expected to account for a considerable share of the world production of these minerals in both the near and more distant future. (See Table IV.)

Heavy-mineral deposits have also been found on a number of coasts in other regions of the Far East and these are also exploited locally (Fig. 4). Titaniferous iron sands, for example, have been reported from the south coast of Java and Bali (Indonesia), in the beach sand of northern Taiwan and on the coasts of Luzon (Philippines) and of Japan. In Korea, placer deposits of gold, monazite, ilmenite and mixed heavy minerals are found, mostly in the western and southern coastal areas (Kim, 1970). The marine ilmenite placer deposits of Travancore, Malabar coast, southern India, have a heavy-mineral content of over 70%. Beside ilmenite, magnetite, rutile, zircon, garnet and locally monazite also occur in this region.

In various locations on the coasts of North America, attempts have been made to produce gold and other heavy minerals, amongst others, offshore of the state of Oregon, in the waters surrounding Alaska (U.S.A.) and offshore of Nova Scotia (Canada) (Fig. 4). Monazite, ilmenite, rutile and zircon, for instance, have been found on ancient beaches of Florida, and black sands in the beach and dune sands of Cape Cod and vicinity (Pepper, 1958; Trombull and Hathaway, 1968).

On the beaches of Brazil in South America, especially along the coasts of Estado do Espirito Santo and Estado de Bahia, monazite, ilmenite, rutile, etc., occur.

In Europe, tin placer (cassiterite) has been recovered from beach sand in western Cornwall for many years (Hosking and Ong, 1964). According to Choffat (1913), at the beginning of the nineteenth century gold-containing coast sands were produced on the coast or Portugal between the mouth of the Tagus River and Cabo Espichel. In the North Sea, placer deposits have been found on the coasts of northwestern Germany (especially ilmenite and zircon) (Lüttig, 1974). The diamond deposits off the coast of South West Africa are Late Pleistocene — Recent unconsolidated sediments resting on Precambrian bedrock. These deposits were first found onshore and later also offshore in wave-cut platforms, submarine cliffs, gullies and submerged river valleys (Nesbitt, 1967; L.G. Murray et al., 1970). The mining operations in the sea have ceased as from April 1971 (Nowak, 1973).

Many shoreline placer deposits of economic value have been discovered accidentally.

However, the search for nearshore and offshore accumulations of heavy minerals requires a very close examination of the geology of the shelf and bordering land areas. Only in this way is it possible to obtain good results, as is discussed in the following example.

For some considerable time, the occurrence of placer deposits has been reported in some places in beach and dune sands on the coasts of Mozambique (SE Africa), north of the mouth of the Zambesi River, particularly in the region of Pebane (Figs. 3, 4). The crystalline basement complex of the African continent, in which the heavy minerals occur finely distributed as accessory components and notably in the granite rocks, extends north of the mouth of the Zambesi, almost of the coast and only separated from the ocean by a small zone of younger deposits. Together with the eroded materials from this basal complex, the heavy minerals have been transported by the rivers to the coast where they could deposit along the beach due to the longshore drift which runs predominantly parallel to the coast. It is because of the fact that during the Pleistocene Glacial ages large regions of the East African shelf were part of the adjacent mainland, that offshore occurrence of black sands in shelf deposits between the mouth of the Zambesi and Ilha de Mozambique on 15°S latitude can be assumed to be present (W. Scott, 1970b, p. 25).

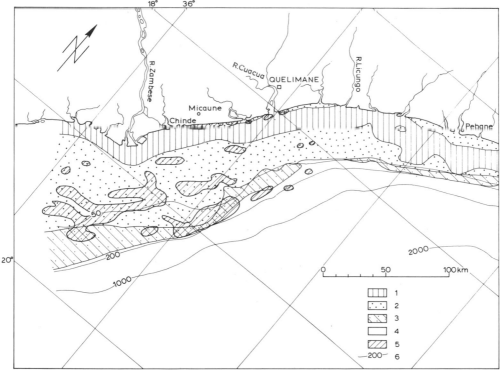

Fig. 8. Placer of titanium ores. Shelf of Mozambique, Indian Ocean. *1*: Shallow-water mud (75−100 weight-percentage <63 μ); *2*: terrigenous sand; *3*: biogenous sands and relict sediments (partially cemented); *4*: deep-sea mud; *5*: accumulation of heavy minerals >3%; *6*: water depth in meters. (From Putzer and Von Stackelberg, 1973.)

In order to recognize better the possible local occurrences of black-sand accumulations on the shelf, geological studies were also carried out on the coast before and during examination of the shelf region, with the object of clarifying the Recent geological development of these coastal areas with placer deposits.

Thanks to this simultaneous geological work on the coast and on the shelf, which was associated with seismic studies, the German research vessel "Valdivia" (W. Schott, 1974) was able, in 1971, to discover sands with a heavy-mineral content of >3% in Pleistocene terrigenous and biogenic relict sediments in the middle and outer shelf region surrounding the mouth of the Zambesi (Fig. 8). Together they cover an area of about 800 km^2. The main component of the usable heavy minerals is ilmenite of commercially acceptable quality (46% TiO_2, 50% Fe_2O_3 and 0.09% Cr_2O_3) (Beiersdorf, 1972; Putzer and Von Stackelberg, 1973). The main task of the second research cruise of the "Valdivia" in 1973 was to investigate the extent in depth of these black-sand accumulation by means of vibratory coring. This operation produced further promising results concerning the distribution of the heavy minerals (U. von Stackelberg, personal communication, 1974).

Minerals originating from chemical concentration

Among the mineral neoformations developed due to chemical concentration on the ocean floor, phosphorite concretions, manganese nodules and hydrothermal metallic deposits are particularly interesting for the production of mineral raw materials.

Phosphorite concretions[1]. Phosphorite concretions are widespread on the sea floor — Agulhas Bank of South Africa, the coast of SW and NW Africa, the coast of California, Blake Plateau off Florida, west coast of South America, the coast of Australia, etc. Phosphorites occur mostly as larger nodules, and locally as sand and mud, especially in the region of the outer shelf, the upper continental slope and on the tops and flanks of submarine banks, and predominantly they are found at water depths less than 400 m. Consolidated phosphatic horizons of Tertiary age have also been found locally on the sea floor. The nodules have quite a variety of shapes; some of them are angular or slightly rounded (Fig. 9), others are nodules the size of a fist and composed of many small nodules, the orderless growth of which produces an irregular pitted appearance. These nodules are commonly bored or inhabited by organisms (bryozoans, serpulid worms, and sponges). Frequently phosphate-containing internal molds of bivalves, gastropods, brachiopods, corals, serpulids, etc., also occur in the nodules (J. Murray and Renard, 1891; J. Murray and Philippi, 1908; Dietz et al., 1942; Emery, 1960; Mero, 1965; D'Anglejan, 1967; United Nations, 1970; Tooms et al., 1971; Parker and Siesser, 1972; Romankevich and Baturin, 1972; etc.). As long ago as 1873 the British "Challenger" expedition had dredged the first samples of phosphorite concretions from the Agulhas Bank off South Africa. Generally, sea-floor phosphorite contains 20–30% P_2O_5 (Table V).

---

[1] For a treatment of an ancient phosphorite deposit, see Chapter 11 by Cook, Vol. 7.

Fig. 9. Phosphorite nodules from the shelf of NE Africa between Cap Guardafui and Socotra Island. "Meteor"-station 102 (11°38'N 52°52'E), water depth between 190 and 290 m (from W. Schott, 1970a).

TABLE V

Average composition of some phosphorite nodules from the Californian offshore area (in weight percentage of dry samples) (after Mero, 1965)

45.0% calcium oxide (CaO)
28.1% phosphor pentoxide (P_2O_5)
1.5% metal oxide (Me_2O_3)*
4.4% carbon dioxide (CO_2)
3.1% fluorine (F)
1.1% organic matter
6.2% insoluble in HCl
10.6% remaining portions, largely magnesium carbonate ($MgCO_3$), silicon dioxide (SiO_2) and water (H_2O)

* Me for iron, aluminium, titanium and manganese. The nodules can also contain small quantities of strontium, barium, uranium, thorium and rare earths.

The formation of submarine phosphorite deposits particularly requires the presence of major quantities of organic matter. Essentially this was already recognized by Murray and Renard (1891) during the "Challenger" investigations. Nearly all the known phosphorites occurring in larger quantities lie on the present sea floor in areas of the oceans where under particular oceanographic conditions cold nutrient-rich and therefore organic-rich water wells up to the ocean surface, and transportation of detritus from the mainland is limited.[1]

A typical example of this is the occurrence of phosphorite nodules off the coast of Somalia which, as far as can be established, was first discovered in 1964 by the German research vessel "Meteor", at Station 102, east of Cape Guardafui (Figs. 9, 10, Table VI) (W. Schott, 1970b). This area is influenced by the monsoon. As the wind direction changes semi-annually, so also does the direction of the offshore surface currents. During summer in the Northern Hemisphere cold, deep water upwells to a greater extent along the Somalian coast, since the NEE-directed strong Somalia current runs slightly offshore (G. Schott, 1935). In this cold upwelling water high values of nitrate and nitrogen (>10 μg atoms per l) from May to October are found at the surface of the sea between Cape Guardafui and 8°N; at the same time the phosphate content in the surface water is >1 μg atoms per l between 11.5° and 8.5°N (Wyrtki, 1971). Just below the nitrate- and phosphate-rich surface water in this region, the phosphorite nodules have been found. The cold upwelling water can differ in temperature as much as 7°C from the subtropical warm surface water of the Somalia current. Suddenly occurring, greater fluctuations in the distribution of these water masses which differ in temperature and nutrient content may lead to a whole-sale mortality in the environment (Brongersma-Sanders, 1948; Foxton, 1965), and consequently intensifying the conditions favourable for the formation of phosphorite on the sea floor. As the result of the close relationship between the formation of phosphorite on the sea floor and a stronger development of the organisms in the surface water, the phosphate content generally fluctuates with the increase and decrease

TABLE VI

Analyses of two phosphorite nodules from the northeastern African shelf between Cape Guardafui and Sokotra Island; dredged on the "Meteor"-station 102 (11°38′N 52°52′E, waterdepth 190–290 m) (Analyses carried out by Dr. V. Marchig, Hannover.)

	CO_2 (%)	SiO_2 (%)	P_2O_5 (%)	MnO (p.p.m.)	MgO (%)	Fe (%)
Nodule 1 (inner part)	8.84	3.83	24.68	13	0.81	8.23
Nodule 1 (surface)	7.53	1.93	20.31	14	1.63	22.80
Nodule 2 (inner part)	9.50	3.88	24.52	25	1.03	5.83
Nodule 2 (surface)	9.06	4.36	32.18	20	1.25	4.83

---

[1] A general treatment of the importance of organic matter in the origin of ores is given in Chapter 5 by Saxby, Vol. 2.

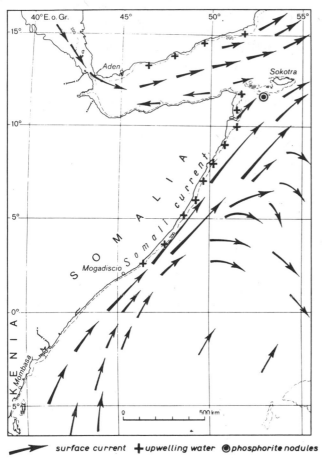

surface current ➤ ✛ upwelling water ⊚ phosphorite nodules

Fig. 10. Surface current and cold upwelling water during summer (after G. Schott, 1935) and "Meteor"-station 102 with phosphorite nodules (from W. Schott, 1970b).

of organic carbon (C_{org}) in the sediments (Marchig, 1972; Setty and Rao, 1972). (See also Chapter 11 by Cook, Vol. 7.)

Phosphorite sands have been reported from the East Coast of the U.S.A. (off North Carolina, Georgia and Florida), off Baja California and off the east coast of Australia (New South Wales). Phosphorite mud occurs, e.g., off the Malabar coast (south India) and in the Bass Strait (south Australia). Phosphate-rich weathered material from the adjacent mainlands probably also plays an important role in the formation of phosphorite sands and muds.

Manganese concretions[1]. Already 100 years ago the British "Challenger" expedition of

---

[1] For detailed discussions on Recent marine and freshwater Mn-deposits see respective chapters in this multi-volume publication. Chapter 9 by Roy, Vol. 7, describes ancient Mn-ores.

1872–1876 discovered the first ore found on the deep-sea floor in the form of manganese nodules. Manganese concretions on the sea floor mostly occur as nodules, which usually range from 0.5 to 25 cm in diameter, as well as in small grains and as crusts up to 15 cm thick (e.g. J. Murray and Renard, 1891; Menard, 1964; Mero, 1965; Arrhenius, 1966; Bender et al., 1966; Cronan and Tooms, 1969; Grasshoff, 1970; Kerl, 1970; Seibold, 1970, 1973; Sapozhnikov, 1971; Schweisfurth, 1971; Bezrukov and Andruschenko, 1972, 1973; Frazer and Arrhenius, 1972; Horn, 1972; Horn et al., 1972, 1973a, b, c; Heye and Beiersdorf, 1973; Meyer, 1973b; Seabed Assessment Program, 1973; Hammond, 1974).

In the Pacific, Indian and Atlantic oceans manganese nodules occur mainly at great depths (>4,000 m). Small grains of manganese concretions appear commonly in deep-sea sediments and manganese-containing crusts have also been found in shallow-water areas. The main components of the nodules are manganese and iron oxide. Furthermore, there are not insignificant quantities of nickel, copper, cobalt, molybdenum, zinc, and rare earths (Table VII). The contribution of these additional components fluctuates: it depends on the water depth and, according to recent observations, also apparently on the facies of the clayey deep-sea sediments. Higher values of nickel, copper and cobalt seem to appear particularly in siliceous deposits; in the Pacific Ocean such values can amount to 2.0, 1.6 and 2.3%, respectively, but the finds recorded up till now in the Indian and Atlantic oceans have a generally lower content. The content of the main components also show regional differences. In the Pacific Ocean the manganese/iron ratios in the nodules are generally less than 1 along the continents. Here the iron content is high, because iron derived from weathered material from the land precipitates earlier from the seawater than the manganese.

Further west of South America, in the Gulf of California and off the southeast coast of Japan, on the contrary, the content of manganese is high. On the average, the manganese/iron ratios in these areas are about 30. The high manganese content close to the southeast coast of Japan can be attributed to submarine volcanic activity (Menard, 1964; Mero, 1965; Arrhenius, 1966; Horn et al., 1972).

TABLE VII

Average percentages (dry-weight basis) of some chemical elements in manganese nodules of the Pacific Ocean (after Mero, 1965)

24.2%	manganese (Mn)
14.0%	iron (Fe)
0.99%	nickel (Ni)
0.53%	copper (Cu)
0.35%	cobalt (Co)
0.67%	titanium (Ti)

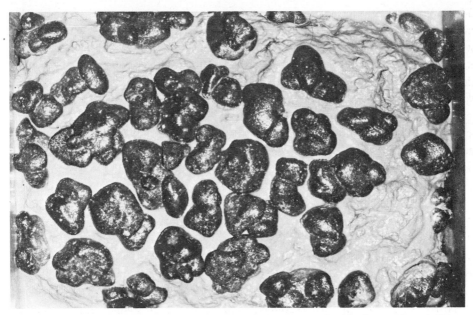

Fig. 11. Manganese nodules on the ocean floor taken by box grab, Pacific Ocean, about 1,000 nautical miles southeast of Hawaii Island, water depth about 5,500 m. RV "Valdivia" VA04/1972 (the area shown is 20 × 30 cm). (Photo Metallgesellschaft AG, Abt. "Meerestechnik".)

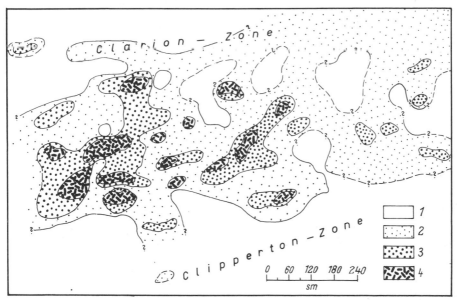

Fig. 12. Manganeses nodules, Pacific Ocean, about 600 nautical miles southeast of Hawaii Island, water depth about 5,500 m. Preliminary trend analysis of nodule distribution and their population density. *1*: no nodules; *2*: <20% distribution; *3*: 20–50% distribution; *4*: >50% distribution. (From Schultze-Westrum, 1973.)

Manganese nodules are mainly found on the surfaces of the deep-sea floor (Fig. 11). but also in deeper layers. On the sea floor they are distributed very irregularly, in some places they lie closely packed, whereas a short distance away only few specimens occur or they are entirely lacking. The form and size of nodules can also vary over a small area. A detailed investigation ca. 600 nautical miles southeast of Hawaii clearly illustrates this irregularity (Fig. 12) (Schultze-Westrum, 1973). The area, covering about $1 \cdot 10^6$ km^2, is located between the Clarion and Clipperton fracture zones in the "Abyssal Hill Region" (Menard, 1964, p. 34), and here the nodules have higher contents of valuable metals (the sum of nickel, copper and cobalt is ca. 3%). Therefore, in recent years (1972–1974) this area has been subjected to more thorough morphological, geologic and geophysical investigation by the German research vessel "Valdivia" to gain a better insight into the occurrence of manganese nodules and their formation on the deep-sea floor (water depth of between 4,500 and more than 5,500 m). Here, the nodules lie on radiolarian ooze (= "Siliceous ooze" facies). Within the facies autochthonous and allochthonous sediments occur. The age of autochthonous sediments increases from south to north. In the southern parts an age of Early Miocene/Late Oligocene has been observed and in the northernmost area, Eocene. The allochthonous sediments may have been formed particularly by bottom currents and slumping (Beiersdorf and Wolfart, 1974). Seismic reflection studies have demonstrated that seismically there is a clear difference between the siliceous-ooze facies with manganese nodules in this area and the calcareous ooze facies south of the Clipperton Fracture Zone (Dürbaum and Schlüter, 1974). For more results of the "Valdivia" cruises see: Beiersdorf and Bungenstock (1973), Heye and Beiersdorf (1973), Hinz and Schlüter (1973), Meyer (1973b), Morgenstein (1973), Friedrich and Plüger (1974), Hartmann and Müller (1974).

The manganese nodules have a concentrically layered build-up like carbonate ooids (Fig. 13). Where their nuclei are recognizable, they are composed of fossil remains, volcanic glass or other solid material. According to observations in the Pacific Ocean, the formation of manganese nodules on the deep-sea floor began, in some places at least, already in the Tertiary, for the teeth of sharks which became extinct in the Tertiary have been found in them. The chemical formation of the nodules usually progresses very slowly; according to recent age determinations, nodules in the deep-sea grow, on the average, only some millimeters per million years. The problem of the irregular distribution of the manganese nodules on the sea floor is still unsolved, likewise it is uncertain by what factors their form and size are influenced.

Despite many and sometimes very detailed examinations, the formation of manganese nodules in the deep sea is still not satisfactorily elucidated. Very possibly the manganese transported from the mainlands into the sea and the manganese derived from submarine volcanic activity both participate in the build-up of the manganese concretions. Microbiological processes probably also play a part, to some extent. Which criteria are decisive for the formation of nodules is uncertain. The state of our present knowledge is clearly reflected in Wogman et al. (1974), who state on page 1:

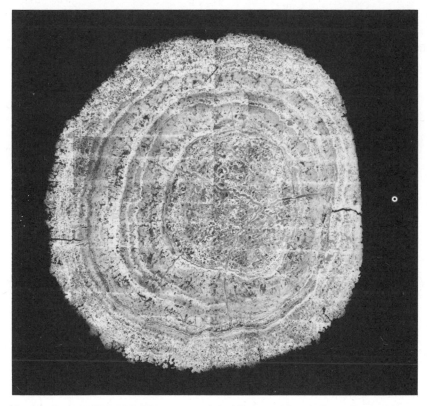

Fig. 13. Cross-section of a manganese nodule (diameter 18 cm) from the central Pacific (water depth about 5,100 m). RV "Valdivia" VA08/1972, Station 68 DK. (Photo Metallgesellschaft AG, Abt. "Meerestechnik".)

"An understanding of the relationships between the internal and external structures of the manganese nodules and their associated mineralogical and chemical compositions is essential to explain the sources of ferromanganese accumulations on the sea-floor. In general, at the present time there is no clear-cut method by which to evaluate the significance of the supply of elements directly from "normal" seawater, directly from the sea-floor sediments, or indirectly from hot submarine emanations, or as a result of terrestrial or marine weathering of pre-existing rocks in the formation of ferromanganese accumulations. The mechanism or mechanisms concentrating the valuable metals such as nickel, cobalt and copper at economic levels are equally unknown. . ."

(For detailed information see the appropriate chapters on recent marine and freshwater ferromanganese deposits, i.e. Chapters 7 and 8 by Glasby/Read and Callender/Bowser, respectively, in Vol. 7.)

Uranium in marine sediments[1]. The heavy metal uranium occurs only in small quantities

[1]For a discussion on the origin of uranium in a continental or terrestrial setting, precipitated by subsurface fluids within a fluvial complex, see Chapter 3 by Rackley, Vol. 7.

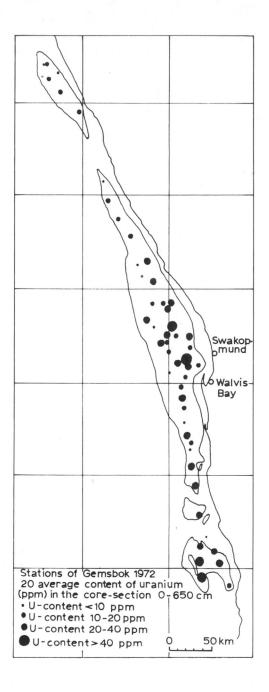

Fig. 14. Distribution of uranium in diatomaceous mud off southwest Africa (from Meyer, 1973a).

in seawater (0.003 mg/l, see p. 247; e.g. Koczy, 1956; Koczy et al., 1957; and Goldberg, 1963). In Recent marine deep-water ocean sediments, the uranium content is generally only 1—3 p.p.m. In fossil marine black shales containing more than 2% of organic carbon, this content can be greater. The Cambrian alum shales of Sweden, for example, have an average uranium content of 0.0168% (Bates and Strahl, 1958). According to Brongersma-Sanders (1948, 1951), black shales originate either in closed sea basins with poorly aerated bottom water or in sea areas in which hypertrophic water conditions prevail. Such water conditions appear during an overproduction of plankton as the result of very good nutrient relationships in surface water, e.g., in areas of upwelling water. Through the oxidation of the sunken plankton which dies off in large quantities, the oxygen content of the bottom water in such areas can be totally depleted. Thus, under conditions of strong enrichment and preservation of plankton remains in the sediment, uranium can be concentrated more strongly, not only in phosphorite concretions of the upwelling areas, but also in the mud of the ocean floor. This is clearly demonstrated by investigations made in the upwelling region off the coast of SW Africa (Calvert and Price, 1970, 1971; Baturin et al., 1971; Meyer, 1973a; Veeh et al., 1974).

Exceptionally organic-rich, nearshore diatomaceous Holocene muds have been recorded in this region on the inner shelf between about 19°20' and 25°10'S in water depths of less than 120 m (Fig. 14). The maximum thickness was about 15 m. Examination of 180 samples revealed that the uranium content of this mud fluctuates between 5 and 93 p.p.m.; the average was 21 p.p.m., which is higher than in the normal Recent sediments. In addition, the concentrations of nickel, zinc, copper and lead in these inner-shelf sediments are relatively high compared with recent sediments in shallow water or shelf sediments in general.

Therefore, these results correspond with the observations made in fossil black shales. It is not clear, however, whether offshore of SW-Africa it is a question of a purely chemical segregation of uranium and the other heavy metals in the sediment which is due to the peculiar water conditions in that region. According to Brongersma-Sanders (1965, 1969), the stronger enrichment of metals can also be caused by a high plankton production. Such a high plankton production in upwelling water can also lead to a greater accumulation on sea floors of numerous metals which are concentrated in living plankton. Several elements which are dissolved in trace quantities in seawater are enriched by a factor of more than 10^5 by organisms; among them are V, Fe, Sn, Zn, Pb, Cu, J, Ni, Co, Ti, Zr, Rb, Ba, U, Mn (Bäcker and Schoell, 1974, p. 2). In the region of high heat flow on the East Pacific Rise, sediments appear which contain, on the crest of the Rise, more uranium than on the flanks. Fisher and Boström (1969, p. 64) suggest that the origin of this uranium is probably volcanic. According to Rydell et al. (1974), the leaching of basalt by seawater was probably an insufficient source for the high concentrations of uranium found in one sediment core on this Rise (see also p. 281).

Hydrothermal sedimentary metallic deposits[1] . Newly-formed minerals related to volcanic activity have been reported from various sea areas. Warm acidic iron-containing volcanic spring waters either well up from the sea-floor subsurface and mix with the salt water at the sea floor, or they come into the sea from the coasts. It depends on the concentration or rate of the thermal inflows, whether the precipitates from these spring waters precipitate as a metallic deposit on the sea floor or mix with the normal sea sediments.

White et al. (1963) differentiated between chloride- and sulphate-rich spring waters. The chloridic waters mostly have a temperature of about 100°C. Somewhat higher contents of metals, such as copper (Cu), nickel (Ni), zinc (Zn), lead (Pb), silver (Ag), among others, have been determined in several springs (Harder, 1964).

A small occurrence of ferric hydroxide segregation in nearshore shallow water has been known for some considerable time off the Santorini archipelago (Greece) in the Aegean Sea. In several bays of the young volcanic island Palaea Kameni, within the caldera of the Santorini Islands, brownish iron-rich sediments are accumulated by submarine thermal activity in nearshore water. Among the main components of these sediments, siderite, vivianite, ferric hydroxide, ferrous hydroxide and possible ferrous silicate have been recorded. Copper, zinc, cobalt, vanadium, chromium and molybdenum were found only in trace quantities. The formation of these iron sediments is thought to have resulted from the leaching of volcanic calc-alkaline rocks by hot acid solutions. According to in situ investigations, the iron bacterium *Gallionella ferruginea* participates in the precipitation[2] of the reddish-brown ferric hydroxide sediments (Puchelt, 1973; Puchelt et al., 1973). The iron-rich sediments of the Santorini Islands are an important recent example of submarine exhalative-sedimentary iron-ore formation.

Likewise, in the bay of "Porto di Levante" of Vulcano Island, the southernmost of the volcanic islands in the Aeolian Archipelago (Lipari Islands), in the Tyrrhenian Sea, hot and cold fumaroles are found on the beach and nearshore to a water depth of 35 m. The examination of certain trace elements in the shallow-water sediments of this bay has revealed some interesting results concerning the different ways of mixing between the hydrothermal segregations which originate from the fumarole inflows and the detrital material of the Tyrrhenian seawater. The sediment of this bay is generally coarse-grained nearshore and fine-grained in the centre of the bay. The sediment fraction below 40 μ is mostly a mixture of fine particles of glass and feldspar. Manganese, copper, nickel and cobalt are concentrated in the sediments of the deeper part of the bay (water depth more than 50 m) (Fig. 15). The contents of these elements decrease towards the shoreline. The contents of chromium, vanadium, strontium and barium, on the other hand, are larger in

---

[1] As a result of the great genetic significance of volcanic-exhalative processes, the reader will find discussions on both Recent and ancient near-surface volcanic activities in numerous chapters, such as those by Degens/Ross, Vokes, Sangster/Scott, Ruitenberg, King, Gilmour, Mookherjee, Evans, Quade, Solomon, Maucher, and Lambert.

[2] For a detailed discussion on bacterial processes, see Chapter 6 by Trudinger, Vol. 2.

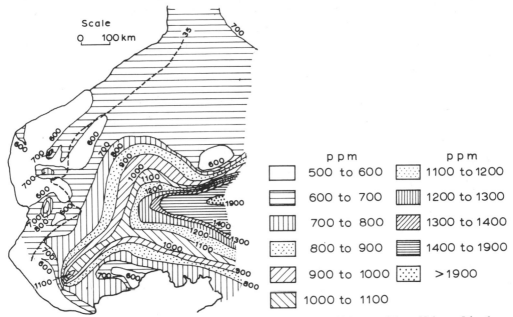

Fig. 15. Distribution of manganese in the sediments of the "Porto di Levante" bay, Vulcano Island, Aeolian Archipelago, Tyrrhenian Sea. 35 = water depth (m). (From Valette, 1973.)

the fumarole sectors of nearshore water (compare Fig. 16 with Fig. 15). Lead zinc, and boron have been found in the deeper water and in the areas of hot fumaroles.

According to these different distributions, Valette (1973) has divided the trace elements in the sediments in the bay into three categories: (1) deep-water concentrates; (2) fumarolian concentrates; and (3) mixed concentrates.

Two factors especially influence the horizontal distribution of trace elements in the sediments of the "Porto di Levante" bay: on the one hand, the bathymetry of the sea, and on the other, the volcanic phenomena near the shoreline, but fumarolian activity plays the main part.

Transport of dissolved iron into the Sea of Okhotsk is caused by hydrothermal processes of the Ebeko Volcano on the Island of Paramushir (Kuril Islands). A small thermal flow transports, amongst other materials, 35 tonnes per day of dissolved iron into the sea. At the mouth of the small river, the precipitation of the suspension takes place at a considerable distance from the coast (Selenov, 1958).

The most interesting discovery of ore deposits was made, more than ten years ago, on the floor of the Red Sea (Fig. 4) and as such it will be discussed somewhat more extensively in this general review of submarine Recent mineral resources. The young metal-containing sediments (Late Pleistocene–Holocene) in this region may be mainly of hydrothermal origin and may be connected with younger volcanic activity.[1]

---

[1] For a detailed summary, see Chapter 4 by Degens and Ross, Vol. 4.

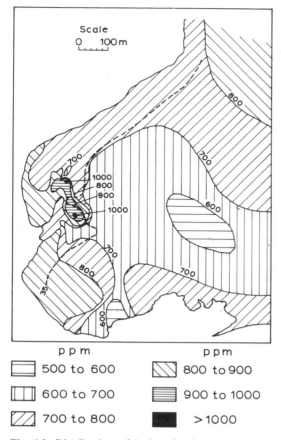

ppm
☐ 500 to 600   ⟋ 800 to 900
▥ 600 to 700   ▤ 900 to 1000
▧ 700 to 800   ■ >1000

Fig. 16. Distribution of barium in the sediments of the "Porto di Levante" bay, Vulcano Island, Aeolian Archipelago, Tyrrhenian Sea. 35 = water depth (m). (From Valette, 1973.)

The Red Sea forms part of one of the greatest tectonic lineaments in the earth's crust; this can be traced over a distance of about 6,500 km from the East African Rift Valley (approximately 19°S) over the Red Sea, the Wadi-Araba—Jordan—Bekaa—Ghab Graben Zone up to the fold mountain chains of southern Turkey (about 37°N). In the area of the Red Sea, this major tectonic fault zone crosses the Precambrian consolidated rock masses of the Nubian—Arabian Shield. The formation of the Graben, on the flanks and within the Graben, has been accompanied by a heavy Late Tertiary to Subrecent volcanic activity, which continues up into historical times. In the deepest parts of the Red Sea, the central graben, in which the sea floor reaches a maximal depth of 2860 m, olivine—tholeiite basalts without modal pyroxene are distributed (Schneider and Wachendorf, 1973).

The hydrothermal deposits are located on the sea floor in some particular deeps of this central graben. Mostly they occur to the west of the Saudi-Arabian port of Jeddah.

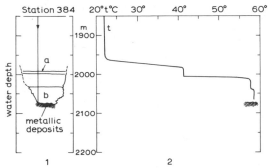

Fig. 17. Hot brines (*b*) and hydrothermal-sedimentary metallic deposits in the Atlantis-II Deep, Red Sea ("Meteor"-station 384). *1*: Echo-sounder record with three deep scattering layers (*a*); *2*: temperature (*t*) record in the bottom water. (After Dietrich and Krause in Degens and Ross, 1969.)

Bathymetric measurements of the sea floor and physicochemical studies of the bottom water in this area of the Red Sea by research vessels of various nations first gave indications of this occurrence. The normal bottom water of the Red Sea has a temperature of ca. 22°C with a salt content of 40.5‰. In some deeps of the sea floor echo-sounder measurements below a water depth of ca. 2000 m revealed a distinct layering of the bottom water. These scattering layers of thermal and saline origin separate water bodies of different density. In these sinks the water temperature can run up to 60°C and the salinity to about 326.5‰ (Fig. 17). On the bottom of these depressions containing hot brines, the hydrothermal sediments were discovered in the form of Recent heavy-metal deposits. The first finds were made in the Atlantis-II Deep and in the Discovery and the Chain Deeps (Fig. 18).

The results of the first period of study (1964–1966) of this interesting occurrence are discussed in the book "Hot Brines and Recent Heavy Metal Deposits in the Red Sea" (Degens and Ross, 1969), which is composed of excellent contributions from numerous authors. These results gave rise to further activities in the Red Sea, which were undertaken by the German research vessel "Valdivia" in particular during two cruises in 1971 and 1972 and thereby our knowledge about the origin and distribution of the hot brines and the hydrothermal metallic sediments in the Red Sea were considerably increased. An especially detailed study was carried out in the area of the Atlantis-II Deep, site of the greatest occurrence of hydrothermal sediments. Measurements of the tectonically active central zone of the Red Sea Graben, between 26°N and the Hanish Islands in the south, revealed 13 hydrothermally influenced separate deeps. Publications on these more recent investigations are: Hartmann (1971, 1972, 1973), Bäcker and Schoell (1972), Schoell and Stahl (1972), Bäcker and Richter (1973), Baumann et al. (1973), Hackett and Bischoff (1973), Schoell and Hartmann (1973), Bundesanstalt für Bodenforschung (1974), and, in this publication, Degens and Ross's Chapter 4, Vol. 4.

In the area of the central graben, the younger sediments are not very thick. Therefore, the morphological structures of this region are, to a great extent, explained by tectonic and volcanic processes. The steep flanks, for example, of the Atlantis-II Deep are certainly major antithetic faults, which run parallel to the NW—SE direction of the strike of the graben system. The Atlantis-II Deep is divided in several individual basins which are separated from each other, in the middle of the deep, by a hilly zone with shallower sea depth (Fig. 18). This is not only demonstrated by the morphologic measurements but

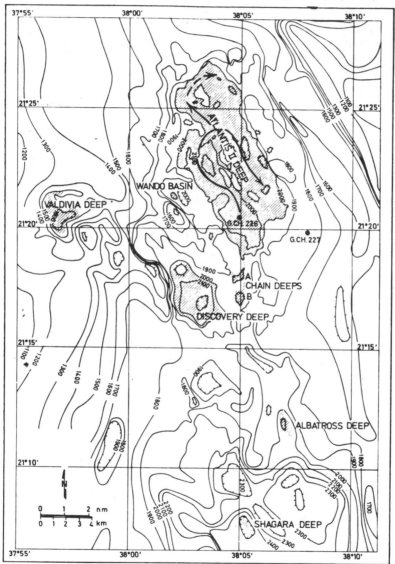

Fig. 18. Atlantis-II Deep and neighbouring brine pools, Red Sea. Contour lines of the water depth in m. The sites of the Glomar Challenger borings (drilled in 1972 in that area) are indicated and the present state of brine distribution in the Atlantis-II Deep. (After Bäcker and Richter, 1973; brine distribution after Schoel and Hartmann, 1973.)

also by the distribution of the deepest brine layer.

In the Atlantis-II Deep the hot brine is clearly layered; this has already been proved by echo-sounder measurements (Fig. 17). The lower 60°C brine, with an average 156.5‰ Cl, is separated by a thermic boundary layer from the overlying 50°C brine (49.8°C, 82‰ Cl). In 1972, this density layer between both brine bodies at a water depth of 2044 m consisted only of a small transition zone. The salinities show that the upper 50°C brine originates from mixing of the 60°C brine with the normal bottom water of the Red Sea. This is also clearly demonstrated by the scattering layer between the 50°C brine and normal bottom water; it is widely drawn apart. In 1972 the upper boundary layer was at a depth of 1,992 m (Fig. 17).

The temperature of the brine body has increased since the first observations; in the lower brine from 56.5°C in 1966 to 60.0°C, in the upper brine from 44.3°C to 49.8°C. The warmest brine of 60°C occurs in the central part of the SW basin. According to bathy-sonde temperature profiles, this brine rises from the bottom up to the scattering layer between the two brine bodies at 2,044 m depth. It expands laterally into the eastern and northern basins with a lesser temperature decrease (up to 58.2°C) (Fig. 18). The sites of brine outflow in the SW basin could not only be deduced from the maximum temperatures in the brine and in the sediment (in this region up to 62.3°C), but also they could be indicated as open channels in the sediment (Fig. 19). This hot brine is muddy and light-brown coloured, due to a higher iron and manganese content; an increase in copper, cobalt, zinc, nickel and lead contrary to the contents of normal seawater, has also been determined. Examination of the suspended matter in the brines showed that their contents of iron, manganese, copper and zinc are generally lower than those in the brine itself. In the deeper parts of the 60°C brine of the SW basin, these contents in the suspended matter are higher than those in the corresponding zones of the other basins. This is also an indication that today the youngest 60°C brine in the SW basin rises from the subsurface. The brines are a mixture of normal Red Sea bottom water with salts from the Miocene evaporite series, which is widely distributed in the Red Sea region. Miocene evaporite has been located in the subsurface of the sea floor, in the close vicinity of the Atlantis-II Deep by borings carried out by the "Glomar Challenger" (see Fig. 18; see further p. 279).

The sediments of the Atlantis-II Deep contain nearly all known sediment facies which derive from sedimentary-hydrothermal deposit formations. Metal oxide, sulphide and carbonate participate in the build-up of the hydrothermal deposits. After the first research period, the ore deposits have been shown to have, on the average, the following composition: 29% iron(Fe); 3.4% zinc (Zn); 1.3% copper (Cu); 0.1% lead (Pb); 54 g silver (Ag) per tonne and probably 0.5 g gold (Au) per tonne. These contents fluctuate within the stratified ore deposits. For example, zinc can be stratiformly enriched up to 8.9% and copper up to 3.6%. During the second research period (1971–1972), the contents of valuable metals determined were on the the average higher. The colour of the hydrothermal deposits varies greatly: from black to white and yellow, to blue-grey or red, it

Fig. 19. Atlantis-II Deep, Red Sea. Open fissure in the oxidic-anhydritic zone in the SW basin of the deep. About 4 mm long anhydrite crystals and sulfides with a smaller grain size cover the walls of the fissures (from Bäcker and Richter, 1973, fig. 4).

depends on the predominating metal compounds (Fig. 20). Iron oxide is yellow to ochre, metal sulphide violet grey, metal carbonate yellowish to pink. In the ore deposits, sulphates often occur, mainly as anhydrite. The sediments are predominantly very fine grained; 95% of the upper parts of the ore deposits consists of particles smaller than 0.06 mm.

In spite of strong vertical and lateral changes in the facies, the following lithostratigraphic units could be distinguished between the overlying and underlying strata, i.e., between the brine and the basaltic underlayer:

amorphous silicate zone   3–4 m thick
upper sulphitic zone   ca. 4 m thick
middle oxidic zone   1–10 m thick
lower sulphitic zone   2.5–4 m thick
detrital oxidic-pyritic zone   1.3–6 m thick

About 25,000 years ago, the first newly formed hydrothermal minerals at this depth started with the detrital oxidic-pyritic zone. The boundary between the Upper Pleistocene and Holocene lies within this zone. The Holocene sediments are on the average 25 m

Fig. 20. Suakin Deep, Red Sea. Sediment core (length of one box 1 m). Dark beds: hydrothermal metalliferous influenced deposits. Light beds: detrital-organogenic sediments. RV "Valdivia" VA03/1972, Station 332 K, W Basin. (Photo H. Bäcker, Preussag).

thick. The sulphide zones are the sedimentologically most important facies for economic geology. They contain the valuable metals copper, zinc, and cadmium. The separation of the two sulphitic zones by the middle oxidic zone indicates that the hot brine, from which the ore mud segregated, was temporarily strongly mixed with the more oxygen-rich normal bottom water of the Red Sea. Only in this way could the building-up of oxidic metal compounds take place.

During the second research campaign, an increased sedimentation of metal sulphides has been determined than during the 1966 cruise, particularly in the SW basin of the Atlantis-II Deep. This is attributed to an increased thermal activity, i.e., a greater supply of hot brines. This occurrence in the Atlantis-II Deep is thus a hydrothermal-sedimentary deposit that is now forming. On an average, at least 100 cm of sediment accumulates during 1000 years in this region.

In their habitus, these hydrothermal sediments of the Atlantis-II Deep recall the submarine hydrothermal-sedimentary ore deposits of Rammelsberg near Goslar/Harz (Germany) and the pyrite − sphalerite − barite ore of Meggen/Westfalen (Germany) (Ehren-

TABLE VIII

Comparison of the hydrothermal-sedimentary deposits of the Atlantis II and Chain deeps with the ore deposits of Rammelsberg/Harz (adapted from table 1 of Bäcker, 1973)

	Atlantis-II Deep (Chain Deep)	Rammelsberg
Geological age	Holocene/Pleistocene (0 – 25,000 years)	Middle Devonian
Major tectonic unit	graben	marginal ridge and basin
Morphology	lenticular and bedded (in deepest parts of the graben)	lenticular and bedded
Water depth (m)	1990 – 2200	500 – 1000?
Pre-sedimentary volcanism	+	+
Synsedimentary volcanism	+	+
Type of volcanism	tholeiitic–basaltic	acidic tuff
Main facies of the precipitate	sulphide, silicate, limonite, hematite, manganite, Mn-siderite, anhydrite	silica-sulphide (Kniest), sulphide, barite-sulphide (Grauerz)
Thickness (m)	10–25	10–15
Type of sedimentary host rock	detrital-organogenic (marl)	detrital (mostly shaly)
Elements introduced	Si, Fe, Mn, Ca?, Zn, Cu, Mg (Pb, Ag, Cd, Mo, Ba)	Si, Zn, Fe, Ba, Pb, Cu, Mn(Sb, As, Cd, Ag)
Lamellar bedding	+	+
Turbidites, sedimentary breccia	+	+
Post-sedimentary mineralization	+	+
Impregnation	o	+

+ = present, o = not observed.

berg et al., 1954; Kraume, 1955; Hannak, 1968). This is clearly demonstrated by the comparative studies of Bäcker (1973) (see Table VIII). Some of the differences between the modern and ancient deposits may be the result of hydrothermal, metamorphic, and tectonic effects on the fossil deposits. James (in Degens and Ross, 1969, p. 525) also compared the Red Sea iron-rich deposits with older ironstone and iron formations. Knowledge about the Recent volcanic iron segregations at Santurini and Vulcano (see p. 270) also contribute significantly to explaining the Lahn–Dill type iron ore in Germany (Harder, 1964; Lippert, 1970; Chapter 6 by Quade, Vol. 7 of the present publication).

An American estimate, based on the results of the first research period (1964–1966) has placed the value of zinc, copper, lead, silver and gold in the hydrothermal-sedimentary metal deposits on the sea floor of the Atlantis-II Deep at about $2.5 \cdot 10^9$ (Bischoff and Manheim, in Degens and Ross, 1969, p. 535). Another estimate by Hackett and Bischoff (1973) amounts to $2.33 \cdot 10^9$. These values do not include estimated exploitation costs. The second research campaign (1969, 1971–1972) indicated on the average higher contents of valuable metals. It is not yet possible to give a final judgement of the potentiality of these occurrences of the metalliferous mud as copper and zinc deposits.

The southernmost discovery of hot brine and hydrothermal deposits is located in the Suakin Deep, about 100 nautical miles south of the Atlantis-II Deep, where the greatest water depth in the Red Sea (2,850 m) has been determined. The Suakin Deep, which is also a tectonic graben structure, consists of two basins. The graben has – at least locally – a basaltic basement below a thin sedimentary bed. The sediments of the West Basin show only slight hydrothermal influences. Light olive-grey marls of detrital organogenic origin prevail (Fig. 20). In the East Basin, ferromanganous and sulphidic ore deposits alternate with the standard detrital sediments of the Red Sea. Zinc and copper are enriched in the sulphidic layers as is the case in the hydrothermal sediments of the Atlantis-II Deep. The rate of sedimentation of these young deposits (Holocene–Late Pleistocene: Würm–Wisconsin) amounts to 15–20 cm per 1,000 years in the West Basin and 35 cm per 1,000 years in the East Basin (Baumann et al., 1973).

The northernmost discovery of brine has been made in the Oceanographer Deep at 26°17.2'N with a water depth of >1466 m. In the Vema Deep at 23°52.0'N, a sediment core has indicated the presence of an almost pure goethite deposit, 45 cm thick, under 4.60 m of calcareous sediments mixed with precipitations of iron oxides (mostly goethite). This bed of goethite has a higher content of some heavy metals (e.g., zinc, tin, molybdenum) than those reported for marine sediments. This northernmost occurrence of hydrothermal goethite deposits is older than that of the Atlantis-II Deep: Early Wisconsin (Würm), 90–100,000 years B.P. (Herman and Rosenberg, in: Degens and Ross, 1969).

Sites 225 and 227 of the Deep Sea Drilling Project, both drilled east of the Atlantis-II Deep (at distances of about 16 and 5 km from the Deep, respectively) recovered dark muds and shales in and overlying the evaporite sequence. These muds and shales are occasionally enriched in vanadium, molybdenum, copper and iron. At Site 228, near the Suakin Deep, on the western side of the Red Sea axial valley, black shales in a brecciated shale-anhydrite sequence also have a higher content of copper, zinc and vanadium (one sample showed 5% zinc) (Whitmarsh et al., 1974)[1]. Such heavy metal-rich muds with a relatively high content of organic material are also well known in other evaporite formations, for instance, the Zechstein formation (Upper Permian) in Germany (Richter-Bernburg, 1941). It is not yet clear whether there are any possible relationships between these metal-rich shales of the evaporite series and the metalliferous hot brines in the Red Sea deeps[2].

---

[1] Dark to blackish-grey bituminous clayey sediments are obviously widespread in the Miocene evaporite sequence of the Red Sea. In November 1952 the present author observed similar sediments of ca. 4 m thickness in the outcropping salt diapir of Salif between anhydrite and rock salt in the coastal plain of Yemen (Tihama). The outcropping evaporite of the Djebel Quizan in southern Tihama in Saudi Arabia contains some intercalations of gray marl between gypsum, which also include a thin bed of black shale (Richter-Bernburg and W. Schott, 1954).

[2] Editor's Note: A solution to this problem is of fundamental importance in determining the precise mechanisms involved in the dissolution, transportation and reprecipitation of metals in the subsurface of sedimentary-volcanic rock piles. For related problems, see Wolf (1976).

Such mineral occurrences are not restricted to the central graben of the Red Sea. On the western flank of the Red Sea Graben between 24° and 25°N, hydrothermal impregnations of lead, copper and zinc ores appear in Miocene sediments off the Egyptian coast. Their origin is related to Tertiary volcanic activities. The mineralization started shortly after the deposition of the enveloping sediments (Miocene or post-Miocene) (Said, 1962). On the eastern side of the graben lead, copper and zinc ores have also been found in Oligocene, Miocene and Plio-Pleistocene sediments. They were discovered in the Saudi-Arabian coastal plain at ca. 25°30'N, near Umm Lajj (Dadet et al., 1970). The direction of the trend in both mineral occurrences (NNW–SSE) is parallel to the main tectonic feature of the Red Sea Graben.

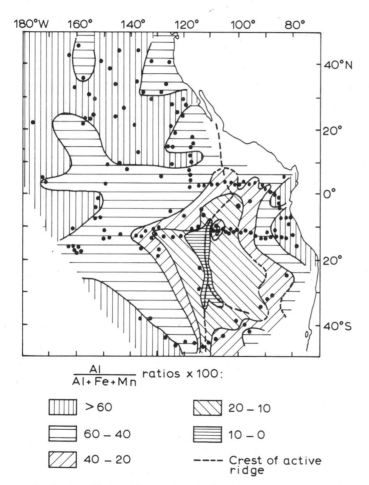

Fig. 21. Eastern Pacific. Distribution of aluminium-poor sediments (for definition of the symbols see legend). Dashed lines approximately indicate the position of the crest of the East Pacific Rise (from Boström and Peterson, 1969).

According to all these observations the Recent, Subrecent and Late Pleistocene hydro-thermal metallic sediments in the central graben of the Red Sea are located within a greater zone of ore mineralization which is correlated with the tectonic origin of the Red Sea Graben.

In the deep parts of the world's oceans and particularly in areas of ocean ridges, which are characterized by volcanic activity and high heat flow, the occurrence of hydrothermal sedimentary deposits can also be counted upon. There are already some indications of this. For example, on the East Pacific Rise between 0 and $35°S$ precipitates from hydrothermal exhalations have been observed. These sediments are abnormally poor in aluminium and titanium, rich in iron (about 18% Fe), manganese (about 6% Mn), boron, arsenic, cadmium, vanadium and chromium as compared to pelagic sediments in general (Fig. 21) (Boström and Peterson, 1966, 1969; Boström et al., 1969). Extensive metalliferous sediments were also discovered east of the East Pacific Rise at ca. $10°S$; they lie between this Rise and the Galapagos Rise, in the Bauer Deep at a water depth of ca. 4,000 m. This is a dark-brown mud, rich in colloidal material and particularly rich in iron and manganese hydroxides (up to 18% Fe and up to 6.5% Mn) with high concentrations of copper, nickel and zinc (Dash et al., 1971; Bischoff and Sayles, 1972; Seibold, 1973). The Bauer Deep is the only area of metalliferous sediment remote from the crest of a mid-ocean ridge. It is characterized by high, but variable, heat flow. The origin of these metalliferous deposits in the Bauer Deep is still an unsolved problem, though there are several hypotheses (see Anderson and Halunen, 1974; see also p. 282).

The JOIDES Deep Sea Drilling Project has also found some evidence of hydrothermal enrichment in older sediments beneath the ocean floor (Gealy, 1971). For example: in the northwest Atlantic Ocean, on the northeastern flank of the Bermuda Rise (Site 9, water depth 4,981 m), iron-rich clay containing 11% of hematite together with rhodochrosite, siderite and ferromanganese nodules, lies at a depth of 835.5 m below the sea floor just above the basalt contact. The age of this clay is probably Maastrichtian (Peterson et al., 1970). At Site 105 on the hills of the lower continental rise, southeast of New York (water depth 5,245 m), multicoloured silty clays contain abundant iron and manganese oxide minerals, siderite, pyrite, rhodochrosite. This facies of Late Cretaceous?– Eocene? age is probably also of hydrothermal origin. It has been found between 250 and 295 m below the sea floor (Hollister et al., 1972).

Above the basalt at Sites 37, 38 and 39 in the East Pacific Ocean, a basal amorphous iron-manganese oxide sediment facies has been discovered (Von der Borch and Rex, 1970). These sites have been drilled in the area of the Mendocino and Murray Fracture Zone (above $140°W$). The following facies have been observed:

At Site 37 (water depth 4,682 m), 25–30 m below the sea floor. Age:?

At Site 38 (water depth 5,134 m), 33–48 m below the sea floor. Age: upper part of Lower Eocene.

At Site 39 (water depth 4,929 m), 11–17 m below the sea floor. Age: Lower Eocene.

East of the Marquesas Islands (French Polynesia) amorphous iron oxide-rich sediments interbedded with calcareous nannoplankton oozes have been observed. This facies, immediately overlying the basalt, occurs in the basal zones of:

Site 74 (water depth 4,431 m), about 98–102 m below the sea floor. Age: Upper–Middle Eocene.

Site 75 (water depth 4,181 m) 74–82 m below the sea floor. Age: Lower Oligocene.

At both sites there is a very sharp boundary between the overlying nannoplankton ooze and those sediments which were enriched in metals by submarine hydrothermal exhalations (Von der Borch et al., 1971).

In the northern part of the North East Pacific basin, southeast of the Hawaiian Islands, the basal sediments at Sites 159–162, just above the basaltic basement, are also ferruginous (Cronan, 1973). They are enriched in iron as well as in manganese. Some deposits also contain higher concentrations of Zn, Ni, Co, Cu and Pb. The average composition of eleven ferruginous sediments from Sites 159, 160 and 162 is: 17.5% Fe; 4.52% Mn; 535 p.p.m. Ni; 83 p.p.m. Co; 917 p.p.m. Cu; 145 p.p.m. Pb; 358 p.p.m. Zn. The geological age of these ferruginous deposits is:

Site 159: possible latest Oligocene to earliest Miocene.

Site 160: Early Oligocene.

Site 161, 162: Middle Eocene.

Above the basaltic basement of the Bauer Deep (SE Pacific), the early Miocene–Quaternary sediments, about 115 m thick, contain metalliferous components. They were recovered at Site 319 at a water depth of 4,296 m (Hart et al., 1974; see also p. 281).

In the Wharton Basin of the eastern Indian Ocean, Sites 212 and 213 (water depth 6,243 m and 5,611 m, respectively) produced an iron oxide-rich clay facies in the basal sediments above the basalt (Pimm, 1974). This facies is about 30 and 40 m thick, respectively, at these sites, but only the lowermost sample contained more than 10% total iron at each site. At Site 212 the geological age of this facies is probably close to the Early/Late Cretaceous boundary, whereas at Site 213 it is Late Paleocene. An iron/manganese-rich basal deposit of Early Paleocene age overlying basalt was recovered in the southern Madagascar Basin in the Indian Ocean (Site 245, water depth 4857 m).

The iron-rich basal sediments established by the JOIDES Program in older geological formations overlying basaltic rocks can certainly be attributed to volcanic activity (Warner and Gieskes, 1974), but it is not clear whether or not the mineralizing solutions are original components of the magma. This is also not yet explained in the case of some metalliferous sediments on the present ocean bottom (Boström, 1970; Corliss, 1971; Piper, 1973; Smith, 1974). Boström (1973) has summarized our present knowledge of these deposits.

In blocks of solid rock dredged from oceanic rift zones, some ore components have been observed in places; these are partly to be explained by hydrothermal processes (Udintsev and Dimitriev, 1970; Dimitriev et al., 1971). Some basalts and dolerites from the Carlsberg Ridge (Indian Ocean), for instance, contain about 5% copper and 400 g/tonne of zinc. Cassiterite, pyrite, chalcopyrite, etc., have been found in serpentinized peridotites in this region. Strongly crushed and serpentinized ultrabasic rocks (harzburgites and lherzolites) representing blocks of mantle substratum have been dredged from the sea floor of the Mid-Atlantic Ridge in the region of the Azores Islands. In these rocks an increased amount of lead, zinc, copper, silver, mercury, chrome, vanadium, cobalt, nickel and sometimes tin and beryllium has been found. For example, one sample of serpentinized harzburgite contained 1800 g/tonne of Zn; 121 g/tonne of Cu; 6 g/tonne of Sn; 10 g/tonne of Pb; and 0.260 g/tonne of Hg.

All these observations illustrate that metalliferous deposits can be found in sediments of Cretaceous, Tertiary and Quaternary age, respectively, beneath and on the ocean sea floor and in solid rocks of oceanic rift zones.

SUMMARY AND CONCLUSIONS

Inorganic mineral resources are to be found in ocean water, on the ocean floor and deeper beneath the sea bottom. This general review has dealt with the resources of the oceans and the ocean floors, whereas the (inorganic) resources occurring in the deeper subsurface of the ocean floors are only discussed with regard to the recent finds of the Deep Sea Drilling Project. Mineral resources in bedrock formations of the shelf regions are either submarine extensions of discoveries on the adjacent mainlands or they appear in tectonic units which correspond to those of the mainlands. Of the elements which are dissolved in seawater, only hydrogen and oxygen (in the form of freshwater), sodium and chlorine (together as salt) and magnesium and bromine are utilized at present as mineral resources. Already in ancient times, the human needs for salt were predominantly covered by the evaporation of seawater in artificially made ponds within sea bays.

The inorganic mineral resources of the ocean floor consist of locally enriched minerals or of minerals originating from chemical concentration. Sand, gravel and heavy-mineral deposits belong to the locally enriched minerals, and they occur especially in nearshore areas and on the shelf. The materials of these mineral resources derive mainly from the detrital masses of the mainlands, and this material is subjected to a strong sorting by ocean currents and wave action. Sand and gravel are already being exploited along many coasts. The heavy-mineral deposits (placer deposits) may be subdivided according to their occurrence or their specific gravity, respectively, into: dunes placer, beach placer, placer on sand banks, submarine deep leads, placer caused by submarine erosion, and into heavy heavy minerals, light heavy minerals and gems. The most common minerals of economic value in placer deposits are zircon, ilmenite, rutile, monazite, magnetite, garnet, epidote, topaz, tourmaline, corundum, gold, platinum, cassiterite and clinozoisite. They are widespread in the shelf regions of the three oceans, but the economically most valuable deposits are in SE Asia and on the coasts of Australia.

Phosphorite concretions, manganese concretions and hydrothermal sedimentary metallic deposits belong to the mineral (inorganic) resources originating from chemical concentration. Phosphorite concretions predominantly occur in the areas of the outer shelf, the upper continental slope and on the tops and flanks of submarine banks; they are widespread. Especially the presence of major quantities of organic matter is required for their genesis. This is particularly the case where cold, nutrient-rich and therefore organic-rich, upwelling water rises to the sea-surface. Manganese concretions in the form of nodules may be found above all on the deep-sea floor at water depths greater than 4000 m. Besides manganese and iron oxides they contain smaller quantities of nickel, copper, cobalt, molybdenum, zinc, and other metals. The contribution of these additional components fluctuates and it seems to depend on the depth of the ocean and on the facies of clayey deep-sea sediments on which the nodules are located. In spite of numerous and sometimes very detailed and very recent studies, the problem of the origin of the manganese nodules on the deep-sea floor is not yet uncontestably explained. The manganese which was transported from the continents into the sea and the manganese derived from submarine volcanic activity may have participated in the formation of the manganese concretions, in which microbiologic processes also exert an influence. Uranium occurs only in small quantities in the deep-sea sediments, though in closed sea basins with poorly aerated bottom water, or in sea areas with a higher plankton production, the uranium content may increase.

Hydrothermal sedimentary metallic deposits became known in connection with volcanic activity from the shallow and deep sea and mostly it is iron compounds which are involved. However, traces of copper, zinc, cobalt, vanadium, chromium and molybdenum also often occur in such sediments. The most interesting ore deposits of this type have been found in deeps in the Red Sea, the bottom water of which chiefly consists of hot brines with a higher iron and manganese content. In the Atlantis-II Deep it is a case of a hydrothermal sedimentary deposit *in statu nascendi,* which can be correlated with fossil sedimentary iron occurrences; a further remarkable find of iron-rich sediment has been made in the Bauer Deep to the east of the East Pacific Rise, an area which is characterized by high heat flow as are several crests of mid-ocean rises. Borings carried out by the DSDP in the deep-sea bottoms of the three oceans have also revealed some evidence of hydrothermal enrichment in sediments of Cretaceous and Tertiary age on the basis of the sedimentary sequence overlying basaltic rocks.

Some economic aspects remain to be discussed. In conclusion, the demand for mineral raw materials has increased considerably due to the strong growth of industrialization in most countries of the world; it is a well-known fact that in the near or more remote future the needs for some mineral raw materials will no longer be satisfied only by continental occurrences even though in the last decennia the higher prices of raw materials have made it possible to extract and to process deposits of lower-grade ores, by improving the required technologies. Thus, we shall soon be forced to utilize, to a greater extent than before, the mineral (inorganic) resources of the oceans and the ocean floors (Mero, 1965, 1971, 1972a, b; Brooks and Lloyd, 1968; Inst. Int. Recht, University of Kiel, 1970; UNO, 1970; Romanowitz, 1971; Wirtschaftsvereinigung Industrielle Meerestechnik, 1971; Moore, 1972; Ward, 1972; Meiser and Müller, 1973; Weinstein, 1973; Schott, 1974).

Some examples: The present annual consumption of freshwater per head of population for domestic purposes, agriculture and industry is estimated at: 1,500 l for tropical Africa; 500,000 l for western Europe; 1,000,000 l for the U.S.A.

It is supposed that in the next 10 years the demand for water in the highly industrialized countries will increase by ca. 50% and in some developing countries by as much as 500%. These needs can be covered only by freshwater production from seawater. It is estimated that the production of freshwater in desalination plants of ca. $1.6 \cdot 10^6$ m^3 per day at the beginning of 1970, will be increased to ca. $5 \cdot 10^6$ m^3 per day in 1980.

The production of sand and gravel from submarine deposits will also increase considerably. These raw materials are available in sufficient quantities on the sea floor of the shelf regions.

The increasing demand for heavy minerals of the most widely differing composition will lead — even more so than formerly — to the exploitation of shelf occurrences, even though these are mostly "disseminating ores". The technology necessary for the exploitation of these occurrences must be improved so as to make their mining economically justified. That such deposits can be relied upon is illustrated by the submarine exploitation of tin deposits in the Indonesian shelf (see p. 256) and the discoveries of titanium ore deposits on the shelf off Mozambique (see p. 259).

Due to its phosphate content, phosphorite is not only sought after as a fertilizer, but it is also utilized in the chemical and pharmaceutical industry. Beside this, it can also contain strontium, barium, magnesium, uranium, thorium and rare earths. According to Mero (1965), $3 \cdot 10^{11}$ tonnes of phosphorite are available on the ocean shelves. Phosphate occurrences are widespread on the mainlands, but Weinstein (1973) assumes that, in spite of the annually increasing production, the reserves of phosphate on land will not cover the demand in the second half of the 21st century. By that time at the latest, submarine deposits will have to be mined.

Manganese nodules are widespread in the deep-sea basins of the oceans. Mero (1965) estimated that there are $1,656 \cdot 10^9$ tonnes on the sea floor of the Pacific Ocean alone. These manganese nodules are of economical interest, due to their additional components of valuable metals. Possibly they will become useful as a long-term supply of copper, nickel and cobalt, but economic mining of the manganese nodules in the ocean at depths of more than 4,000 m will, however, only be possible if a use can be found for the very great quantities of manganese which will then become available.

Due to their content of valuable metals, the hydrothermal metallic deposits in the Red Sea are probably also of economic interest for the more remote future and perhaps also the metalliferous sediments in the Bauer Deep of the South East Pacific. On the grounds of the present knowledge about the occurrence of hydrothermal deposits in the oceans, it is not beyond possibility that similar metal-containing sediments of the Red Sea type will also be found in other ocean areas.

The manganese nodules and the hydrothermal deposits of the Red Sea have aroused the economic interest of different industrialized countries and commercial research has already produced interesting results. However, the technology necessary to exploit these mineral raw materials from great depths and the refining methods have yet to be improved considerably. To some extent the first usable results have already been reported. (Manganese nodules: Mero, 1972a, b; Hänig, 1973; Kauczor et al., 1973; Meyer-Galow et al., 1973; Ulrich et al., 1973; Hydrothermal deposits of the Red Sea: Clement et al., 1973; Supp et al., 1973.) Up to the present, the developments of the technology of mining and processing manganese nodules have required an expenditure of about US $100 million, but economic exploitation of the nodules is still uncertain (Hammond, 1974).

According to present knowledge about the mineral (inorganic) resures of the oceans and ocean floors, it appears possible that in the near or more remote future the submarine resources which are today not yet used will be rendered economically productive, when the unresolved problems are cleared up by further research activities. It is therefore the duty of the present and next generations to dedicate themselves intensively to these tasks so that the demand for mineral (inorganic) resources can continue to be covered also in the future.

REFERENCES

Anderson, N.R. and Halunen Jr., J., 1974. Implications of heat flow for metallogenesis in the Bauer Deep. *Nature,* 251: 473–475.

Andrée, K., 1920. *Geologie des Meeresbodens, II.* Bornträger, Leipzig, 689 pp.

Arrhenius, G., 1966. Pelagic sediments. In: M.N. Hill (Editor), *The Sea, 3.* Wiley, New York, N.Y., pp. 655–727.

Bäcker, H., 1973. Rezente hydrothermal-sedimentäre Lagerstättenbildung. *Erzmetall,* 26 (11): 544–555.

Bäcker, H. and Richter, H., 1973. Die rezente hydrothermal-sedimentäre Lagerstätte Atlantis-II-Tief im Roten Meer. *Geol. Rundsch.,* 62 (3): 697–741.

Bäcker, H. and Schoell, M., 1972. New deeps with brines and metalliferous sediments in the Red Sea. *Nat. Phys. Sci.,* 240 (103): 153–158.

Bäcker, H. and Schoell, M., 1974. Anreicherungen von Elementen zu Rohstoffen im marinen Bereich. *Chem. Ztg.,* 98: 299–305.

Bates, Th.F. and Strahl, E.O., 1958. Mineralogy and chemistry of uranium-bearing black shales. *United Nations, Peaceful Uses of Atomic Energy,* pp. 407–411.

Baturin, G.N., Kochenow, A.V. and Senin, Y.M., 1971. Uranium concentration in recent ocean sediments in zones of Rising Currents. *Geochem. Int.,* 8 (2): 281–286.

Baumann, A., Richter, H. and Schoell, M., 1973. Suakin Deep: Brines and hydrothermal sediments in the deepest part of the Red Sea. *Geol. Rundsch.,* 62 (3): 684–697.

Beiersdorf, H., 1972. Erkundung mariner Schwermineralvorkommen. *Meerestech. Mar. Technol.,* 3 (6): 217–223.

Beiersdorf, H. and Bungenstock, H., 1973. Reflexionsseismik bei der Manganknollen-Exploration mit dem Rohstoff-Forschungsschiff "Valdivia". *Meerestech. Mar. Technol.,* 3: 77–78.

Beiersdorf, H. and Wolfart, R., 1974. Sedimentologisch-biostratigraphische Untersuchungen an Sedimenten aus dem zentralen Pazifischen Ozean. *Meerestech. Mar. Technol.,* 6: 192–198.

Bender, M.L., Ku, T.L. and Broecker, W.S., 1966. Manganese nodules: their evolution. *Science,* 151: 325–328.

Bezrukov, P.L. and Andrushenko, P.F., 1972. Ferro-manganese nodules from the Indian Ocean. *Proc. U.S.S.R. Acad. Sci. Geol. Ser.,* 7: 3–20.

Bezrukov, P.L. and Andrushenko, P.F., 1973. Geochemistry of ferro-manganese nodules from the Indian Ocean. *Proc. U.S.S.R. Acad. Sci. Geol. Ser.,* 9: 18–37.

Bischoff, J.L. and Sayles, F.L., 1972. Pore fluid and mineralogical studies of Recent marine sediments Bauer Depression Region of East Pacific Rise. *J. Sediment. Petrol.,* 42 (3): 711–724.

Boström, K., 1970. Submarine volcanism as a source for iron. *Earth Planet. Sci. Lett.,* 9 (4): 348–354.

Boström, K., 1973. The origin and fate of ferromanganoan active ridge sediments. *Stockholm Contrib. Geol.,* XXVII: 149–243.

Boström, K. and Peterson, M.N.A., 1966. Precipitates from hydrothermal exhalations on the East Pacific Rise. *Econ. Geol.,* 61: 1258–1265.

Boström, K. and Peterson, M.N.A., 1969. The origin of aluminium-poor ferromanganoan sediments in areas of high heat flow on the East Pacific Rise. *Mar. Geol.,* 7 (5): 427–447.

Boström, K., Peterson, M.N.A., Ioensuu, O. and Fisher, D.E., 1969. Aluminium-poor ferromanganoan sediments on active oceanic ridges. *J. Geophys. Res.,* 74 (12): 3261–3270.

Brongersma-Sanders, M., 1948. The importance of upwelling water to vertebrate Paleontology and oil geology. *Verh. K. Ned. Akad. Wet. Afd. Natuurk. Tweede Sec., Deel XLV,* 4: 1–112.

Brongersma-Sanders, M., 1951. On conditions favouring the preservation of chlorophyll in marine sediments. In: *Proc. World Pet. Congr., 3rd, the Hague, 1951, Section I,* pp. 401–413.

Brongersma-Sanders, M., 1965. Metals of Kupferschiefer supplied by normal sea water. *Geol. Rundsch.,* 55: 365–375.

Brongersma-Sanders, M., 1969. Permian wind and the occurrence of fish and metals in the Kupferschiefer and Marl slate. In: *Proc. Inter-University Geol. Congr. 15th, Leicester, 1967, Spec. Publ.,* 1: 61–71.

Brooks, D.B. and Lloyd, B.S., 1968. Mineral economics and the oceans. In: E. Keiffer (Editor), *Proc. Symp. on Mineral Resources of the World Ocean. Univ. Rhode Island, Occas. Publ.,* 4: 23–30.

Bundesanstalt für Bodenforschung, 1974. *"Valdivia" VA 01/03 Rotes Meer–Golf von Aden. Wissenschaftliche Ergebnisse I. Interner Bericht.* Bundesanstalt für Bodenforschung, Hannover, 165 pp.

Calvert, S.E. and Price, N.B., 1970. Minor metal contents of recent organic rich sediments off South West Africa. *Nature,* 227 (5228): 593–595.

Calvert, S.E. and Price, N.B., 1971. Recent sediments of the South West African shelf. In: *The Geology of the East Atlantic Continental Margin. ICSU/SCOR Working Party 31, Symp., Cambridge, 1970, Rep. No. 70/16,* pp. 175–185.

CCOP/ECAFE, United Nations, 1967. *Report of the Third Session of the Committee for Co-ordination of Joint Prospecting for Mineral Resources in Asian Offshore Areas,* E/CN11/L186: 152 pp.

CCOP/ECAFE, United Nations, 1968a. *Report of the Fourth Session of the Committee for Co-ordination of Joint Prospecting for Mineral Resources in Asian Offshore Areas and Report of the Third Session of its Technical Advisory Group,* E/CN11/L190: 150 pp.

CCOP/ECAFE, United Nations, 1968b. Regional Geology and Prospects for Mineral Resources on the Northern Part of the Sunda Shelf. *Economic Commission for Asia and the Far East – Committee for Co-ordination of the Joint Prospecting for Mineral Resources in Asian Offshore Areas, Technical Bulletin,* 1: 129–142.

Choffat, P., 1913. Rapports de géologie économique, I. Sur les sables aurifères marins, d'Adiça et sur d'autres dépôts aurifères de la côte occidentale de la péninsule de Setubal. *Comm. Serviço Geol. Portugal,* IX: 5–26.

Cissarz, A. and Baum, F., 1960. Vorkommen und Mineralinhalt der Zinnerzlagerstätten von Bangka (Indonesien). *Geol. Jahrb.,* 77: 541–580.

Clement. M., Imhof, R. and Mertins, E., 1973. Beitrag zur Aufbereitbarkeit von hydrothermalen Erzschlämmen. *Interocean '73,* 1: 404–416.

Corliss, J.B., 1971. The origin of metal-bearing submarine hydrothermal solutions. *J. Geophys. Res.,* 76 (33): 8128–8138.

Cronan, D.S., 1973. Basal ferruginous sediments cored during leg 16 of the Deep Sea Drilling Project. In: *Initial Report of the Deep Sea Drilling Project, XVI.* U.S. Government Printing Office, Washington, D.C., pp. 601–604.

Cronan, D.S. and Tooms, J.S., 1969. The geochemistry of manganese nodules and associated pelagic deposits from the Pacific and Indian Oceans. *Deep-See Res.,* 16: 335–359.

Cruickshank, M.J., 1974. Mineral resources potential of continental margins. In: C.A. Burk and C.L. Drake (Editors), *The Geology of Continental Margins.* Springer, New York, N.Y., pp. 965–1000.

Dadet, P., Marchesseau, J., Millon, R. and Motti, E., 1970. Mineral occurrences related to stratigraphy and tectonics in Tertiary sediments near Umm Lajj, eastern Red Sea area, Saudi-Arabia. *Philos. Trans. R. Soc. London, Math. Phys. Sci.,* 267: 99–106.

D'Anglejan, B.F., 1967. Origin of marine phosphorites off Baja California, Mexico. *Mar. Geol.,* 5: 15–44.

Dash, E.J., Dymond, J.R. and Heath, G.R., 1971. Isotopic analysis of metalliferous sediment from the East Pacific Rise. *Earth Planet. Sci. Lett.,* 13 (1): 175–180.

Degens, E.T. and Ross, D.A. (Editors), 1969. *Hot Brines and Recent Heavy Metal Deposits in the Red Sea.* Springer, New York, N.Y., 600 pp.

Dietrich, G., 1963. *General Oceanography – an Introduction.* Interscience, New York, N.Y., 588 pp.

Dietz, R.S., Emery, K.O. and Shepard, F.P., 1942. Phosphorite deposits on the sea floor off southern California. *Bull. Geol. Soc. Am.,* 53: 815–848.

Dimitriev, L., Barsukov, V. and Udintsev, G., 1971. Rift-zones of the oceans and the problem of ore-formation. *Int. Assoc. Genesis Ore Deposits, Soc. Min. Geol. Jap. Spec. Issue,* 3: 65–69.

Dürbaum, H.-J. and Schlüter, H.-U., 1974. Möglichkeiten der Reflexionsseismik für die Mangan-Knollen-Exploration. *Meerestech. Mar. Technol.,* 6: 188–192.

Dunham, K.C., 1967. Economic geology of the continental shelf around Britain. In: *The Technology of the Sea and the Sea Bed,* 2: 328–341 (U.K. Atomic Energy Authority).

Dunham, K.C., 1970. Gravel, sand, metallic placer and other mineral deposits on the East Atlantic continental margin. In: *The Geology of the East Atlantic Continental Margin. ICSU/SCOR Working Party 31, Symp., Cambridge, 1970, Rep. No. 70/13,* pp. 79–85.

Dunham, K.C. and Sheppard, J.S., 1970. Superficial and solid mineral deposits of the continental shelf around Britain. In: *Commonwealth Min. Metall. Congr. 9th.* Inst. Min. Metall., London, pp. 1–23.

Ehrenberg, H., Pilger, A. and Schröder, F., 1954. Das Schwefelkies–Zinkblende–Schwerspatlager von Meggen (Westfalen). *Beih. Geol. Jahrb.,* 12: 1–352.

Emery, K.O., 1960. *The Sea off Southern California a Modern Habitat of Petroleum.* Wiley, New York, N.Y., 366 pp.

Emery, K.O., 1966. Geological methods for locating mineral deposits on the ocean floor In: *Trans. Mar. Tech. Soc. Conf., 2nd, Exploiting the Ocean,* pp. 24–43.

Emery, K.O., 1968. Relict sediments on continental shelves of world. *Bull. Am. Assoc. Pet. Geol.,* 52: 445–464.

Emery, K.O. and Noakes, L.C., 1968. Economic placer deposits of the continental shelf. *Economic Commission for Asia and the Far East – Committee for Co-ordination of the Joint Prospecting for Mineral Resources in Asian offshore areas, Technical Bulletin,* 1: 95–111.

Ewing, J.I., Worzel, J.L. and Ewing, M., 1962. Sediments and oceanic structural history of the Gulf of Mexico. *J. Geophys. Res.,* 67: 2509–2527.

Ewing, M., Worzel, J.L., Beall, A.O. et al., 1969. *Initial Reports of the Deep Sea Drilling Project, I.* U.S. Government Printing Office, Washington, D.C., 672 pp.

Fisher, D.E. and Boström, K., 1969. Uranium-rich sediments on the East Pacific Rise. *Nature,* 224: 64–65.

Fowler, H.W. and Fowler, F.G., 1964. *The Concise Oxford Dictionary of Current English.* Oxford, 421 pp., 5th ed.

Foxton, P., 1965. A mass fish mortality on the Somali Coast. *Deep-Sea Res.,* 12: 17–19.

Frazer, J.Z. and Arrhenius, G., 1972. World-wide distribution of ferromanganese nodules and element concentrations in selected Pacific Ocean Nodules. *Tech. Rep. No. 2, NSF-GX 34659,* 4 pp.

Friedrich, G. and Plüger, W., 1974. Die Verteilung von Mangan, Eisen, Kobalt, Nickel, Kupfer und Zink in Manganknollen verschiedener Felder. *Meerestech. Mar. Technol.*, 6: 203–206.

Gardner, D.E., 1955. Beach-sand heavy-mineral deposits of eastern Australia. *Bur. Min. Res. Aust.*, 28: 103 pp.

Gealy, E., 1971. Results of the Joides Deep Sea Drilling Project 1968–1971. In: *Proc. World Pet. Congr., 8th, Moscow*, 2: 337–348.

Girelli, A., 1965. Fresh water from the sea. In: *Proc. Int. Symp. Milan*. Pergamon, London.

Goldberg, E.D., 1963. The oceans as a chemical system. In: M.N. Hill (Editor), *The Sea, 2*. Wiley, New York, N.Y., pp. 3–25.

Grasshoff, K., 1970. Chemie der Manganknollen. In: G. Dietrich (Editor), *Erforschung des Meeres*. Umschau, Frankfurt, 141–149.

Haber, F., 1927. Das Gold im Meerwasser. *Z. Angew. Chem.*, pp. 303–314.

Hackett, J.P. and Bischoff, J.L., 1973. New data on the stratigraphy, extent and geologic history of the Red Sea geothermal deposits. *Econ. Geol.*, 68 (4): 553–564.

Hänig, G., 1973. Drucklaugung von Manganknollen im schwefelsauren Medium. *Interocean '73*, 1: 432–444.

Hammond, A.L., 1974. Manganese nodules. I: Mineral resources on the deep seabed; II: Prospects of deep-sea mining. *Science*, 183 (4124): 502–503; 183 (4125): 644–646.

Hannak, W., 1968. Die Rammelsberger Erzlager. In: *Zur Mineralogie und Geologie der Umgebung von Göttingen. Aufschluss, Sonderheft*, 17: 62–74.

Harder, H., 1964. Untersuchung rezenter vulkanischer Eisenausscheidung zur Erklärung der Erze vom Lahn-Dill-Typus. *Ber. Geol. Ges. DDR*, 9: 469–473.

Hart, S.R., Ade-Hall, J.M. et al., 1974. Leg 34, Oceanic basalt and the Nazca Plate. *Geotimes*, 19 (4): 20–24.

Hartmann, M., 1971. Bericht über geochemische Untersuchungen in den Hydrothermallaugenbecken am Boden des Roten Meeres. *Geol. Rundsch.*, 60: 244–256.

Hartmann, M., 1972. Sound velocity data for the hot brines and corrected depth of the interfaces in the Atlantis II Deep. *Mar. Geol.*, 12 (5): M 16–M 20.

Hartmann, M., 1973. Untersuchung von suspendiertem Material in den Hydrothermallaugen des Atlantis-II-Tiefs. *Geol. Rundsch.*, 62 (3): 742–754.

Hartmann, M. and Müller, P., 1974. Geochemische Untersuchungen an Sedimenten und Porenwasser. *Meerestech. Mar. Technol.* 6: 201–202.

Heye, D. and Beiersdorf, H., 1973. Radioaktive und magnetische Untersuchungen an Manganknollen zur Ermittlung der Wachstumsgeschwindigkeit bzw. zur Alterbestimmung. *Z. Geophys.*, 39: 703–726.

Hinz, K. and Schlüter, H.U., 1973. Ergebnisse reflexionsseismischer Messungen der "Valdivia"-Fahrt "Manganknollen I" im äquatorialen Pazifik. *Interocean '73*, 1: 217–222.

Hollister, Ch.D., Ewing, J.I., Habib, D., Hathaway, J.C., Lancelot, Y., Luterbacher, H., Paulus, C., Poag, C.W., Wilcoxon, I.A. and Worstell, P., 1972. *Initial Reports of the Deep Sea Drilling Project, XI*. U.S. Government Printing Office, Washington, D.C., 1077 pp.

Horn, D.R. (Editor), 1972. *Ferromanganese Deposits on the Ocean Floor*. Office for the International Decade of Ocean Exploration, National Science Foundation, Washington, D.C., 293 pp. (papers from a conference).

Horn, D.R., Horn, B.M. and Delach, M.N., 1972. Ferromanganese deposits of the North Pacific Ocean. *Tech. Rep. No. 1, NSF-GX 33616*, 78 pp.

Horn, D.R., Delach, M.N. and Horn, B.M., 1973a. Metal content of ferromanganese deposits of the oceans. *Tech. Rep. No. 3, NSF-GX 33616*, 51 pp.

Horn, D.R., Horn, B.M. and Delach, M.N., 1973b. Ocean manganese nodules metal values and mining sites. *Tech. Rep. No. 4, NSF-GX 33616*, 57 pp.

Horn, D.R., Horn, B.M. and Delach, M.N., 1973c. Factors which control the distribution of ferromanganese nodules and proposed research vessel's track North Pacific. *Tech. Rep. No. 8, NSF-GX 33616*, 20 pp.

Hosking, K.F.G., 1971. The offshore tin deposits of southeast Asia. *Economic Commission for Asia and the Far East — Committee for Co-ordination of the Joint Prospecting for Mineral Resources in Asian offshore areas, Technical Bulletin,* 5: 112–129.

Hosking, K.F.G. and Ong, P.M., 1964. The distribution of tin and certain other heavy metals in the superficial portions of the Gwithian/Hayle Beach of West Cornwall. *Trans. R. Geol. Soc. Cornwall,* XIX (5): 352–390.

Ingham, F.T. and Bradford, E.F., 1960. *The Geology and Mineral Resources of the Kinta Valley, Perak.* Geol. Surv. Malaya, Kuala Lumpur, 347 pp. (District Mem. 9).

Institut für Internationales Recht, 1970. Die Nutzung des Meeresgrundes ausserhalb des Festlandsockels (Tiefsee). In: *Symp. 1969, Veröff. Inst. Int. Recht Univ. Kiel.* Hansescher Gildenverlag, Hamburg, 64: 257 pp.

International (European) Symposium on Fresh water from the Sea, 1973. 1. – 4. Symposium (preprints) (4. (last) Symposium), Heidelberg 1973.

Kauczor, H.W., Junghanss, H. and Roever, W., 1973. Nassmetallurgische Aufarbeitung metallhaltiger Lösungen aus dem Manganknollen-Aufschluss. *Interocean '73,* 1: 469–473.

Keiffer, E. (Editor), 1968. Mineral resources of the World Ocean. In: *Proc. Symp. Naval War College, Newport, Rhode Island, 1968,* 108 pp.

Kerl, J.F., 1970. Eigenschaften, Vorkommen und Entstehung von nickel-, kupfer- und kobalthaltigen Manganknollen des Meeresbodens. *Erzmetall,* 1: 1–10.

Kim, W.J., 1970. Placer deposits of detrital heavy minerals in Korea. *Tech. Bull. ECAFE,* 3: 127–136.

Koczy, F.F., 1956. Geochemistry of the radioactive elements in the ocean. *Deep-Sea Res.,* 3: 98–103.

Koczy, F.F., Tomic, E. and Hecht, F., 1957. Zur Geochemie des Urans im Ostseebecken. *Geochim. Cosmochim. Acta,* 11: 86–102.

Kraume, E., 1955. Die Erzlager des Rammelsberges bei Goslar. *Beih. Geol. Jarhb.,* 18: 394 pp.

Krishnan, M.S., 1956. *Geology of India and Burma.* Higginbothams Ltd., Madras, 555 pp.

Layton, W., 1966. Prospects of offshore mineral deposits on the eastern seaboard of Australia. *Min. Mag.,* 115 (5): 344–351.

Lippert, H.J., Hentschel, H. and Rabien, A., 1970. *Erläuterungen zur geologischen Karte von Hessen 1: 25 000, Bl. Nr. 5215 Dillenburg,* 410–423 (Eisenerze), 2nd ed.

Lotze, F., 1957. *Steinsalz und Kalisalze, I.* Gebr. Borntraeger, Berlin, 465 pp., 2nd ed.

Love, L.G., 1967. Sulphides of metals in Recent sediments. In: *Proc. Inter-University Geol. Congr., Leicester, 15th, Spec. Publ.,* 1: 31–60.

Ludwig, G. and Vollbrecht, K., 1957. Die allgemeinen Bildungsbedingungen litoraler Schwermineralkonzentrate und ihre Bedeutung für die Auffindung sedimentärer Lagerstätten. *Geologie,* 6 (3): 233–277.

Lüttig, G., 1974. Seifenlagerstätten an der niedersächsischen Küste. *Glückauf,* 110 (5): 169–171.

MacDonald, E.H., 1971. Detrital heavy mineral deposits in eastern Asia (General introduction and country reports). *Economic Commission for Asia and the Far East—Committee for Co-ordination of Joint Prospecting for Mineral Resources in Asian Offshore Areas (CCOP), Tech. Bulletin,* 5: 13–111.

Marchig, V., 1972. Zur Geochemie rezenter Sedimente des Indischen Ozeans. *"Meteor"-Forschungsergebnisse C,* 11: 1–104.

McKelvey, V.E. and Wang, F.F.H., 1969. *World Subsea Mineral Resources.* U.S. Geol. Surv., Washington D.C., Miscell. Geol. Investig. Map 632.

Meiser, H.J. and Müller, E., 1973. Manganknollen – eine weitere Quelle für die Deckung des zukünftigen Rohstoffbedarfs? *Meerestech. Mar. Technol.,* 4 (5): 145–150.

Menard, H.W., 1964. *Marine Geology of the Pacific.* McGraw-Hill, New York, N.Y., 271 pp.

Mero, J.L., 1965. *The Mineral Resources of the Sea.* Elsevier, Amsterdam, 312 pp.

Mero, J.L., 1971. Oceanic mineral resources and current developments in ocean mining. In: *Colloque International sur l'Exploitation des Oceans, Bordeaux, 1971, Thème IV, Tome 1* (G2-03), 1–39.

Mero, J.L., 1972a. Recent concepts in undersea mining. *Min. Congr. J.,* 58 (5): 43–48; 54.

Mero, J.L., 1972b. The future promise of mining in the ocean. *CIM-Bulletin,* 65 (270): 21–27.

Meyer, K., 1973a. Uran-Prospektion vor Südwestafrika. *Erzmetall*, 26 (7): 313–317.

Meyer, K., 1973b. Surface sediment- and manganese nodule facies, encountered on R.V. "Valdivia" cruises 1972/73. *Meerestechn. Mar. Technol.*, 4 (6): 196–199.

Meyer-Galow, E., Schwarz, K.H. and Boin, U., 1973. Sulfatisierender Aufschluss von Manganknollen. *Interocean '73*, 1: 458–468.

Moore, J.R., 1972. Exploitation of Ocean Mineral Resources – Perspectives and Predictions. In: *Proc. Int. Congr. History Oceanog.*, 2nd, R. Soc. Edinb., Section B, 72: 193–206.

Morgenstein, M. (Editor), 1973. Papers on: *The Origin and Distribution of Manganese Nodules in the Pacific and Prospects for Exploration*. Symposium organized by "Valdivia" Manganese Exploration Group and the Hawaii Institute of Geophysics, Honolulu, Hawaii, July 1973.

Murray, J. and Philippi, E., 1908. *Die Grundproben der "Deutschen Tiefsee-Expedition"*. Deutsche Tiefsee-Expedition 1898/99, Band X, 206 pp.

Murray, J. and Renard, A.F., 1891. Report on deep-sea deposits based on the specimen collected during the voyage of H.M.S. "Challenger" in the years 1872–1876. *Rep. Sci. Res. Deep-Sea Deposits, Chall. Exp.*, 525 pp.

Murray, L.G., Joynt, R.H., O'Shea, D.OC., Foster, R.W. and Kleinjan, L., 1970. The geological environment of some diamond deposits off the coast of South West Africa. In: *The Geology of the East Atlantic Continental Margin, ICSU/SCOR Working Party 31 Symposium, Cambridge, 1970. Rep. No. 70/13*, pp. 119–141.

Nesbitt, A.C., 1967. Diamond mining at sea. *Proc. First World Dredging Conf. New York*, 1967, pp. 697–725.

Newell, R.A., 1971. Characteristics of the stanniferous Alluvium in the southern Kinta Valley, west Malaysia. *Geol. Soc. Malaysia*, 4: 15–37.

Nowak, W.S.W., 1973. Rise and fall of the marine diamond industry, 1–3. *Offshore Services*, Oct.: 43/46; Nov.: 68/70; Dec.: 33/34.

Osberger, R., 1968. Über die Zinnseifen Indonesiens und ihre genetische Gliederung. *Z. Dtsch. Geol. Ges.*, 117: 749–766.

Osberger, R. and Romanowitz, C.M., 1967. How the off-shore Indonesian tin placers are explored and sampled. *World Min.*, 20 (12): 52–58.

Parker, R.J. and Siesser, W.G., 1972. Petrology and origin of some phosphorites from the South African continental margin. *J. Sediment. Petrol.*, 42: 434–440.

Pepper, J.F., 1958. Potential mineral resources of the continental shelves of the western hemisphere. In: *An Introduction to the Geology and Mineral Resources of the Continental Shelves of the Americas. U.S. Geol. Surv. Bull.*, 1067: 43–65.

Peterson, M.N.A. et al., 1970. *Initial Reports of the Deep Sea Drilling Project, II*. U.S. Government Printing Office, Washington, D.C., 501 pp.

Pimm, C.A., 1974. Mineralization and trace element variation in deep-sea pelagic sediments of the Wharton Basin, Indian Ocean. In: *Initial Reports of the Deep Sea Drilling Project, XXII*. U.S. Government Printing Office, Washington, D.C., pp. 469–476.

Piper, D.Z., 1973. Origin of metalliferous sediments from the East Pacific Rise. *Earth Planet. Sci. Lett.*, 19 (1): 75–82.

Puchelt, H., 1973. Recent Iron Sediment Formation at the Kameni Islands, Santorini (Greece). In: *Ores in Sediments, IUGS, Ser. A*, 3: 227–245.

Puchelt, H., Schock, H.H. and Schroll, E., 1973. Rezente marine Eisenerze auf Santorin, Griechenland. *Geol. Rundsch.*, 62 (3): 786–812.

Putzer, H. and Von Stackelberg, U., 1973. Exploration von Titanerzseifen vor Moçambique. *Interocean '73*, 1: 168–174.

Richter, G., 1941. Geologische Gesetzmässigkeiten in der Metallführung des Kupferschiefers. *Arch. Lagerstättenforsch., N.S.*, 73: 61 pp.

Richter-Bernburg, G. and Schott, W., 1954. *Geological Researches in Western Saudi Arabia*. Dir. Gen. Mineral Resources, Djiddah, S. Arabia (unpublished report).

Riley, J.P. and Chester, R., 1971. *Introduction to Marine Chemistry*. Academic Press, London, 465 pp.

Romankevich, Ye.A. and Baturin, G.N., 1972. Composition of the organic matter in phosphorites from the continental shelf of southwest Africa. *Geochem. Int.,* 9 (3): 464–470.

Romanowitz, C.M., 1971. On shore alluvial mining results as a guide to future offshore mining. *World Min.,* 24 (2): 52–59; 24 (3): 46–52 (European ed.).

Ryabinin, A.L., Romanov, A.S., Khatanov, Sh. et al., 1974. Gold in waters of the World Ocean. *Geokhimiya,* 1: 158–182 (in Russian).

Rydell, H., Kraemer, T., Bostroem, K. and Ioensun, O., 1974. Postdepositional injections of uranium-rich solutions into East Pacific Rise sediments. *Mar. Geol.,* 17 (3): 151–164.

Said, R., 1962. *The Geology of Egypt.* Elsevier, Amsterdam, 377 pp.

Sapozhnikov, D.G., 1971. On manganese content in sea water (in connection with the problem of manganese deposit formation). In: Proc. of the IMA–IAGOD Meetings '70. *Soc. Min. Geol. Jap. Spec. Issue,* 3: 484–488.

Sayles, F.L. and Bischoff, J.L., 1973. Ferromanganoan sediments in the equatorial East Pacific. *Earth Planet. Sci. Lett.,* 19: 330–336.

Schlee, J., 1968. Sand and gravel on the continental shelf off the northeastern United States. *U.S. Geol. Surv. Circ.,* 602.

Schneider, W. and Wachendorf, H., 1973. Vulkanismus und Graben-Bildung im Roten Meer. *Geol. Rundsch.,* 62 (3): 754–773.

Schoell, M. and Hartmann, M., 1973. Detailed temperature structure of the hot brines in the Atlantis II Deep Area (Red Sea). *Mar. Geol.,* 14: 1–14.

Schoell, M. and Stahl, W., 1972. The carbon isotopic composition and the concentration of the dissolved anorganic carbon in the Atlantis II Deep brines/Red Sea. *Earth Planet. Sci., Lett.,* 15: 206–211.

Schott, G., 1935. *Geographie des Indischen und Stillen Ozeans.* C. Boysen, Hamburg, 413 pp.

Schott, W., 1970a. Das Meer und der Meeresboden als mineralische Rohstoffquelle. In G. Dietrich (Editor), *Erforschung des Meeres.* Umschau, Frankfurt, pp. 167–181.

Schott, W., 1970b. Möglichkeiten der Nutzung mineralischer Rohstoffe aus dem Meeresboden und dem Meeresuntergrund. *Interocean '70,* 1: 23–28.

Schott, W., 1974. Untermeerische mineralische Rohstoffe in den Ozeanen. *Naturwissenschaften,* 61 (5): 192–199.

Schott, W. and Von Stackelberg, U., 1965. Über rezente Sedimentation im Indischen Ozean, ihre Bedeutung für die Entstehung kohlenwasserstoffhaltiger Sedimente. *Erdöl Kohle, Erdgas, Petrochem.,* 18: 945–950.

Schultze-Westrum, H.-H., 1973. Fahrtrouten und Stationsdichte des Forschungsschiffs "Valdivia" in Abhängigkeit zur Manganknollenverteilung. *Meerestech. Mar. Technol.,* 4 (5): 163–169.

Schweisfurth, R., 1971. Manganknollen im Meer. *Naturwissenschaften,* 58: 344–347.

Seabed Assessment Program, 1973. (Wogman, N.A., Chave, K. et al. (Chairmen). *Phase I Report, Inter-University Program of Research on Ferromanganese Deposits of the Ocean Floor.* Int. Decade of Ocean Exploration, Washington, D.C. (unpublished).

Seibold, E., 1970. Der Meeresboden als Rohstoffquelle und die Konzentrierungsverfahren der Natur. *Z. Chem. Ing. Tech.,* 42 (23): A2091–A2102.

Seibold, E., 1973. Rezente submarine metallogenese. *Geol. Rundsch.,* 62 (3): 641–684.

Seibold, E., 1974. *Der Meeresboden, Ergebnisse und Probleme der Meeresgeologie.* Springer, New York, N.Y., 183 pp., Hochschultext.

Selenov, K.K., 1958. Über die Belieferung des Ochotskischen Meeres mit gelöstem Eisen durch die Hydrothermen des Ebekovulkans (auf der Insel Paramušir). *Dokl. Akad. Nauk S.S.S.R.,* 120 (5): 1084–1092 (In Russian).

Setty, A.P. and Rao, C.M., 1972. Phosphate, carbonate and organic matter distribution in sediment cores off Bombay–Saurashtra Coast, India. In: *Proc. Int. Geol. Congr., XXIV Session, Sec.* 8: pp. 182–191.

Shigley, C.M., 1968. Sea water as a raw material. In: E. Keiffer (Editor), *Proc. Symp. on Mineral Resources of the World Ocean. The University of Rhode Island, Occas. Publ.,* 4: 45–50.

Smith, P.J., 1974. Origin of metalliferous sediments. *Nature,* 251: 465 pp.

Supp, A.Ch., Nebe, R. and Schöner, H., 1973. Physikalische Aufbereitung von Thermalschlämmen aus dem Roten Meer. *Interocean '73,* 1: 417–431.

Sverdrup, H.U., Johnson, M.W. and Fleming, R.H., 1946. *The Oceans, their Physics, Chemistry and General Biology.* Prentice-Hall, New York, N.Y., 1087 pp.

Tooms, J.S., Summerhays, C.P. and McMaster, R.L., 1971. Marine geological studies on the north-west African margin Rabat-Dakar. In: *The Geology of the East Atlantic Continental Margin, 4. Africa ICSU/SCOR Working Party 31 Symposium, Cambridge 1970,* pp. 9–25.

Trombull, J.V.A. and Hathaway, J.C., 1968. Dark mineral accumulation in beach- and dune sands of Cape Cod and vicinity. *U.S. Geol. Surv. Prof. Pap.,* 600 (B): B 178–B 184.

Udintsev, G.B. and Dimitriev, L.V., 1970. The ultrabasic rocks of the ocean floor. In A.E. Maxwell (Editor), *The Sea,* 4 (1): 521–573.

Ullmanns Encyklopädie der Technischen Chemie. W. Foerst (Editor), vol. 4: 842 pp., 1953; vol. 12: 800 pp., 1960; vol. 18: 792 pp., 1967; Verlag Urban u. Schwarzenberg, München/Berlin/Wien.

Ulrich, K.H., Scheffler, U. and Meixner, M.J., 1973. Aufarbeitung von Manganknollen durch saure Laugung. *Interocean '73,* 1: 445–457.

United Nations, 1970. Mineral resources of the sea. *UN Publ.,* E.70.II.B.4: I–VI; 1–49.

United States Department of the Interior, 1965. *Saline Water Conversion Report.* U.S. Department of the Interior, Office of Saline Water.

Untung, M., 1968. Results of a sparker survey for tin ore off Bangka and Belitung Islands, Indonesia. *ECAFE, CCOP Report of 4th session,* pp. 61–67.

Valette, J.N., 1973. Distribution of certain trace elements in marine sediments surrounding Vulcano Island (Italy). In: *Ores in Sediments. IUGS, Ser. A,* 3: 321–337.

Van Overeem, A.J.A., 1960. The geology of the Cassiterite Placers of Billiton, Indonesia. *Geol. Mijnb.,* 39: 444–457.

Van Overeem, A.J.A., 1970. Offshore tin exploration in Indonesia. *Inst. Min. Metal. Trans. Sec. A,* 79: A81–A85.

Veeh, H.H., Calvert, S.E. and Price, N.B., 1974. Accumulation of uranium in sediments and phosphorites on the South West African shelf. *Mar. Chem.,* 2 (3): 189–202.

Veenstra, H.J., 1970. Sediments of the southern North Sea. In: *The Geology of the East Atlantic Continental Margin, ICSU/SCOR Working Party 31 Symposium, Cambridge, Institute of Geological Science,* Rep. No. 70/15, pp. 13–23.

Von Buschman, O., 1909, 1906. *Das Salz, dessen Vorkommen und Verwertung in sämtlichen Staaten der Erde.* (Bd. I, 1909, Bd. II, 1906), Leipzig.

Von der Borch, C.C. and Rex, R.W., 1970. Amorphous iron oxide precipitates in sediments cored during leg 5, Deep Sea Drilling Project. In: *Initial Reports of the Deep Sea Drilling Project, V.* U.S. Government Printing Office, Washington, D.C., pp. 541–544.

Von der Borch, C.C., Nesteroff, W.D. and Galehouse, J.S., 1971. Iron-rich sediments cored during leg 8 of the Deep Sea Drilling Project. In: *Initial Reports of the Deep Sea Drilling Project, VIII.* U.S. Government Printing Office, Washington, D.C., pp. 829–833.

Von Stackelberg, U., 1972. Fazisverteilung in Sedimenten des indisch-pakistanischen Kontinentalrandes (Arabisches Meer). *"Meteor"-Forschungsergebnisse,* C, 9: 1–73.

Ward, J., 1972. Australian resources of mineral sands. *Aust. Min. Ind.,* 25 (1): 12–23.

Warner, Th.B. and Gieskes, I.M., 1974. Iron-rich basal sediments from the Indian Ocean: Site 245, Deep Sea Drilling Project. In: *Initial Reports of the Deep Sea Drilling Project, XXV.* U.S. Government Printing Office, Washington, D.C., pp. 395–403.

Weinstein, R.P., 1973. Mining marine phosphorite – a case study. *Interocean '73,* 1: 243–255.

Wenk, E., 1969. The Physical Resources of the Ocean. *Sci. Am.,* 221 (3): 167–176.

White, D.E., Hem, J.D., and Waring, G.A., 1963. Chemical composition of subsurface waters. In: W. Fleischer (Editor), *Data of Geochemistry,* 6th ed. *U.S. Geol. Surv. Prof. Pap.,* 440-F.

Whitmarsh, R.B., Ross, D.A. et al., 1974. *Initial Reports of the Deep Sea Drilling Project, 23.* U.S. Government Printing Office, Washington, D.C., 1180 pp.

Wirtschaftsvereinigung Industrielle Meerestechnik, 1971. *Meerestechnik in der Bundesrepublik Deutschland*. Düsseldorf, 111 pp.

Wogman, N., Chave, K., Sorem, R. (Co-chairmen), 1974. Workshop on manganese nodule mineralogy and geochemistry methods. *Inter-University Program of Research on Ferromanganese Deposits of the Ocean Floor — Seabed Assessment Program, Tech. Rep.*, 7: 23 pp.

Wolf, K.H., 1976. Ore genesis influenced by compaction. In: G.V. Chilingar and K.H. Wolf (Editors), *Compaction of Coarse-Grained Sediments 2*. Elsevier, Amsterdam, in press.

Wyrtki, K., 1971. *Oceanographic Atlas of the International Indian Ocean Expedition*. National Science Foundation, Washington, D.C.

Zaalberg, P.H.A., 1970. Offshore tin dredging in Indonesia. *Inst. Min. Metall. Trans.*, 79: A 86—A 95.

Reference added in proof

Bäcker, H., 1975. Exploration of the Red Sea and Gulf of Aden during the m.s. "Valdivia" cruises 1971 and 1972. *Geol. Jahrb.*, 13 (in press). (This paper summarizes the operation of "Valdivia" and the subsequent scientific investigations.)

Chapter 7

TYPICAL AND NONTYPICAL SEDIMENTARY ORE FABRICS

OSKAR SCHULZ

(Translated by Ignaz Vergeiner, Innsbruck)

INTRODUCTION

The progress of sedimentology in the past four decades has brought about sometimes fundamental changes in the interpretation of the formation of ore deposits in the science of mineral deposits ("Lagerstättenkunde"). The study of fabrics, in particular (Sander, 1930, 1948–1950, 1970), has given new impulses to sedimentology: its approach, working methods and operational possibilities were shown in detail in 1936 by B. Sander.

Schneider (1953, 1954, 1957) and Taupitz (1953, 1954, 1957) were the first to apply the known facts about "apposition[1]-fabrics" to the science of mineral deposits. They were followed by numerous other authors, for example Maucher (1954, 1957), Schulz (1955, 1957, 1960a,b, 1968) and Siegl (1956, 1957). These applications of the study of fabrics to applied geology consisted of a critical examination of the processes taking place during the formation of Pb-Zn deposits in the Triassic sediments of the Alps. Even now, and in the future, these Mesozoic geosynclinal metal enrichments are going to form a welcome object of study for petrologists and geochemists, as mostly there has been no metamorphism which would have obliterated primary ore apposition fabrics. The sediments still show their mostly well-preserved primary and syndiagenetically modified textures and structures, they permit optimal use of the data, and therefore they also form the core of the considerations contained in this publication.

The knowledge of simple, typical, easily understandable characteristics of the ore sediments, whose unique interpretation cannot be doubted, permits us to extrapolate to more difficult and also more ambiguous features, which have to be regarded in part as outright nontypical. Knowledge of the formation of mineral deposits frequently forms one of the foundations of prospecting for and exploitation of utilizable mineral resources.

---

[1] The technical terms in the study of fabrics coined by Sander (1930, 1936, 1948 and 1950) e.g., apposition, component movement, s-surface, space rhythm, have been translated, defined and examplified in Sander (1970).

Thus, this form of basic research cannot be neglected in its economic significance. Our contribution "Typical and nontypical sedimentary ore fabrics" ought to be seen from this point of view, and it ought to give further impulse to judging the sedimentary ore deposits which are widely distributed in the sedimentary and volcanic rocks of past geologic times. Also, those sedimentary deposits that have undergone metamorphism have to be given consideration, and to understand their secondarily formed textures, a knowledge of the primary and diagenetic (i.e., premetamorphic) characteristics of the rocks is indispensable.

This article quite intentionally emphasizes the Pb-Zn paragenesis, because the related ore minerals sphalerite, galena and iron sulfides as well as the minerals surrounding them (baryte, fluorite, quartz and calcite) have furnished the primary object of the study of sedimentary ore fabrics. The Pb-Zn paragenesis will remain one of the most important ways to make unique statements about sedimentary ore deposition from solutions.

GENERAL REMARKS

First it appears to be important to answer two questions: Which are, in fact, the typical sedimentary ore fabrics? When can apposition fabrics be classified as sedimentary only and not in any other way? It will be shown that quite a few occurrences cannot be called unique or cannot be related to a single cause of formation in the sedimentary cycle. It is the typifiable (i.e., widespread) features which matter, and wrong questions such as: "is the deposit sedimentary or metasomatic?" should not be asked in the first place; for one alternative does not exclude the other, as the processes leading to formation of sediments are very complex. Among the fabric-forming componental movements[1], the "mechanical, chemical and biogenetic" apposition (Sander, 1936), only mechanical ore apposition lends itself to the consideration of the formation of deposits as being typically sedimentary. Therefore, its occurrence will have to be demonstrated whenever unequivocal proof of sedimentary ores is needed, for in this case, only that can be considered typical which is characteristic of the sediment in question and which cannot be observed in other kinds of rocks.

The study of fabrics certainly requires the distinction between mechanical deposition of detritus or of particles which were precipitated from suspension and traction, on one hand, and chemical apposition, on the other hand; the latter occurs independently of gravity by crystallization of the space lattice to crystals, crystal-fibre layers and crystal layers (strata), possibly via the gelatinous state.

For mechanical apposition to occur, there must have existed material before the process of apposition, be it mineral or rock detritus, precipitated crystallites, or only

---

[1] The technical terms in the study of fabrics coined by Sander (1930, 1936, 1948 and 1950) e.g., apposition, componental movement, s-surface, space rhythm, have been translated, defined and exemplified in Sander (1970).

crystal nuclei. The emphasis of our considerations must be on those cases where the first shaping of the zones of formation occurred under the influence of gravity. It is only these cases which yield unequivocal proof of synsedimentary ore apposition. We omit here the obvious cases of placers (alluvial deposits) and clastic deposits.

In the process of mechanical apposition of ore detritus as sand and pebbles, external sediments are formed, which, generally speaking, are not subject to hard-to-understand genetic processes, at least in non-metamorphic regions. Those cases of metal enrichment, however, whose origin is due to precipitation from solutions and apposition caused by gravity deserve our attention.

It is genetically important to answer the following questions: Were the ore minerals subject to *mechanical apposition* (deposition)? Did the apposition occur as polar apposition, e.g. in the form of vertically *graded bedding*, or as *geopetal* apposition? Does the series of strata show *space rhythm* for one or more mineral components? Are there also cases of *nonparallel* apposition such as oblique and *cross beddings*? Do syndiagenetic processes affect the ore minerals together with the surrounding host-rock sediment? Are we dealing with *external* or *internal sediments*? Do *resediments* exist, which would prove the denudation of already ore-containing sediments within the series of strata?

Admittedly, knowledge of the process of apposition by itself does not tell us the source of the solutions from which apposition had occurred: weathering solutions, exhalations, hydrothermals; and even less about the possible connections between metal ions and a magma reservoir. For clarification, other structural data must be obtained, e.g. the mineral and trace-element parageneses, that is geochemical data and other geo-scientific aspects.

Again and again we see that fabrics with a relatively low metal content may be genetically interpreted more easily than fabrics with a high ore content. This is because high ore content causes recrystallization and formation of coarse crystalline ore in many minerals, while low content permits subtly differentiated details of the zones of forma-tion to stand out. The latter, however, are the most important for petrologists, and in this case the discrepancy between scientific research and economics becomes apparent. In any case, it is advisable for the geoscientist to start his observations at a place where ore minerals make only minor contributions to the sediment structure, and where the recog-nition and interpretation of the fabric-forming componental movements is made easier.

BEDDING, FINE BEDDING, RHYTHM

A simple diagnostic feature of sedimentary formation of deposits is the usually strati-fied, in a general sense conformable arrangement of an ore body (Fig. 1). This is not, however, by itself a unique, reliable, infallible feature.

The term "stratification" as used in the study of fabrics (but not in all geological definitions) presupposes an inhomogeneous parallel-fabric. Accordingly sediments, whose

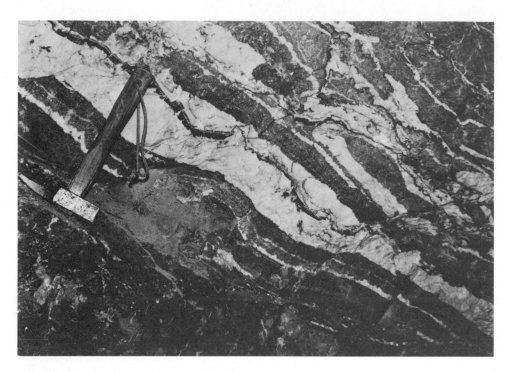

Fig. 1. Stratiform ore body with coarse-crystalline sphalerite and "Schalenblende" (grey, dark grey), as well as calcite (white) in Wetterstein Limestone (light grey). Lafatsch Mine, Karwendel range, Tyrol.

anisotropy lies solely in containing statistically parallel lamellae (clay lamellae, haematite, baryte lamellae), or columnar components where a preferred orientation has been produced morphologically (tiny quartz columns, rutile needles), form a homogeneous parallel-fabric, but not a stratum.

In order to extend our considerations, it is anticipated here that *stratified appearance* (not stratification) can come about in ore bodies through selective metasomatic processes (mimetic metasomatism), as well as through internal and external chemical sedimentation. Some material layerings are indeed formed by the process of stratification, e.g. sinter fabrics, botryoidal fabrics, colloform fabrics (Fig. 2). Here a real chemical precipitation takes place through partly rhythmic precipitation.

Furthermore, crystallite fabrics can cause stratified structure (e.g. parallel-fabric) through change of grain size. They may have formed primarily as such, or they may have attained their shape only by secondary crystallizations. All these layered fabrics with a more or less parallel structure are common in sediments, but they are ambiguous. Fabrics laid down chemically do not tell us very much; they yield some valuable additional data on the formation of the respective ore deposits, and a rounding of the overall picture, but they are useless for deciding whether and when a sedimentary ore deposit has been formed. Diffusion processes, of the type of the "Liesegang" rings, can also produce

Fig. 2. Colloform-layered fabric of Schalenblende (various greys) in the ZnS ore body of Fig. 1. Galena black, calcite white. Plain section, 1-cm scale with mm-divisions.

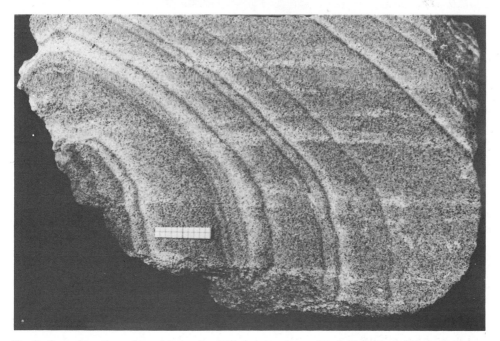

Fig. 3. Secondary formation of layers by diffusion processes with precipitation of Fe-hydroxide unconformably with respect to the stratification. The bedding (horizontal) is shown by fine bedding rich in clay, embedded in arenite rich in limonite. Post-Cretaceous cavity sediment, Bleiberg mining complex, Gailtal Alps, Carinthia. Plain section, 1-cm scale with mm-divisions.

patterns *similar to stratification* (Fig. 3). Such a secondary layering is not, however, formed by apposition on a solid surface, but frequently by rhythmic precipitation from diffusing solutions in the capillary space of sediments, or by seemingly stratiform, stationary mineral transformations, which are determined by diffusing solutions.

The favorable, unequivocal structures are those with a *polar* and if possible repeated series of strata, such as, e.g., vertical sorting of grains (graded bedding) or the repeated sequence of increase and decrease in ore mineral content with height in the individual layers (Figs. 4,5). Such polar series of strata make the interpretation of the formation as selective metasomatism an untenable assertion.

Fig. 4. Polar fabric formed by space-rhythmic apposition of sphalerite (fine-grained, light grey) in calcilutite. Repeated sequences of strong increase and decrease in ore content with height. Wetterstein Limestone, Lafatsch Mine, Tyrol. Plain section, 1-cm scale with mm-divisions.

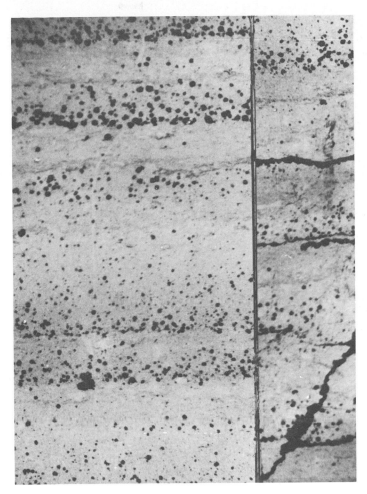

Fig. 5. Rhythmic polar apposition of galena (black) causes graded bedding in the Wetterstein Dolomite (Middle Triassic) of the Mežica deposit, Karawanken, Yugoslavia. Clay minerals grey, dolomite–microsparite white-grey. Thin section, nicols –, sample height 11 mm. On the right edge of the picture recrystallized PbS layers, syndiagenetic dislocation and unconformable PbS precipitate.

Another case of unidirectional, polar apposition, may also be put to optimal use. This is the case in which one and the same interface (e.g. the plane of apposition) is lined with ore grains of different sizes side by side, and where there is in addition, if possible, a characteristic coating by closely joining fine bedding or by laminae which clearly fill in the relief. Syndiagenetically grown crystals also cause similar polar fabrics (Fig. 6).

If need be, the evaluation may concentrate on the fine-clastic sediments, that is ore-containing sands (ore arenites) and ore-containing muds (ore lutites). In particular, terms like galena-arenite, galena-lutite, sphalerite-arenite, sphalerite-lutite, chalco-arenite, chalco-lutite, etc., may be used.

Fig. 6. Top: Polarity by mechanical apposition of resedimentated detritus of varying size onto the same plane. ZnS-rhythmite with dolomicrite, quartz, fluorite and clay minerals in Raibler Schichten (Karnian). Bleiberg-Kreuth mining complex, plain section.

Bottom: Syndiagenetically sprouted fluorite crystal causes polarity in the overlying finely bedded sphalerite—calcite—lutite by static pressure. Wetterstein Limestone, Lafatsch Mine. Plain section, 1-cm scale with mm-divisions.

Ore components of anisotropic form, that is columnar and tabular minerals, give some help in assessing stratiform fabrics. In the case of free mechanical apposition they were probably either put in a preferred orientation morphologically already in the process of apposition onto the s-surface[1], or they experienced this statistical orientation ("Einregelung") by early diagenetic compaction processes while the sediment was still in a plastic state (Fig. 7).

The explanations given so far will now be discussed in more detail and complemented using some examples.

---

[1] The technical terms in the study of fabrics coined by Sander (1930, 1936, 1948 and 1950) e.g., apposition, componental movement, s-surface, space rhythm, have been translated, defined and exemplified in Sander (1970).

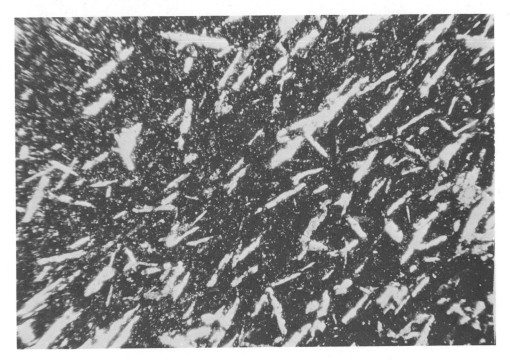

Fig. 7. Morphological production of a preferred orientation ("gestaltliche Einregelung") of baryte plates (white) in the fine bedding (diagonal). Wetterstein Limestone, Bleiberg mining complex. Thin section, nicols +, sample size 0.6 × 0.42 mm.

Pb—Zn—Fe—Ba- ores

Galena-sphalerite-marcasite-pyrite-baryte as mineral paragenesis of ore deposits in the Limestone C. Alps (Kalkalpen) permit a wealth of observations which are valuable in the study of the individual components or of all ore components taken together. Surrounding minerals are occasionally fluorite, quartz, calcilutite, dolomicrite (laid down biogenetically), sparry carbonates, clay minerals and bitumen. The appearance of the minerals existing in the form of single small crystals, often as fine bedding, and their sometimes obvious space rhythm[1] (Figs. 4,5,8) make the conclusion inevitable that either pre-existing complete small crystals were precipitated from a stagnant (isotropic), turbulent (statistically isotropic) or flowing (anisotropic) medium, or that at least crystal phases, in the process of growing or existing as nuclei, were deposited as soon as they experienced the influence of gravity. A choice of the two possibilities may perhaps be made from

---

[1] The technical terms in the study of fabrics coined by Sander (1930, 1936, 1948 and 1950) e.g., apposition, componental movement, s-surface, space rhythm, have been translated, defined and exemplified in Sander (1970).

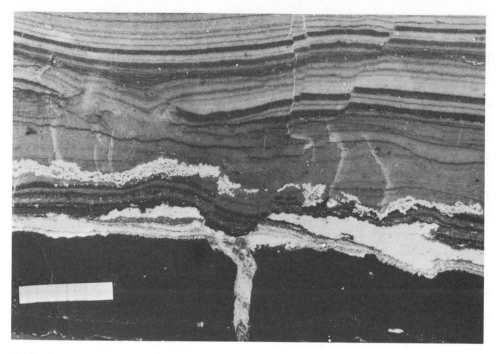

Fig. 8. The rhythmic-finely bedded lutite contains almost all minerals of the paragenesis: sphalerite, galena, marcasite, fluorite, baryte, quartz and calcilutite. Bleiberg-Kreuth mining complex. Plain section, 1-cm scale with mm-divisions.

details of the fabric, like structure of the series of sedimentary strata and of the crystals or of the grain boundaries.

The gravitational differentiation in the magma also produces a stratification which thus represents a mechanical process of apposition: e.g., layers of chromite in ultrabasites.

Inclusions in crystals by themselves do not, in my opinion, prove growth in a pelitic medium, but skeleton growth and growth inhibition with formation of fixed contours at crystal boundaries may be taken as sure signs of growth (continued growth) in a colloidal or crystalline medium (Figs. 9,10). Oblique- and cross-bedding with such fabrics are possible and they do not invalidate the above interpretation. Resedimentation, i.e., mechanical denudation of sediments already deposited and their re-apposition in the same sequence, vitally aids the proof of typically sedimentary fabrics, and makes possible a detailed account of the sedimentation processes.

The causes for *space-rhythmic* apposition — not necessarily, but probably derived from time-rhythmic causes — are, as is well known, to be found in the geochemical state of the environment in which the deposition occurs: concentration, temperature, pressure, pH-

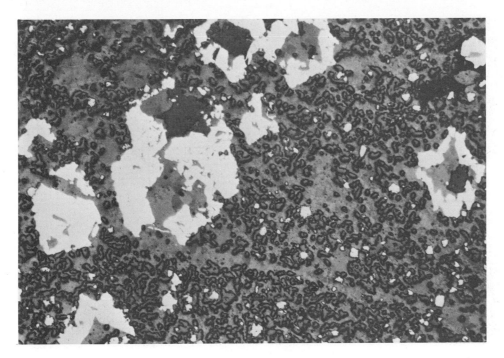

Fig. 9. Crystal skeletons of galena (white) in ZnS-containing quartz-calcilutite. Qartz standing out because of relief polishing. Wetterstein Limestone, Bleiberg mining complex. Polished section, nicols –, sample size 1.5 × 1 mm.

and Eh-values, and others. There is evidence from many finely bedded ore sediments (Fig. 11) that small fluctuations of these quantities resulted in large changes in the sensitive chemical-physical environment. Material content and solubilities as direct causes of the apposition of ore pelites (lutites) and psammites (arenites) may, however, in some cases be influenced by indirect causes, as for example by weathering events on the continent, transport conditions of weathering solutions, magmatic-hydrothermal supply of material into the region of sedimentation, and by suboceanic tectonic events.

From the joint occurrence of the ores with dolomitic carbonate rocks, which is obvious in many deposits, a genetic link was rightly suspected to exist. Salinous conditions may have led to synsedimentary formation of dolomite rock with precipitation of the ore from metal brines, and possibly the influx of Mg is related to the influx of heavy metals and may have caused metasomatic ionic exchange in limestones. Let us be reminded, however, that sedimentary ore parageneses have frequently occurred in limestones, too, and that in this case also calcilutite is built in in finely bedded series of ore-mineral strata (or vice versa) (Fig. 8).

Many data from finely bedded, rhythmic series of strata of Pb-Zn deposits make it probable that during uniform supply of carbonates (precipitation from suspension,

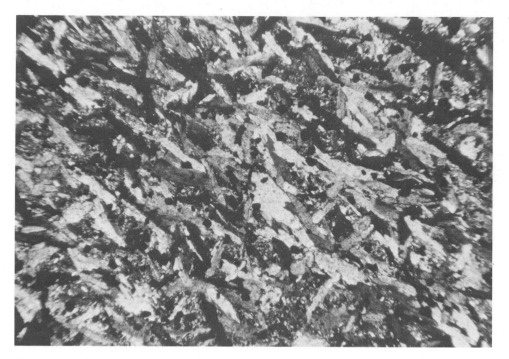

Fig. 10. Mutual growth inhibition exhibited by plate-shaped baryte crystals. Nevertheless preferred orientations (diagonal) may be recognized. From sedimentary ore fine beddings with quartz, fluorite and calcilutite. Wetterstein Limestone, Bleiberg mining complex. Thin section, nicols +, sample size approx. 0.8 × 0.55 mm.

mechanical apposition) time-rhythmic supply of material-induced space-rhythmic apposition. It appears that alternating, directly or indirectly magmatic events are more probable as time-rhythmic agents than are simply pH- and Eh-fluctuations with constant supply of metal ions from weathering solutions. This interpretation is corroborated by fabric data in mine exposures (compare section on geopetal fabrics).

The crystal contours frequently observable in ore minerals and extraneous surrounding minerals often exhibit a mutual growth inhibition (Figs. 9 and 10). From the study of the fabrics of such finely bedded ore bodies, it seems probable that precipitation of crystal nuclei and sinking of sufficiently heavy crystallites onto the plane of apposition (mechanical apposition) was the first componental movement and determined fine bedding and space rhythm. The viscosity of the uppermost sediment layer surely varied from case to case, just as did the supply of crystal nuclei, growing small crystals and fully grown crystals to the zones of formation of the growing sediment. Therefore, the characteristic features of synsedimentary apposition in a narrow sense, which are: alignment into a plane ("Einkippen") of components of anisotropic form into the s-surface during the first process of apposition, or mechanical production of a preferred orientation during

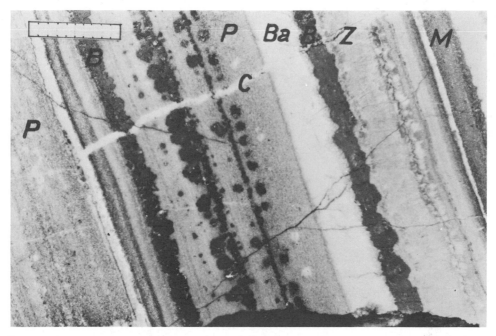

Fig. 11. Synsedimentary finely bedded ore in the Wetterstein Limestone of the Bleiberg deposit with galena (*B*), sphalerite (*Z*), marcasite (*M*), calcite (*C*), baryte (*Ba*) and ore lutite (*P*), the latter consisting of sphalerite, marcasite, baryte, quartz, fluorite and calcilutite. Plain section, 1-cm scale with mm-divisions.

compaction of a fluid still highly susceptible to componental movements (e.g. not yet solidified pelite), are reproduced with varying distinctness; in this way the morphological production of a preferred orientation ("gestaltliche Einregelung") of the crystallites (columns, lamellae) occurs. Only in rare cases will it be possible to discriminate between the two controlling processes, that is in those cases when monoclinic-symmetric apposition can be demonstrated, as it is formed by sedimentation from a flowing medium or on a strongly inclined plane of apposition. This observation is possible in s-parallel fracture faces or faces of thin or polished sections, respectively. With rhombic layering of higher symmetry an inference as to the mechanical process of "Einregelung" is not possible.

Let us be reminded, too, of the (production of a preferred) orientation due to outflow and shearing of still plastic, soft fine layers, which, however, ought to reveal itself by syndiagenetic deformation patterns in the area of concern.

Just as the mechanical apposition of very fine particles — no longer subjected to gravity — is possible onto planes of apposition of any orientation, so preferred orientation into the zones of formation may no longer be expected for fine-pelitic small crystals. This might yet be accomplished by compaction in younger sediments.

Inhomogeneous parallel-fabrics with extremely bulkily deposited ore- and other crystals without preferred orientation, possibly even with mutual growth inhibition, make it necessary to conclude that apposition of nuclei or rather "not fully grown" crystals was mechanical, but that these crystals did not form the complete primary sediment until the stratifying componental movement and subsequent chemical apposition (extension of the space lattice) had occurred. The state of the sediment at this stage of mature growth might well have been that of a finely dispersed lutite (muddy) environment.

Manganese ores

With increasing crystallinity of the deposited mineral it becomes more diffcult to interpret a primary sedimentary structure. Admittedly, no petrologist will doubt the sedimentary origin of rock-salt, potash-salt, anhydrite and gypsum deposits, but the situation changes seriously with the appearance of ore minerals in sedimentary parageneses, and various people have adopted unjustifiably complicated ways of thinking. Again and again colleagues who work on problems of the genesis of deposits have had to recognize the fact that diagnosis becomes increasingly risky with decreasing content of typical sediment (calcilutite, dolomicrite, clays, terrigenous and biogenetic arenites) and with an

Fig. 12. Parallel fine bedding, oblique bedding and syndiagenetic deformation in sedimentary rhodochrosite. Manganese ore deposit Dawin-Alpe, Lechtal Alps, Tyrol. Plain section, 1-cm scale with mm-divisions.

increase of materials extraneous to the sediments. Therefore, it is always recommendable to practise on a well-known material before proceeding to crystallized, that is diagenetically transformed layered fabrics, or to layered fabrics which are already primarily coarse-grained.

It is advantageous to study those metal-containing carbonates which occur as sediment in the pelitic state, with a subsequent tendency towards recrystallization and, therefore, towards increasing grain size. Magnesites, siderites and rhodochrosites could give insight into the genetic history from primarily external apposition and diagenesis to metamorphism.

Fig. 12 and 13 show sections from a rhodochrosite deposit (calci-rhodochrosite and mangano-calcite with Fe-content) of the "Lias-Fleckenkalk" of the Dawin-Alpe near Strengen (Lechtal Alps) with typically sedimentary fabrics. The manganese ore deposit has a thickness of more than two meters, its position is exactly stratiform. With regard to the predominantly very fine-grained manganese-carbonate, tectonic deformations prove to be postcrystalline. The fine bedding and oblique bedding are caused by rhodochrosite grain sizes varying between layers, as well as by varying contents of calcite, clay minerals, framboidal pyrites and recrystallized small pyrite crystals. The largely preserved detailed

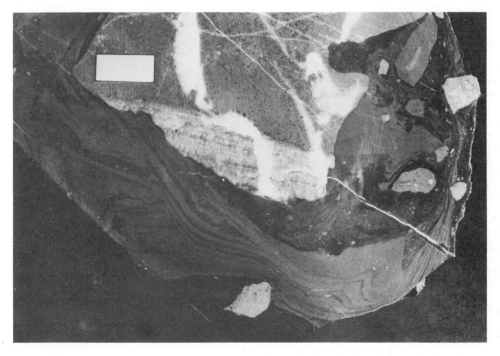

Fig. 13. Subaquatically deformed, finely bedded rhodochrosite-lutite in a sedimentary-brecciated part of the manganese deposit Dawin-Alpe, Lechtal Alps, Tyrol. Plain section, 1-cm scale with mm-divisions.

features of the zones of sediment formation point towards precipitation and mechanical apposition of a carbonate-pelite. It may be seen from the symmetry of the sediment fabric (undisturbed parallel-fabrics, cross bedding, erosion by currents, resedimentation, mechanical orientation) that sedimentation proceeded partly from stagnant and partly from tangentially flowing metal-containing solutions. Immediate subsequent recrystallizations (microsparite) and carbonate-metasomatisms are thereby not excluded. The syn-sedimentary process of mineralization is still further supported by the subaquatic deformations of the manganese lutite (Fig. 13).

Bauxites

In order to complement the topic "stratification", let us add a short note on an alumina-rich rock. The bauxite deposit of Präfingkogel (Sengsen-Gebirge, Austria), which has a prevailing red color and is generally only coarsely bedded, consists mainly of boehmite; in addition, varying quantities of mostly kaolinite, quartz and carbonates occur, as well as Fe-oxides, Fe-hydroxides, pyrite and rutile as coloring pigments. The aluminum-ore deposit, which has been modified tectonically into linearly extended bodies, lies as a sediment on a pre-Gosau (Upper Cretaceous) erosion trough (probably karst). Within the rocks (Hauptdolomit, Norian, Triassic) which form the ancient land surface a cavity is filled with a finely bedded sediment similar to bauxite. The cavity lies a few meters within the footwall of the bauxite deposit and is tectonically oblique, together with its internal sediment, as shown in Fig. 30.

Here the main mineral component is kaolinite with finely bedded layers of rhombohedral dolomite, also with quartz detritus as well as a colouring pigment contribution of haematite changing from layer to layer, lepidocrocite and rutile.

The fabric of the fine bedding was formed by wash-in and mechanical deposition of the pelitic mineral phases and probably corresponds to the mineral paragenesis of the footwall of the bauxite deposit.

Of general interest is the clearly geopetal internal accumulation (compare section on geopetal fabrics) with the parallelism of cavity stratification and surrounding stratification; also of interest is the characteristically polar form of the cavity: a trough, that is, erosion influenced by gravity! If the picture is not misleading, the stratification of the region with dolomite rocks shown here happened to be horizontal at the time of sedimentation in the cavity (= beginning of the formation of the bauxite deposit) despite previous orogenesis. However, the fact that only one surface of cut is visible may be misleading. Also, for this reason we cannot determine the spatial extent of the cavity.

Gels

Colloform fabrics, especially those that occur in sphalerite and wurtzite, haematite (kidney ore) and limonite (kidney form), cannot be considered as typically sedimentary despite their bedded, shell-like, spherical, but also clearly stratiform structure (Figs. 1,2). Therefore, they only play a minor part in a critical search for sedimentary ores. The reason is simple: bedded colloform fabrics are formed by chemical (partly rhythmic) precipitation onto an arbitrary plane of precipitation and are of central interest only in special cases in connection with mechanical apposition, for example when the products of chemical and mechanical apposition show up together in a series of sediment strata. In this way they point up a variety of changes or varying behavior: mechanical resedimentation of the chemically precipitated component, e.g. of the gel fabrics or syndiagenetic cataclastic processes of rigid colloform crusts in soft lutite or breaking of the former gels and further colloform encrusting of these fragments (Fig. 14). Thus it is always essential in such cases to find the colloform structure which, viewed by itself, is genetically ambiguous, in a cycle of clearly mechanical accumulation.

Multivarious kinds of development of colloform fabrics in genuinely sedimentary cycles are exhibited by the Pb-Zn deposits of the Kalkalpen. Their syndiagenetic transformations are also varied (plastic, cataclastic deformations).

Fig. 14. Syndiagenetic reworking and mixing in the ore lutite causes kataclastic processes in lithified crusts of Schalenblende (= syndiagenetic inhomogeneity breccia). Renewed encrusting of the Schalenblende fragments with Schalenblende. Raibler Schichten, Bleiberg-Kreuth mining complex. Plain section, 1-cm scale with mm-divisions.

NONPARALLEL APPOSITION: OBLIQUE BEDDING, CROSS BEDDING

These nonparallel layers reveal very important features in sedimentary ore deposits. Included among them are those series of strata and filling in and levelling the relief, which are distinguished by varying thickness of the laminae. Also to be counted among them are the vector-portraying fabrics which are formed by mechanical apposition from a flow. In this case a demonstrable denudation in the form of disconformities or various kinds of unconformities can affirm the sedimentary character of a series of ore strata even more strongly, as such fine structures would be expected to form only by external or internal, *free* growth of the sedimentary beds or laminae and not by internal metasomatic transport of material. Basically, however, we should at the same time not lose sight of the fact that existing fabrics may be influenced by secondary crystallizations. The figures which demonstrate a selective substitution of baryte by calcite (Fig. 15) or of baryte by fluorite (Fig. 16) are intended to stress the necessity for cautious interpretation and critical, objective study. A flow vector is portrayed and derivable by precise observation also in those cases in which tabular tiny crystals or columnar crystals lying along the direction of the current or slope were stacked on top of each other like roof tiles (= imbrication).

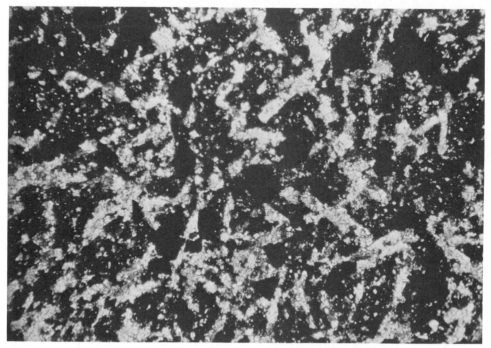

Fig. 15. Pseudomorphisms of calcite to baryte. The baryte plates (light grey to white) in the sphalerite-fluorite-quartz-calcilutite have been completely transformed metasomatically into calcite! Thin section, nicols +, sample size approx. 2.26 X 1.51 mm.

Fig. 16. Pseudomorphisms of fluorite to baryte. The baryte plates have been completely transformed metasomatically into fluorite. Fluorite-marcasite-clay fine beddings. Thin section, nicols −, sample size 10 × 7 mm.

Such features are to be expected when sedimentation occurred onto a rough-ground relief, including internal apposition onto sloping surfaces, e.g. washed-out clefts and cavity systems of various forms and orientations.

Oblique bedding that fills in the relief may be found frequently in the groove- and funnel-shaped cuts, caused by rupture and erosion, in the Triassic sea floor of the Alpine geosyncline (Figs. 17,18). When finely bedded alternating sedimentation onto the ground relief begins, visible angular unconformities result in various places at the interface between overlying and overlain strata. Sedimentation filling in the relief occurs with nonparallel apposition. This detailed fabric, typically sedimentary by itself, becomes even

Fig. 17. Stratiform groove-shaped erosion pattern with geopetal infilling of sedimentary ore. Wetterstein Limestone, Bleiberg mining complex. Layering from lower left to upper right. Finely bedded CaF_2- and SiO_2-rich bottom part (finely bedded black-gray), on top of it $BaSO_4$- and ZnS-rich calcilutite (finely bedded light grey).

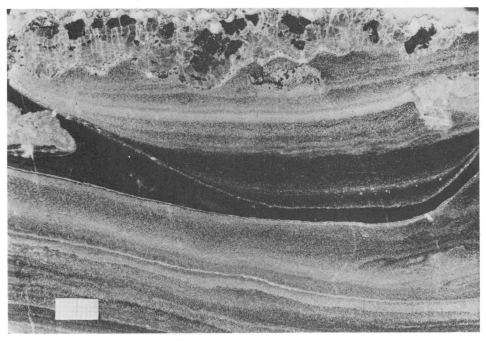

Fig. 18. Erosion on a small scale. Dark fine bedding rich in fluorite and quartz, light fine layers rich in sphalerite. In the upper part of the picture Schalenblende with galena. Wetterstein Limestone, Bleiberg mining complex. Plain section, 1-cm scale with mm-divisions.

Fig. 19. Mechanical, nonparallel apposition of sphalerite, baryte (plate-shaped), fluorite and quartz in oblique bedding. CaF_2-rich parts black-grey, $BaSO_4$- and ZnS-rich fine layers light grey. Plain section, 1-cm scale with mm-divisions.

more significant through the simultaneous portrayal of polar and geopetal ore apposition, for the relief formation and its levelling-out by progressive sedimentation causes polarity in the growing sediment. A sufficient number of such unambiguous examples are useful as top-and-bottom criteria (see section on "Geopetal fabric").

Oblique and cross beddings formed by tangential transport with monoclinic symmetry are worth mentioning as typically sedimentary. Within the cycle of denudating and sediment-forming processes, associated fine-beddings of ore may be interpreted with all details of crystallization and recrystallization of the components involved.

Nonparallel ore apposition is, of course, not restricted to sulfidic ore deposits: examples of baryte- and manganese-ore enrichments are shown in Figs. 12 and 19.

SYNDIAGENETIC DEFORMATIONS

Syndiagenetic (= paradiagenetic) deformations in ore bodies are best usable for solving genetic problems if the closely or remotely surrounding sediment, or parts thereof, were also affected by common deformations. Thus the causal correlation in structure, one zone of formation on top of the other, must stand out. The character of the deformations depends on the deformation velocity on the one hand and on the partial mobility of the material on the other. Sliding of a pelitic sediment portion on a sloping surface can also lead to complete destruction of the material and thus leads to resedimentation. Unsteady, rupturally caused deformations may occur, or steady, folding deformations. Both events may even have taken place in the same environment simultaneously. In that case hetero-mobile processes often act in the sediment, that is, mechanical inhomogeneities appear (Figs. 20,21), inducing the formation of inhomogeneity-breccia. The interplay of crystallization and deformation is one of the most important fabric-forming componental move-

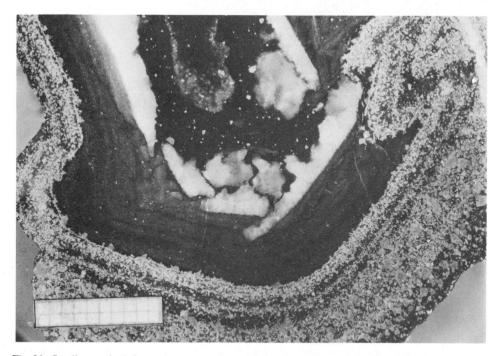

Fig. 20. Syndiagenetic deformation: crumpling of the soft sphalerite-marl-saprolutite, breaking of the rigid calcite fibre layers. Wetterstein Limestone, Lafatsch Mine, Karwendel range. Plain section, 1-cm scale with mm-divisions.

Fig. 21. The mechanical inhomogeneity becomes obvious through syndiagenetic deformation of the sphalerite sediment. Bituminous-clayey calcilutite is crumpled (center of picture and upper right), recrystallized rigid ZnS-crusts in the sphalerolutite are broken. Wetterstein Limestone, Lafatsch Mine. Plain section, 1-cm scale with mm-divisions.

ments, permitting an excellent description of the mineralization process in the many stages of the genesis of sedimentary deposits.

Frequently those transformations are called *early diagenetic events* which started with a more or less mouldable, soft sediment. In this connection many an aggregate of coarse-grained crystals, like sparry carbonate fibre layers, fluorite- and baryte-layers, betray their presence already at the time of deformation. Soft, partially mobile areas undergo plastic deformation, whereas under the same conditions rigid layers will fracture. It is often in this way that the question can finally be resolved as to which aggregates of crystals were

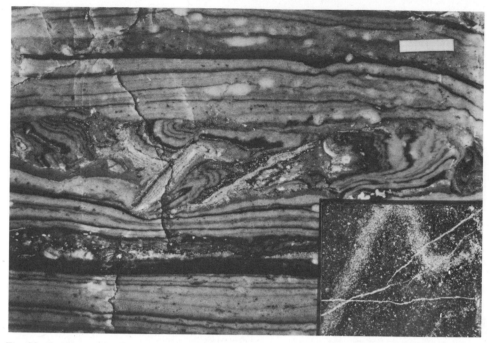

Fig. 22. Syndiagenetic flexural slide and shearing off of an ore lutite highly susceptible to compo-nental movements. The finely bedded sediment contains almost the entire primary mineral paragenesis of the Bleiberg complex. Dark layers rich in fluorite and quartz, light layers rich in sphalerite and calcilutite. Wetterstein Limestone, Bleiberg-Kreuth mining complex. Plain section, 1-cm scale with mm-divisions.
Lower right: thin section (7 X 6 mm) from a surface of flexural slide. Nicols X.

already present at the time of subaquatic deformation. The deforming stirring movement also yields information on the progress of recrystallization and the state of solidification of the sediment (Figs. 20,21,22).

The informative value of the fabric is the larger the more clearly it shows alternating primary fine bedding, because from this latter flexural folds, flexural slips, shear folds or shearings at joints may be recognized (Figs. 13,20,21,22).

Post-crystalline shearing with regard to a certain mineral component or paragenesis with submarine tectonite formation may be pre-crystalline in relation to a second or, generally speaking, later sequence of sedimentation. Because of this, the establishment of an age sequence of mineral generations (i.e., its paragenesis), possibly with varying fabrics, could become feasible (e.g. Fig. 14). However, not all folds of large-scale character and well-formed small-scale foldings are of syndiagenetic origin. An uncertainty of judgement exists particularly in those cases where bitumen and layers of clay are present as fine intermediate layers, increasing the deformability of the sediment even in the solidified state. Errors are quite possible here and therefore various criteria should be used to enable us to date the deformation.

There is no doubt about early diagenetic folding or ruptural displacement if the deformation is restricted to one layer and the above-lying zones of formation show characteristic polar superposition, proving a previously formed ground relief (Fig. 22, 23). Also, limbs of folds cut off by denudation, or generally by mechanical erosion after deformation and resedimentation of the affected portions, give unequivocal proof of the subaquatic nature of the deformation. Important as subtly differentiated patterns of deformation are in ore sediments, their dating may nevertheless be risky, or even their diagnosis may be difficult in cases where ruptural dislocations are involved. Here it is impossible most of the time to distinguish between syn- or postdiagenetic, pre- or syn-tectonic, for only in exceptional cases do criteria exist which permit the geologic se-quence of events to be assessed. Most of the time, however, distinctive marks needed for this decision will be missing or will be so few as to leave the history of their origin still obscure. Here the implications of a right or wrong decision become clear. In these cases it is more useful to interpret a fabric cautiously than to overemphasize the interpretation where it is not supported by suitable distinctive marks.

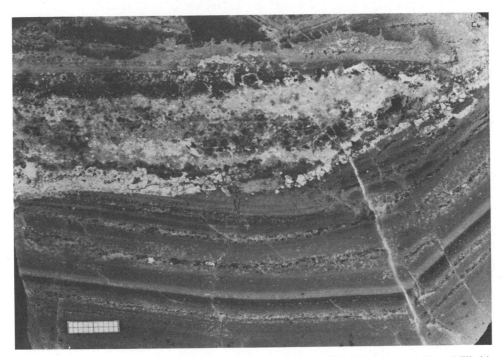

Fig. 23. Ground relief formed by syndiagenetic deformation is partly eroded (center left) and filled in by sedimentation. Lower half of picture: fluorite-calcilutite fine layers. Upper half: coarse-grained sphalerite-fluorite-layers with calcilutite (sphalerite partially transformed into white hydrozincite), overlain by calcilutite with fluorite and sphalerolutite. Wetterstein Limestone, Lafatsch Mine. Plain section, 1-cm scale with mm-divisions.

Fig. 24. Vein unconformable with respect to stratification in Middle-Triassic Wetterstein Limestone, brecciated, with dominating baryte-mineralization (white). Deformation and mineralization turned out to be syndiagenetic. Rudolf Mine, Bleiberg mining complex.

Ore-containing *deformation breccias* occur frequently in deposits, but they are ambivalent: for example, the vein breccias in fissures and disturbed zones (Fig. 24), tube-shaped breccia ore bodies, and also the brecciform, stratiform complexes.

In the first place it is important to distinguish between the two main genetic types of breccias according to Sander (1936, 1948–50, 1970): are they sedimentary breccia or deformation breccia? Furthermore, the distinction must be made between syndiagenetic deformation breccia (in part submarine tectonite) and postdiagenetic, orogenetic-tectonic deformation breccia; especially when the scientific results should be evaluated for purposes of prospecting and exploitation.

When in doubt, one must try a tectonic analysis of the fabric, clarifying the pattern of deformation on the large scale, as well as a sedimentological assessment of the aggregate of layers to elucidate the history of mineralization and the time span of its occurrence. It is the purpose of tectonic analysis to find out whether a brecciform ore body (e.g. cleft- or vein-mineralizations) may be topologically derivable from an analyzable tectonic plan of formation or not. It is not the purpose of this paper to explain the necessary operations. Some specific examples will, however, be chosen, which from a sedimentological point of view contribute to the clarification of mineralization.

The decisive piece of data comes from the geopetal internal sediment (see section on "Geopetal fabrics"), which act as a "geologic bubble level", determining the relationship between the position of the internal-s and external-s as stratum or fine bedding. The time span in which the discontinuity fabric was formed is thereby narrowed down or even fixed. The situation becomes particularly favorable when the mechanical internal sediment is itself an ore sediment and shows perfect agreement with the externally laid down ore lutites in both mineral content and fabric. This is the case, for example, in the Pb–Zn deposit of Bleiberg-Kreuth (Gailtaler Alpen), which will be discussed below, as well as in the sections "Internal or external apposition" and "Nontypical ore fabrics".

Mineralized fissures, which are statistically mutually parallel turn out to be, after tectonic analysis, pre-tectonic relative to the Alpide orogenesis. The mineralization agrees with the paragenesis of the externally formed stratiform ore bodies, but it consists overwhelmingly of chemical internal apposition, mostly attached to a wall with and without metasomatism. There are very rare cases in which finely bedded ore was formed locally as mechanical internal sediment in the fissures and still preserved as remainders (Fig. 25). Internal and external strata lie parallel despite tectonic displacement of layers and agree also in the extent of finish of the components involved. This occurrence proves that as a result of the parallelism of internal and external strata, syndiagenetic or tectonic sloping of the layers had not yet occurred. Thus a time span of Triassic to Cretaceous would result for the finely bedded internal mineralization (beginning of the Alpide orogenesis).

The fact, however, that internal and external finely bedded ores may melt into one another in some ore deposits, that is they are connected (Rainer 1957), proves the synsedimentary structure of the vein systems together with their mineralization. An analogous and equally rare pattern is exhibited in the irregularly tube-shaped, fissure-shaped and lenticular brecciated zones of the so-called "Westschachtscholle" of the Bleiberg-Kreuth deposit. Here a Middle-Triassic carbonate rock complex, not precisely definable in its form and extent, has been multifariously, irregularly and cataclastically destroyed (Fig. 26) and broken up by a network of ruptures down to small scales. This network of joints which runs across layers and across facies areas, forms a preferred path for mineralizing solutions. The minerals of the paragenesis of deposits have been chemically-internally precipitated in the network of joints: partly in tension joints and solution cavities freely attached to a wall, leading to complete healing of the joints, partly with metasomatic ion exchange in the surrounding sediment. In extreme cases mineraliza-

Fig. 25. Ore vein ("Bleiberg type") unconformable with respect to stratification with sphalerite, galena and baryte in Wetterstein Limestone. Narrowly localized geopetal ore lutite in fissures with ZnS, PbS, $BaSO_4$, SiO_2, CaF_2 and calcilutite. Inset in the lower left of the picture: detail from geopetal fissure infilling of 1 cm width. Fissure boundary formed by chemical precipitation of PbS attached to a wall. Right boundary of fissure is missing. Parallelism of internal-s and external-s. Plain section, 1-cm scale with mm-divisions.

tion has practically replaced the carbonate rock, probably in the regions of supply of metal solutions, and sphalerite rocks with a volume far exceeding 1000 m^3 have been formed.

Such networks of mineralizations and ore breccias may be found in numerous ore deposits of varying origin. The distinctive feature of our example from the Bleiberg-Kreuth deposit lies in the fact that here the time span of mineralization can be narrowed down considerably by demonstrating the presence of geopetal ZnS-fine beddings as

Fig. 26. Breccia region darkly colored because of fluorite-quartz-sphalerite-mineralization (syndia-genetic deformation breccia) with irregular contours in light-grey Wetterstein Dolomite. "Westschacht-scholle", Bleiberg-Kreuth mining complex.

internal sediment in solution cavities of the network of joints. The mechanical internal apposition of the sphalero-lutite turns out to be Triassic and therefore the series of ore strata proves to be synsedimentary in a larger sense.

This example should not, however, induce us to consider each ore-containing deformation breccia as syndiagenetic! It is only pointed out here that tectonite may be formed as deformation in the diagenetic sediment stage.

RESEDIMENTATION

There is no better proof for the mechanical part in the process of ore apposition than the evidence for resedimentation. B. Sander has defined resediments as repeated apposi-tions or multiple appositions which come about through denudation and renewed apposi-tion within a series of sediment strata. Increased grain size is a consequence of this, and one will mainly look for sandy and finely conglomeratic resedimented material. It is essential in this connection to identify denuded and re-apposited materials. Therefore, we will only consider mechanical resedimentation and not chemical resedimentation. There is no doubt that chemical resedimentation occurs frequently, but it cannot be identified in repeatedly laid down materials.

Uncertainty about resedimentation can only occur where no lamination has developed, that is where we have an ore sediment unsuitable for sedimentological analysis due to lack of texture, and where, therefore, the possibility exists of confusing it with a sediment which has been reworked syndiagenetically to form a homogeneous "massive" deposit (syndiagenetic inhomogeneity-breccia). When in doubt, it is therefore still necessary to find the plane of apposition onto which the mechanical re-apposition has taken place, in order to identify a resediment. The mechanical denudation of material (fragments, ore sediment detritus) and its renewed sedimentation should have occurred, therefore, during

Fig. 27. Internal sediment rich in minerals in the solution cavity of the cataclastic Wetterstein Dolomite. The internal sediment contains as fine bedding (s) varying amounts of sphalerite, galena, pyrite, marcasite, quartz, fluorite, baryte and dolosparite, as well as resedimentated dolomite rock detritus from the cavity ceiling. "Westschachtscholle", Bleiberg-Kreuth mining complex. Plain section, 1-cm scale with mm-divisions.

Fig. 28. Resedimentary breccia, formed externally on the sea floor by breaking off of (in part already mineralized) rock components after submarine wall formation. Mine exposure, Raibler Schichten, Bleiberg-Kreuth mining complex.

sediment growth. Regressive and progressive processes may also follow each other repeatedly.

The immediate causes for the occurrence of resedimentation are: flows, changes of shape of the sediment boundary, ruptures and drying cracks, that is to say, events which lead to detritus formation, breaking off of particles in small cavities and of coarse blocks in cavities (internal resediment, Fig. 27), as well as tectonic dislocations on the sea floor, perhaps with formation of cliffs. On these walls resedimentary breccia may be formed by falling blocks and crumbling after of small fragments (Fig. 28, and section on "Resedimentary breccias").

GEOPETAL FABRICS

Internal and external geopetal fabrics yield the vertical orientation of the zones of sediment formation. In particular cases even more detailed information about the layering or movement of strata may be obtained. It is common experience that statistical methods are accurate to within ±5%. This kind of accuracy is feasible with mechanical internal apposition in small cavities, in which a finely bedded-s or a sediment level (lutite level) has been portrayed under incomplete internal sedimentation, independently of the environment. The remaining space remains empty or, as is most often the case, becomes filled with crystals which are chemically attached to a wall.

The internal apposition may be formed as a sand trap, that is, by mechanical apposition of detritus which has been washed into the cavities. We may, however, also be dealing with mechanical apposition of tiny crystals after precipitation from the solutions contained in the cavities, whereby in this case also, and as in the case of the external sediments, precipitation is determined by chemico-physical conditions. Carbonaceous infillings of small cavities with dimensions of millimeters are numerous in the algal stromatolites, infillings consisting of ore lutites abound in the remaining spaces of colloform ores, e.g. in schalenblende- and wurtzite-aggregates (Fig. 29). Chemical erosion by solution may also contribute substantially to the formation of small cavities. Its action may be recognized from irregular contours of the small cavities which are discordant relative to the layers of Schalenblende.

Geopetal apposition enables us to obtain evidence as to the genetic history of deposits from internal sediments in cavities of decimeter-, meter- and even larger scales. The finely

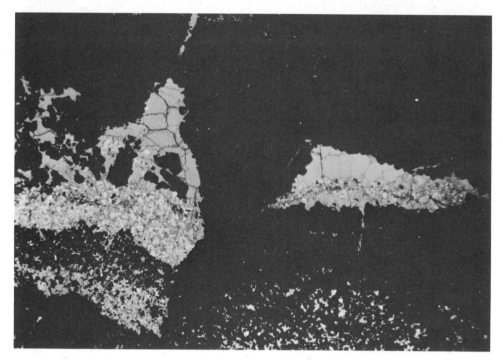

Fig. 29. Geopetal fluorite-quartz-sphalerite-lutite as partial filling in small cavities within Schalenblende. Remaining volume crystallized with sparry fluorite. Raibler Schichten, Bleiberg-Kreuth mining complex. Thin section, nicols ⅹ, sample size 10 × 7 mm.

Fig. 30. Erosion cavity in Hauptdolomit (Norian, Triassic) geopetally filled with finely bedded alternating layers of kaolinite, dolomite and quartz, and with differing pigmentation by Fe-oxide, Fe-hydroxide and Ti-oxide. Internal-s and external-s from lower left to upper right. Bauxite deposit Präfingkogel, Sengsengebirge, Upper Austria. (Compare text p. 310 and below.)

bedded internal sediment in Fig. 22 consists of the same mineral paragenesis which also characterizes the host sediments of the same aggregate of layers, but this is not visible in the exposure. On the small scale we have subaquatic, syndiagenetic crumblings of the fine bedding, which latter has a general orientation parallel to the large-scale stratification and banking structure of the sediment aggregate. The geopetal infilling of an erosional cavity in a dolomite rock (Fig. 30) comes from a bauxite environment. Comparing Figs. 17 and 30, it is very remarkable to see how two genetically quite different fabrics can be so similar in appearance (compare section on bauxites).

A further important, if odd, feature of geopetal ore apposition occurs in the unconformable ore veins of the "Bleiberg type". The vein system (compare section on syndiagenetic deformation) consists of fissures running approximately normal to the stratification with a length of a few hundred meters and a depth of roughly 120 m, or even as much as 200 m. It would be impossible to date the strongly dominant chemical mineralization, consisting of sphalerite, galena, marcasite and pyrite, which is attached to a wall, if geopetal ore lutites in the fissures did not give us information through their fine bedding, acting as a "bubble level". The finding of finely bedded internal mineral apposition in fissures of centimeters to a few meters width (Fig. 25) and observable only very rarely, does indeed prove the syndiagenetic character of the vein mineralization. It does not, however, enable us by itself to determine whether the solvent flow was exclusively ascending or descending. When evaluating the parallelism or angular unconformity, respectively, between the general stratification and the internal pelite level or internal fine bedding in all these cases of geopetal internal ore apposition, one needs to be careful. This is because subaquatic dislocations (affecting, perhaps, only very limited sections) are very well possible, and therefore the interpretation of the geopetal sedimentation must not be made with too narrow a field of view. When in doubt, one will have to try a statistical data treatment. Admittedly, the same is true also for the external geopetal sediments to be discussed below.

The examples of internal geopetal apposition discussed hitherto started from a situation where the ore pelite forms the geopetal fabric, while the possibly remaining space becomes filled with carbonates or other non-ore minerals, which are chemically internally precipitated. The opposite case can also contribute to the genetic clarification of a mineralization, e.g., a small cavity, like bivalve or gastropod shells, whose lower part is covered with ore-free carbonate lutite, thus forming the geopetal sediment, while the remaining space is filled with a mineral involved in the ore paragenesis. This role may, for example, be played by anhydrite, which in the Bleiberg deposit is thought to have originated syndiagenetically in a saline environment.

External geopetal ore apposition is produced by the numerous corrugated ocean-floor surfaces, which are frequently formed by subaquatic erosions and slides. A sediment which fills in and levels troughs and fissures shows very clearly the top and bottom of the zones of formation.

Polarity, that is unidirectional structure in the sediment, is already present because of the submarine relief; the geopetal ore sediment, however, is most easily interpreted when fine bedding is present (Figs. 17,18). In this case, though, it is not advisable to make too much of possible unconformities within a stratigraphic section, as the external apposition may have been influenced by tangential transports (deposition from flows). Possibly apposition with nonparallel bedding will occur on the downslope trough edges, adapting to and filling in the ground relief. Another possible source of misinterpretation should not be forgotten, i.e., the effect of studying an object along one cut only, when it should be studied in two or more planes of projection.

In my opinion, some of the fabrics described under the heading "geopetal sedimentation" allow us to draw certain conclusions regarding the origin of the metal-containing solutions. The frequently observable erosion-shaped extension of lenticular and oblong layers of ore, of small ore channels and ore troughs into the ocean floor, as well as their topological distribution, suggest that corrosive solutions were active at times (compare Fig. 18). These fabrics point towards a hydrothermal origin of the solutions rather than the action of weathering solutions from the continents, especially since indications from the regional geology and paleogeography of the Ostalpen-region also point toward remote associations with Triassic volcanism.

Discussing the genesis of ore deposits, we will find enough cases in which an interpretable geopetal fabric is not sufficient for precise dating. These are the cases in which we must discuss the question of internal or external geopetal fabrics.

INTERNAL OR EXTERNAL APPOSITION?

Conformability of internal-s and host rock only tells us that no displacement of strata occurred before internal apposition took place. This result, however, may determine the date of mineralization, that is, the time span between internal sedimentation and later orogenesis, only within wide limits. Therefore, the interpretation of the internal geopetal ore sediment often does not, by itself, suffice to date the mineralization. The question becomes even more pertinent when the internal or external origin of the geopetal ore sediment is in dispute. Thus it is very important, in most cases even decisive for explaining the genetic context, to answer the question "internal or external sediment?". Frequently we are even dealing with both cases; that is, internal and external mineralization in the same deposit, side by side or one on top of the other in the series of strata.

As an important indication of external ore apposition we may consider gradual transitions of ore fine beddings into ore-free fine beddings or vice-versa, if possible in the continuation of a footwall, hanging-wall and strike-wall (Fig. 31). It remains to be proved, however, that these gradual transitions were not caused by metasomatic ion exchange. Metasomatism often shows a puffy, diffuse structure, which naturally also leads to gradual transitions. Therefore, each case must be considered very carefully.

It is a strong indication of metasomatism if an extraneous mineralization cuts across a finely bedded fabric unconformably with respect to the stratification. (Naturally this metasomatism may have occurred early-diagenetically and may still belong to the synsedimentary mineralization in a general sense.) If, however, we find a finely bedded ore mineral paragenesis placed well within the normal sediment, and if the gradual transitions from ore-fine beddings to no-ore-fine beddings are present also, we can assume that metasomatic mineralization in the sense of mimetic metasomatism is improbable. Such selective metasomatism can be precluded altogether where the finely bedded ore occurs as oblique or cross bedding, where we find resediments of the ore-fine beddings, and where early diagenetic deformations have affected the finely bedded series of ore strata (Figs. 22, 23). A sharply delineated, all-around change of material, possibly even unconformable, is an indication of cavity sedimentation, if not a conclusive one. A reliable diagnosis may be performed if the sediment aggregate was not changed by diagenetic recrystallizations. Another disturbing factor for the diagnosis is the crystallization of ore- and attendant minerals which frequently shows coarse-grained, rough structure and which can also cause a sudden, sharply delineated change of material. For comparison, let us be reminded of layers and lumps of hornstone which for the most part are separated clearly, with sharp boundaries, from the carbonate sediment. Here no one will doubt the synsedimentary existence of silica.

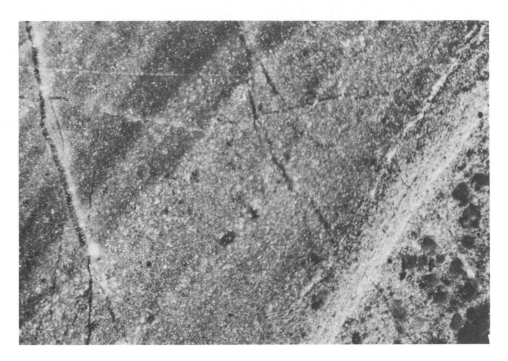

Fig. 31. Gradual transition of sphalerite-rich fine bedding (right, grey-white) to ore-free dololutite fine bedding (left). Gradual vanishing of ZnS-lutite. Coarse ZnS-grains grey-black (lower right). Raibler Schichten, Bleiberg-Kreuth mining complex. Polished section, illuminated obliquely from above, sample size 10 × 7 mm.

In some circumstances cavities, mostly only local ones, may be formed subsequently in ore bodies by syndiagenetic solution-mobilization. In this way similar internal events may happen as with cavity sedimentation in previously "free" spaces.

In any case fragments may break off from the cavity ceiling. The overlying stratum may also break off if, for example, a stratiform ore body which has not yet hardened completely is being overlain as a heteromobile region by an already hardened sediment (e.g. Alpine stromatolite). Such an inhomogeneity of hardness can lead to the breakdown of a layer of rock above a still easily deformable substance.

The repeated placement of fragments into the sediment which are derived from the cavity ceiling is an indication of a cavity infilling.

From the description of sedimentary ore fabrics discussed up to now we can see that while a number of occurrences are considered to be typically sedimentary, an ideal proof for the synsedimentary overall process can only be obtained by several criteria in combination. An improvement of the interpretation will be possible if, apart from the primarily external-sedimentary events and syndiagenetic changes, postdiagenetic crystallizations and tectonic transformations are also included in the overall assessment of the ore deposit. Of course, the changes in oxidation and cementation zones and generally all weathering-transformations and new formations also are part of the sedimentary cycle. These further processes of weathering, transport, dispersion or renewed enrichment of ores, however, are not the subject of this chapter.

Still, a section on nontypical sedimentary ores seems important, in which somewhat odd results will be described, which are ambiguous if viewed as isolated occurrences.

NONTYPICAL SEDIMENTARY ORE FABRICS

In the following we will describe and interpret occurrences of fabrics which, in the course of an overall evaluation of a deposit, prove to be sedimentary, or rather "synsedimentary in a larger sense". These fabrics are, however, unclear, uncertain and frequently not at all interpretable, if viewed by themselves.

Recrystallized aggregates

Those coarse-grained recrystallized aggregates which are related to definitely mechanically laid down ore minerals or their attendant minerals represent a situation which is still relatively easily understood (Fig. 5). The spatial interrelations, once clarified in a deposit, easily permit a rather generous interpretation of the analogous data. Certain ore minerals have a tendency toward recrystallization. Therefore, they may be found only very rarely as single grains in a stratiform position like typical sediments: e.g. galena, chalcopyrite, fahlerz, carbonate ore minerals, baryte and others.

The following features facilitate the analysis of the diagenetic crystallization process: proven syndiagenetic folding and ruptural transformations, that is postcrystalline deformations in a young sediment, moreover resedimentation of areas already recrystallized, and their placement into oblique- and cross-beddings.

There is no doubt that the clarification will also be important for distinguishing between post-diagenetic and orogenic-tectonic crystallizations. Figs. 5 and 11 refer to this topic.

Syndiagenetic network mineralization

In stratiform sedimentary ore deposits, areas mineralized along fractures also appear to be frequent. This mineralization, attractive as it often is economically, as a rule offers little incentive to the scientist because the fabrics are ambiguous. They are, therefore, paid too little attention in the genetic discussion.

Irregular joints and fractures form a network with a mesh size ranging between millimeters and meters. The spatial extent is sometimes not easily recognized. Stratiform network zones prevail, as well as tube- and vein-shaped areas. Sometimes the connection with a genuine stratiform ore body is obvious, sometimes there is certainly no connection. Possibly the mineralization which follows the fractures will make the entire rock formation look brecciated. And indeed, it is a mineralized deformation breccia. The fractural deformation has only taken place in the preexisting sediment aggregate of variable hardness. The time, however, when deformation and mineralization occurred, has to be determined in each case.

There is no foolproof procedure for analyzing the processes. Syndiagenetic to postdiagenetic events must be considered. Therefore, it is necessary to perform either investigations using methods from the study of fabrics and from sedimentology, or tectonic analyses of the axial and planar fabric (bedding, cleavage) and sometimes also analyses of the grain fabric. The emphasis will be on the one or on the other type of investigation and in some cases all of these analyses must be coordinated. They are also the basis for geochemical and geophysical work, if required. The geochemical approach to problems all too often lacks basic information on the fabric structure.

The following features are to be considered as evidence that in the case under discussion the fracture was, in fact, synsedimentary: confinement within a layer, deformation confined to a banking structure, connections with external-sedimentary ore deposits and joint tectonic deformation.

Mineralization of fracture networks may not be accessible to dating if the brecciform ore bodies have an irregular extent. A geopetal internal sediment (compare section "Geopetal fabrics" and Fig. 27), as it happens to be observable in the Bleiberg deposit, offers a possibility for clarification. The healing of the joints of the deformation breccia consists almost exclusively of Schalenblende attached to a wall, with carbonate, some galena, fluorite, quartz, pyrite and marcasite, all laid down chemical-internally. Admittedly, an expansion of the joints, caused by solution, was often connected with this mineralization. In addition we have small to considerable metasomatic ion transports, so that as a whole the original form of the brecciated components was often changed, while the brecciated appearance was even accentuated.

An unusual rarity is the occurrence of geopetal mechanical internal apposition in a solution cavity of the deformation breccia. The parallelism between the internal-s and the bedding of the surrounding sediment tells us that no significant alignment of strata into a plane had occurred in this region before the cavity infilling, but this still leaves a long time span for the process of mineralization. The further fact that the internal ore-fine beddings and ore-fine beddings of the external-sedimentary ore deposits show similar paragenesis and fabric, makes early diagenetic network mineralization probable, although direct connections are lacking.

Finally, a further feature may be used for safe dating: fragments of the ore-containing deformation breccia are found as external resediment in a polymictic breccia, a few dekameters stratigraphically above the series of strata. Therefore syndiagenetic, ruptureforming deformation and syn- to post-deformational mineralization are beyond doubt.

It should be obvious that despite the successful genetic interpretation of this showcase for the study of fabrics, we cannot in addition make a precise statement as to the origin

NONTYPICAL SEDIMENTARY ORE FABRICS

of metal-containing solutions. Basically, in all cases of syn- and postdiagenetic internal mineralization, the possibility of fluidization of the primary, still unconsolidated host sediments and of reworking must be taken into account. This problem is much disputed and its geochemical aspects are still poorly understood for various compounds. In the example described above a primary connection with the overall mineralization could be made likely because of the common occurrence of mechanically laid down internal and external fine beddings (Fig. 27), and large, extensive, secondary transport of material could be excluded. This does not, however, speak against the possibility of local transports by solution and recrystallizations which may be frequently demonstrated.

Resedimentary breccias

In parts of the Bleiberg-Kreuth deposit mineralized and non-mineralized coarse-clastic brecciated areas are exposed. Deformation breccia may be discounted immediately because of the polymictic composition, the very heterogeneous size distribution and the strongly differing orientation of the components relative to each other; furthermore, because of the frequently dense packing with many points and interfaces of contact, as well as mutual imprints without friction detritus. The fact that by far most brecciated fragments consist of the same finely bedded bituminous clay-dolomite sediment as the surrounding sediment, suggests that resedimentation occurred during the Karnian stage of the Middle Triassic.

A particularly striking and genetically important occurrence may be seen in the participation of extraneous light-dolomite rock fragments, originating from a sediment which lies stratigraphically a few dekameters below. This somewhat strange and surprising explanation is confirmed by the fact that some of these extraneous fragments already show the typical sphalerite mineralization, which, therefore, they had already acquired at an earlier stage of the formation of sediments and ore deposits.

As a morphological illustration one can imagine submarine displacements of the sea floor with formation of cliffs and terraces, also with resedimentation caused by collapsing walls which are partly subaquatic and partly perhaps extending above the water level. Such movements, probably accompanied by marine earthquakes, could have been triggered by crustal displacements at preexisting inhomogeneities of the geosynclinal underlying strata.

Syndiagenetic vein breccias, vein mineralizations

Vein mineralizations which are unconformable relative to the stratification, are usually classified not as sedimentary, but as orogenic-tectonic features. This interpretation need not always be correct, and therefore it is advisable to assess vein deposits with caution. This is shown by a case of subaquatic fissure formations on the ocean floor, with depths down to more than 100 m, demonstrated in the Bleiberg deposit. The orientation of the ore-containing fissures runs normal to the banking of the sediment, the width of the vein filling reaches several meters. Frequently we have fissure systems, consisting of some or many parallel ore veins cutting through several layers, and associated with them are also small joints from millimeters to only centimeters wide. The veins are spatially confined to a certain complex of layers, but individually they penetrate it to varying depths and towards the top they run into strata-bound ore bodies of varying stratigraphic orientation. In the immediate vicinity of the mineralized veins cataclastic areas and in the veins themselves breccia zones have also frequently been formed. Although these vein mineralizations are typically mineralizations of large planar extent, it is not unusual to find brecciated regions which are similar to ore bodies found in the so-called network mineralization. This is no accident, for both fabrics may be derived from syndiagenetic-ruptural events, as is shown here.

We must suspect the same causes of deforming events to be active which have also led to the brecciation of sediment banks and irregular cataclastic zones with network mineralization. Associations with the submarine floor displacements (see section on re-sedimentary breccias) are also probable. Sloping of the sea floor in the linearly extended sedimentation trough of the "Bleiberg special facies" could have favored tectonic fracture with formation of tension joints.

Submarine tectonics, therefore, has in one case led to mainly vertical clefts in the sediment, but in the other case, to breaking of the sediment plate, over the depth of the banking, into a giant mosaic. Transitions between the two types occur in the groove- to tube-shaped brecciated zones of irregular spatial extent. The state of the fissure walls, the contours of the brecciated zones and of the fragments are indications that unsteady ruptural deformations affected a more or less lithified rock. Naturally, the sections of the veins close to the surface, which lead into the sediment surface, form an exception.

The mineralization comprises the entire mineral paragenesis of the deposit with ore and attendant minerals; the apposition, however, was overwhelmingly chemical, that is independent of gravity, attached to a wall and closing joints. In part strong metasomatism has also taken place and has changed the cleft rims and fragment contours.

The very rare discovery of finely bedded mechanical apposition of ore lutite in fissures permits that stunning explanation of fundamental importance, according to which the all-encompassing vein system and the many small accompanying ruptures were still formed suboceanically, i.e., while the camgit (= Ca-Mg rock, Sander, 1936) sediment and its stratabound ore deposits were still continuing to grow. The internal ore lutite, that is, may be evaluated as geopetal sediment due to its fine bedding, and this fine-bedding-s agrees in orientation with the layering which is interrupted by the veins.

The occurrence of the geopetal sediment would not be accessible to interpretation if the present-day sediment complex were lying horizontally. The tectonic steep slope of the limestone-dolomite layers together with their unusual internal sediment enables us to make the elegant statement on the date of formation of the fissure system.

Finally, the complex of genetic problems is completely resolved due to the fact that the mineral content and fabric of the internal, cleft-filling lutite agree with the externally sedimentated ore-fine beddings, and that in several cases veins were found to run into the former sea floor, with a corresponding transition from the strata-bound sedimentary ore to the vein material (Rainer, 1957). Accordingly, several cycles of fracture deformations and mineralization phases may be demonstrated. The ore veins were sheared in the course of the Alpide orogenesis.

From these unambiguous data we may draw the conclusion that all the Bleiberg ore veins which are unconformable relative to the stratification were formed during the Triassic and were mineralized in several cycles during the Middle Triassic together with the formation of the strata-bound ore deposits.

At this point, the rarity and the rather accidental discovery of such internal fabrics suggest that there exist many synsedimentary vein mineralizations in sedimentary ore deposits as well as in deposits which cannot be identified as sedimentary. In that case the following features could at least give reason for suspecting veins to be synsedimentary: their spatial confinement to certain stratigraphic horizons and the pretectonic formation of the mineralized vein system which may be checked by tectonic analysis.

CONCLUDING REMARKS

In most of the preceding sections an attempt was made to describe typical sedimentary ore fabrics as they have been preserved in favorable cases, proving synsedimentary ore apposition. Not all relevant data were thereby given sufficient consideration. For example, biodetritus in ore sediments is not mentioned; this is a rare, but certainly important aid in genetic classification of an ore precipitate (Fig. 32). It will not always be possible to furnish one or another characteristic piece of evidence. All too easily primary fabrics are affected by diagenetic transformation. This is considered entirely normal for sediments, but it is often an insurmountable obstacle for a critical genetic interpretation of the deposit.

In some difficult cases we will already count it a success if we can designate the deposit as "synsedimentary in a larger sense". These are the cases in which ores formed externally on the free sediment surface cannot be found, in which the overall picture may, however, be classified as syndiagenetic.

For these cases the collective term "synsedimentary in a larger sense" is applicable and intelligible, for it comprises not only the primary fabrics of the external zones of sediment formation, but also events in the upper or lower regions of the sea floor. In individual cases and depending on the thickness of the ore-containing complex, terms like "strata-bound, layer-bound, banking-bound, horizon-bound" and possibly also "stratigraphically bound to geologic periods" may be chosen.

Considering the fact that sedimentary genesis already causes a great deal of change and destruction in sedimentary ore fabrics — a fact which considerably complicates the inter-

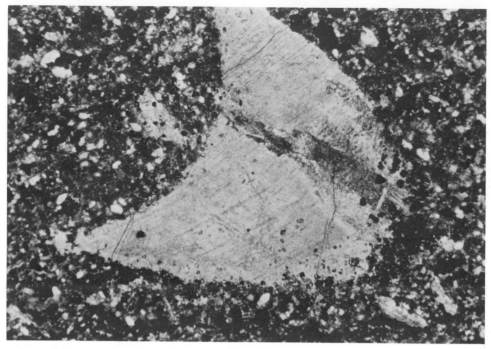

Fig. 32. Biodetritus (fragment of Ophiura) in ore lutite, with ZnS, SiO_2, CaF_2 and calcilutite. Wetterstein Limestone, Bleiberg mining complex. Thin section, nicols +, sample size 2.5 × 1.7 mm.

pretation — the hope of achieving a reliable interpretation of a sedimentary deposit subjected to metamorphism appears to be minimal. However, the original sedimentary material content has already been made plausible in a variety of metamorphic ore deposits. Of course, we know that ore minerals are liable to react very differently, but nevertheless many a remaining fabric from typically sedimentary structures has survived weak to moderate metamorphism.

Apart from the search for relics from the sedimentary genetic cycle, the analysis of metamorphic ore deposits has to pay special attention to common tectonic deformations of deposit and surrounding rock, to the relation between crystallization and deformation, to patterns of preferred orientation of the grain fabric ("Tectonite-Regelung") of all mineral components, and to the stability ranges of the individual minerals.

The investigation of deposits by way of the analysis of fabrics will always be worthwhile in explaining the genesis, but also as a basis for the well-directed application of geochemical methods which will increasingly contribute to successful scientific interpretation.

337

ACKNOWLEDGEMENT

This work was in part sponsored by the Austrian "Fonds zur Förderung der wissen-
schaftlichen Forschung", partly it is related to the I.G.C.P. (International Geological
Correlation Programme).

REFERENCES

* Amstutz, G.C. (Editor), 1964. *Sedimentology and Ore Genesis*. Elsevier, Amsterdam, 170 pp.
* Amstutz, G.C. and Bernard, A.J. (Editors), 1973. *Ores in Sediments*. Springer, Berlin, 350 pp.
* Amstutz, G.C., Ramdohr, P. and Park, W.C., 1964. Diagenetic behaviour of sulphides. In: G.C.
   Amstutz (Editor), *Sedimentology and Ore Genesis*. Elsevier, Amsterdam, pp. 65–90.
* Arnold, M., Maucher, A. and Saupe, F., 1973. Diagenetic pyrite and associated sulphides at the
   Almadén Mercury Mine, Spain. In: G.C. Amstutz and A.J. Bernard (Editors), *Ores in Sediments*.
   Springer, Berlin, pp. 7–19.
* Bogacz, K., Dźulynski, S., Harańczyk, C. and Sobczyński, P., 1972. Contact relations of the ore-
   bearing dolomite in the Triassic of the Cracow-Silesian region. *Ann. Soc. Géol. Pol.*, 42 (4):
   347–372.
Chilingar, G.V., Bissell, H.J. and Wolf, K.H., 1967. Diagenesis in carbonate rocks. In: G. Larsen and
   G.V. Chilingar (Editors), *Diagenesis in Sediments*. Elsevier, Amsterdam, pp. 179–322.
Cros, P. and Lagny, Ph., 1969. Paléokarsts dans le Trias moyen et supérieur des Dolomites et des Alpes
   Carniques occidentales. Importance stratigraphique et paléogéographique. *Sci. Terre*, 14 (2):
   139–195.
* Germann, K., 1973. Deposition of manganese and iron carbonates and silicates in Liassic marls of the
   northern Limestone Alps (Kalkalpen). In: G.C. Amstutz and A.J. Bernard (Editors), *Ores in Sedi-
   ments*. Springer, Berlin, pp. 129–138.
Lebedev, L.M., 1967. *Metacolloids in Endogenic Deposits*. Plenum Press, New York, N.Y., 298 pp.
Maucher, A., 1954. Zur "alpine Metallogenese" in den bayerischen Kalkalpen zwischen Loisach und
   Salzach. *Tschermaks Miner. Petrogr. Mitt.*, 4: 454–463.
Maucher, A., 1957. Die Deutung des primären Stoffbestandes der kalkalpinen Pb-Zn Lagerstätten als
   syngenetisch-sedimentäre Bildung. In: *Entstehung von Blei-Zinkerzlagerstätten in Karbonatgestei-
   nen, Tagung 1956–Berg-Hüttenmänn. Monatsh.*, 9: 226–229.
* Mengel, J.T., 1973. Physical sedimentation in Precambriam cherty iron formations of the Lake-
   Superior type. In: G.C. Amstutz and A.J. Bernard (Editors), *Ores in Sediments*. Springer, Berlin,
   pp. 179–193.
Mlakar, J. and Drovenik, M., 1971. Strukturne in genetske posebnosti idrijskega rudišča. *Geologija,
   Ljubljana*, 14: 67–126.
* Padalino, G., 1973. Ore deposition in karst formations with examples from Sardinia. In: G.C. Amstutz
   and A.J. Bernard (Editors), *Ores in Sediments*. Springer, Berlin, pp.209–220.
* Perna, G., 1973. Fenomeni carsici e giacimenti minerari. *Atti Seminar. Speleogenesi–Le Grotte d'Italia*,
   4: 71–72.
* Popov, V.M., 1973. On the anisotropy of ore-bearing series in stratiform deposits. In: G.C. Amstutz
   and A.J. Bernard (Editors), *Ores in Sediments*. Springer, Berlin, pp. 221–225.
Rainer, H., 1957. Diskussionsbeitrag. In: *Entstehung von Blei-Zinkerzlagerstätten in Karbonatgesteinen,
   Tagung 1956–Berg-Hüttenmänn. Monatsh.*, 9: 235–237.
* Ramdohr, P., 1969. *The Ore Minerals and their Intergrowths*. (English translation.) Pergamon Press,
   London, 1174 pp.
Sander, B., 1930. *Gefügekunde der Gesteine*. Springer, Wien, 352 pp.
Sander, B., 1936. Beiträge zur Kenntnis der Anlagerungsgefüge (Rhytmische Kalke und Dolomite aus

der Trias). *Tschermaks Miner. Petrogr. Mitt.*, 48: 27–139. (Translated into English by E.B., Knopf, 1951: Contributions to the Study of Depositional Fabrics. Rhythmically Deposited Triassic Limestones and Dolomites–*Am. Assoc. Pet. Geologists*, Tulsa, Okla.)

Sander, B., 1948–1950. *Einführung in die Gefügekunde der geologischen Körper, I, II*. Springer, Vienna, I 215 pp.; II 409 pp.

Sander, B., 1970. *An Introduction to the Study of Fabrics of Geological Bodies*. Pergamon Press, London, 641 pp.

* Schadlun, T.N., 1973. On the origin of "Kies"-ore and Pb-Zn deposits in sediments. In: G.C. Amstutz and A.J. Bernard (Editors), *Ores in Sediments*. Springer, Berlin. pp. 267–273.

Schneider, H.J., 1953. Neue Ergebnisse zur Stoffkonzentration und Stoffwanderung in Blei-Zink-Lagerstätten der nördlichen Kalkalpen. *Fortschr. Mineral.*, 32: 26–30.

Schneider, H.J., 1954. Die sedimentäre Bildung von Fluss-Spat im Oberen Wettersteinkalk der nördlichen Kalkalpen. *Abh. Bayer. Akad. Wiss. Math.- Naturwiss. Kl.*, 66: 1–37.

Schneider, H.J., 1957. Diskussionsbeitrag. In: *Entstehung von Blei-Zinkerzlagerstätten in Karbonatgesteinen, Tagung 1956–Berg-Hüttenmänn. Monatsh.*, 9: 238–240, 246, 248, 256, 242–244.

* Schneider, H.J., 1964. Facies differentiation and controlling factors for the depositional lead-zinc concentration in the Ladinian geosyncline of the Eastern Alps. In: G.C. Amstutz (Editor), *Sedimentology and Ore Genesis*. Elsevier, Amsterdam, pp. 29–45.

Schulz, O., 1955. Montangeologische Aufnahme des Pb-Zn Grubenrevieres Vomperloch, Karwendelgebirge, Tirol. *Berg-Hüttenmänn. Monatsh.*, 9: 259–269.

Schulz, O., 1957. Diskussionsbeitrag. In: *Entstehung von Blei-Zinkerzlagerstätten in Karbonatgesteinen, Tagung 1956–Berg-Hüttenmänn. Monatsh.*, 9: 241–242.

Schulz, O., 1960a. Die Pb-Zn-Vererzung der Raibler Schichten im Bergbau Bleiberg-Kreuth (Grube Max) als Beispiel submariner Lagerstättenbildung. *Carinthia* 2 (22): 1–93.

Schulz, O., 1960b. Beispiele für synsedimentäre Vererzungen und paradiagenetische Formungen im älteren Wettersteindolomit von Bleiberg-Kreuth. *Berg-Hüttenmänn. Monatsh.*, 1: 1–11.

* Schulz, O., 1964. Lead-zinc deposits in the Calcareous Alps as an example of submarine-hydrothermal formation of mineral deposits. In: G.C. Amstutz (Editor), *Sedimentology and Ore Genesis*. Elsevier, Amsterdam, pp. 47–52.

Schulz, O., 1968. Die synsedimentäre Mineralparagenese im oberen Wettersteinkalk der Pb-Zn-Lagerstätte Bleiberg-Kreuth (Kärnten). *Tschermaks Miner. Petrogr. Mitt.*, 12 (2/3): 230–289.

* Schulz, O., 1973. Wirtschaftlich bedeutende Zinkanreicherung in syndiagenetischer submariner Deformationsbreccie in Kreuth (Kärnten). *Tschermaks Miner. Petrogr. Mitt.*, 20: 280–295.

* Schulz, O., 1975. Resedimentbreccien und ihre möglichen Zusammenhänge mit Zn–Pb-Konzentrationen in mitteltriadischen Sedimenten der Gailtaler Alpen (Kärnten). *Tschermaks Miner. Petrogr. Mitt.* (in press).

* Shrock, R.R., 1948. *Sequence in Layered Rocks*. McGraw-Hill, New York, N.Y., 507 pp.

Siegl, W., 1956. Zur Vererzung der Pb-Zn-Lagerstätten von Bleiberg. *Berg Hüttenmänn. Monatsh.*, 5: 108–111.

Siegl, W., 1957. Diskussionsbeitrag. In: *Entstehung von Blei-Zinkerzlagerstätten in Karbonatgesteinen, Tagung 1956–Berg-Hüttenmänn. Monatsh.*, 9: 237–238.

Taupitz, K.C., 1953. Die verschiedene Deutbarkeit von "metasomatischen" Gefügen auf "telethermalen" Blei-Zink-Lagerstätten. *Fortschr. Mineral.*, 32: 30–31.

Taupitz, K.C., 1954. Erze sedimentärer Entstehung auf alpinen Lagerstätten des Typs "Bleiberg". *Z. Erzbergbau Metallhüttenw. (Erzmetall)*, 7 (8): 1–7.

Taupitz, K.C., 1957. Diskussionsbeitrag. In: *Entstehung von Blei-Zinkerzkagerstätten in Karbonatgesteinen, Tagung 1956–Berg-Hüttenmänn. Monatsh.*, 9: 241, 247, 248, 253.

* Zimmermann, R.A. and Amstutz, G.C., 1964. Small-scale sedimentary features in the Arkansas barite district. In: G.C. Amstutz (Editor), *Sedimentology and Ore Genesis*. Elsevier, Amsterdam, pp. 157–163.

* Zimmermann, R.A. and Amstutz, G.C., 1973. Intergrowth and crystallization features in the Cambrian mud volcano of Decaturville, Missouri, U.S.A. In: G.C. Amstutz and A.J. Bernard (Editors), *Ores in Sediments*. Springer, Berlin, pp. 339–350.

* Papers not referred to, but also recommended, on the topic of sedimentary ore fabrics.

Chapter 8

RHYTHMICITY OF BARITE–SHALE AND OF Sr IN STRATA-BOUND DEPOSITS OF ARKANSAS

RICHARD A. ZIMMERMANN

INTRODUCTION

The problem of ore genesis as related to textural, structural and other morphologic features also includes the problem of rhythmicity and the origin of banded features in sedimentary rocks. Our recent papers on the rhythmicity of barite and shale in Arkansas (Zimmermann and Amstutz, 1961, 1964a, b, 1965) and Zimmermann (1965, 1967, 1969, 1970) appear to be the earliest on this subject concerning the Arkansas Stanley Shale deposits of Mississippian age, or perhaps any other strata-bound barite province.

The barite forms sedimentary layers in the Arkansas deposits and similarly banded barite also occurs in deposits in sedimentary rocks in the Shoshone Range near Battle Mountain, Nevada and in Meggen, Germany (Zimmermann and Amstutz, 1964c; and Zimmermann, 1969, 1970). As can be observed elsewhere in numerous cases of banded materials in nature (varves, glacier ice, wood, etc.), sharp as well as gradational contacts occur between bands. In the Chamberlain Creek syncline barite deposit near Malvern, Arkansas, as well as in other deposits where barite alternates with shale, two or at the most, three different bands are involved in the rhythm: in the Arkansas deposits, the bands consist simply of fine-grained barite and quartz plus shale layers, sometimes with a greater concentration of calcite in the upper part of the barite layers. Pyrite is concentrated in the shale layers and gypsum may occur as small grains along the margins of the shale and barite layers. The sharp contact lies between the barite layer below and the shale layer above due to loading of clay during sedimentation and diagenesis. (The shale grades upward into barite and this grades into calcite, as shown in Fig. 1.a.) A specimen can, therefore, be oriented anywhere in the field, whether in the outcrop or in float specimens, with respect to top and bottom.

In comparison, stratiform sulfides from a deposit like the Sullivan mine, in British Columbia, Canada also show a type of rhythmic deposition with strong variation between sulfides alone. In the Sullivan mine, the lowest of the "B" Band Triplets (a section of about 25 cm thickness) shows the sharp contact lying at the base of thin sphalerite bands which grade upward to bands of galena, followed by a thin band of pyrite. The picture is not as simple as the barite–shale rhythms because various sulfides are also present in

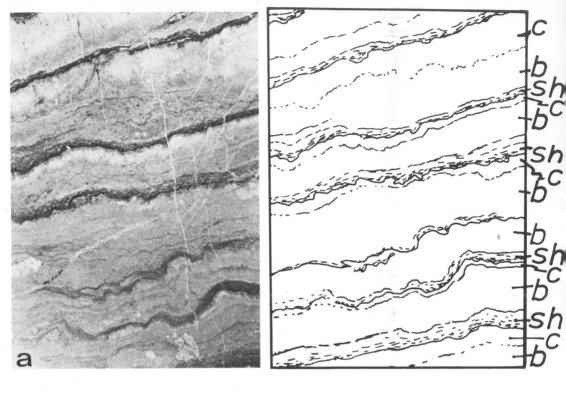

Fig. 1.a. Thin section. Banded barite (*b*, grey in photo) and shale (*sh*, dark laminated in photo) with calcite (*c*, white in photo) in upper part of barite layers. b. Nearly vartically dipping bedded barite and shale. Barite layers thicken from left to right in the outcrop or from bottom to top in the section.

layers (sphalerite, galena and iron-sulfide layers). Also, the lowest of the "B" Band Triplets does not consist of a constant repetition of the rhythm given above.

STRATA-BOUND BARITE AND SHALE IN THE CHAMBERLAIN CREEK SYNCLINE, ARKANSAS

On classifying rhythmic types according to ABAB or AB(C)AB(C) [where (c) may or may not be present] or ABAABA where, in each type, the same letter designates the recurrence of the same layer, we would find also rhythmic variations in thicknesses of both the barite and shale layers of the ABAB type of rhythm. This is illustrated in Fig. 1b where barite—shale bands, in groups of about 30 cm thickness, exhibit a gradual thickening of barite layers from bottom to top.

Taking into account the entire deposit, the lower part contains fissile barite—shale (ABAB type) whereas higher in the section the individual barite layers thicken to form bulging lenses in places. Shale layers 2—4 cm thick recur in various places still higher in the section. Near the middle and upper part of the section, the type of banding illustrated in Fig. 1b occurs.

The rhythmic layers of shale and sandstone which overly the Arkansas Novaculite are interlayered with the barite and also overly it, thus giving an insight into the sedimentary processes going on along with barite formation. The black-shale beds alternate with barite and the 4—10 cm thick siltstone layers in occurring places in the barite section, together with the overlying sandstone—shale sequence, can all be included together to form a large-scale rhythm of clastics. Fig. 2 illustrates in graphical form thick beds of black shale alternating with thin sandstone beds in the lower part of the sequence, forming groups of beds (1 and 2); these groups are interbedded with sandstone units of thicknesses similar to those of the sandstone—shale groups. (In Fig. 2, brackets and numbers are placed beside groups of sandstone—shale.) Higher in the section, thick shale units alternate with thick sandstone units, as shown by the groups of beds bracketed as 3 and 4 (Fig. 2). Each of the shale units becomes progressively thicker up section (bracket 5). Still higher in the section, the shales are again interbedded with sandstone (brackets 5, 6 and 7). The sandstone units now become thicker and, therefore, form thick groups of alternating sandstone and shale. The sandstone—shale groups are separated by thick siltstone units (between 7 and 8). Still higher in the section the groups become even thicker (bracket 9).

The graph (Fig. 2) of average shale thickness vs. number of sandstone units in the various sandstone—shale alternating groups shows the following features: (1) the thickness of the shale units decreases with increasing number of sandstone units within a sandstone—shale alternation group; and (2) in the case of groups 7, 8 and 9, with each group being stratigraphically higher in the section, the thickness of the shale units decreases steadily upwards. The first point could indicate that a greater number of sandstone units, with thinner shale between them, is the result of more frequent interruptions in normal shale deposition in sandstone units.

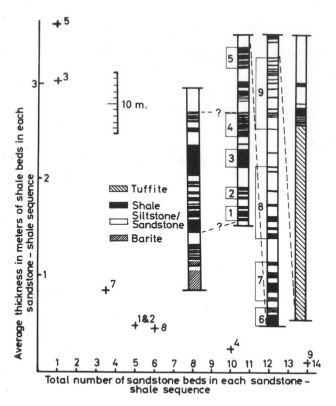

Fig. 2. Rhythmically layered shale–sandstone section in the Chamberlain Creek syncline measured along the west wall of the barite open pit. Brackets along the columnar section enclose sandstone–shale alternations which are also illustrated graphically.

The sandstone beds always grade upward into shale and on a larger scale (as is the case with alternations *8* and *9*) there is, to some extent, a steady upwards decrease in sandstone and overlying shale thicknesses.

The entire sandstone–shale sequence (Fig. 3) is overlain by an olive-green, massive-layered rock identified as a tuffite (Zimmermann, 1965).

With the exception of the sandstone lenses known to occur in the barite–shale section, all of the barite appears to have been deposited in a low-energy environment with slow deposition of shale; pyrite and organic matter in the black shale indicates a stagnant environment with formation of hydrogen sulfide. The load casts and sharp contacts often found at the base of the shale bands, and the gradation upward of shale to barite indicate cyclic deposition of either clay or barite; perhaps rapid deposition of clay took place, followed by increased deposition of barite, with maximum deposition of barite at a time when clay was not being deposited; finally, rapid deposition of clay again took place. It is possible that the barite deposition was continuous, with frequent interruptions of clay. A 1-mm layer of shale would correspond to about a 1-cm deposit of clay. Since the barite is

Fig. 3. Sequence of sandstone (gray) and shale (black) below the tuffite section in the Chamberlain Creek syncline.

but little compacted, the barite and clay layers may have been of equal thickness at the time of deposition.

It seems that two possibilities as to the mode of formation, involving sedimentary and diagenetic processes, might be considered to explain the origin of the layering between barite and shale in the Stanley Shale deposits of Arkansas. Bearing in mind the shale—barite contact relationships, Fig. 4,A illustrates successive deposition of clay and barite. As

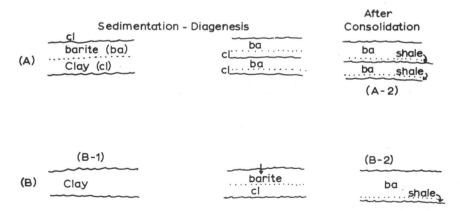

Fig. 4. Formation of barite layers by (A) direct precipitation of barite followed by compaction of clay, and (B) deposition of barite contemporaneously with compaction of clay.

the clay undergoes compaction, pore solution separates out and precipitates barite in a more or less separate layer (above or below), thus producing a sort of diagenetic differentiation into separate layers (Fig. 4,B). In case A, barite is deposited as a sediment (with, of course, later crystallization or recrystallization); in case B, crystallization of barite from saturated barium-sulfate solution moving out of the clay layer takes place. There may be, of course, transitional cases between both processes.

The pyrite which occurs in both the shale and barite beds reaches amounts of 10% (Fig. 5,E) in the shale, and sulfur bacteria very likely played a role in its formation. The very consistently similar thickness of several meters of the barite bands, each usually with thin intermediate shale seams of a thickness of 1 or 2 cm, can be interpreted in the following manner. It is possible that precipitation during similar time intervals, such as annual rhythms, is responsible for the bands. From interval to interval, however, the rate and type of biological activity, responsible for the chemical precipitation, may have varied according to the seasons throughout the year, but were similar when the same

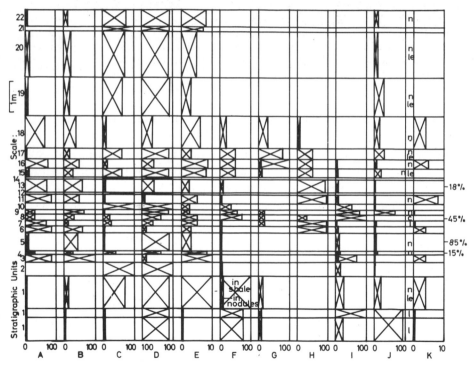

Fig. 5. Microscopic data from the barite–shale section near the upper part of the barite deposit in the Chamberlain Creek syncline (Section VII, Zimmermann, 1965). A: % quartz and coarse detritals; B: quartz grain size in microns; C: % argillaceous material; D: 100% sandstone–100% shale (only clastics); E: % pyrite; F: % pyrite spheres; G: size of pyrite spheres in microns; H: % pyrite crystals; I: size of pyrite crystals in microns; J: % barite; K: % calcite. In the column to the right of column J the letters indicate the following: l: barite in layers; le: barite in lenses; n: barite in nodules; c: barite in crystals.

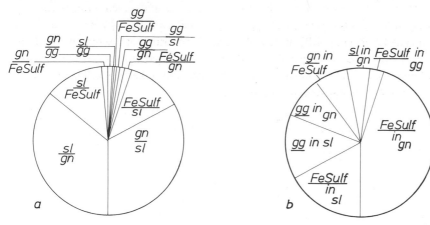

Fig. 6a. The ratios of 403 layers (sulfide and gangue) in contact, and Fig. 6b. 58 observations of sulfide in gangue or in another sulfide from the lowest of the "B" Band Triplets, a bed of about 25 cm thickness, of the Sullivan deposit, British Columbia, are here illustrated. Note that the galena is not only commonly overlain by sphalerite, but the sphalerite likewise is commonly overlain by galena. We have an alternating deposit of galena and sphalerite with other sulfide layers and gangue and/or rhythms superposed on this scheme. Note also that the contact between iron sulfides and sphalerite is likewise common and agrees with our observation that the sphalerite—galena—iron sulfide (bottom to top) rhythm is overlain by sphalerite of a new rhythmic band. *FeSulf*: iron sulfides, here pyrite and pyrrhotite; *gn*: galena; *sl*: sphalerite; *gg*: gangue.

seasons of different years are compared. Thus, beds of similar thicknesses with nearly identical internal features were the result. A greater than normal abundance of sulfate in seawater and sediment pore waters may also have played a role in precipitating low-barium concentrations which might have been available, but it is still uncertain whether these substances were supplied by erosion or by volcanism in accumulations sufficient to form large commercial barite deposits.

Fig. 5 (A—K) shows microscopic data from units of one of the sections measured in the upper part of the barite sequence in the Chamberlain Creek syncline. Fig. 5,B shows in many cases that the quartz grain size in adjacent beds is larger in the thicker bed. Over the entire section there is no consistent relationship between bed thicknesses and grain size; this is probably due to high abundances of shaly materials in thick, silty shale beds (units *5, 19, 20*) so that the bed thickness no longer show a relationship with the sizes of the relatively few quartz grains present.

Graded bedding from sandstone to shale is clearly shown in Fig. 5 (A to D) in the graph of 100% sandstone to 100% shale. The five graded sequences involve units *1* and *2*; *3—5*; *6—10*; *11—12*; *13—14*, respectively. Most of the remainder of the section consists of silty shale and does not clearly show sandstone grading to shale.

Fig. 5 (E—K), which gives relationships between barite, pyrite, and calcite, clearly shows that a high barite content occurs with low pyrite, and that low barite occurs together in the same bed with a higher pyrite content. This relationship exists in the

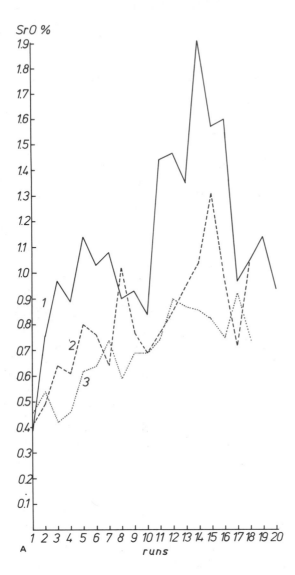

Fig. 7. X-ray fluorescence analyses for Sr of barite nodules of different sizes from the Chamberlain Creek syncline barite deposit of Arkansas. The nodules range from 10 mm in diameter (curve 1) to 2 mm in diameter (curve 10). Along the horizontal axes the runs 1–20 (A), 1–21 (B) and 1–14 (C) are labeled; run No. 1 refers to the first recovered powders after milling (outermost shells of nodule).

Meggen and Rammelsberg deposits in Germany (see Ehrenberg et al., 1954; Kraume, 1955) and probably exists in many other layered barite deposits. In Meggen, the inner part of the bedded deposit consists of pyrite and the sulfides of zinc and lead, while the outer or peripheral part of the deposit consists of barite. At Rammelsberg, in the Neues Lager, both barite and pyrite occur together. The pyrite is more abundant at the base of

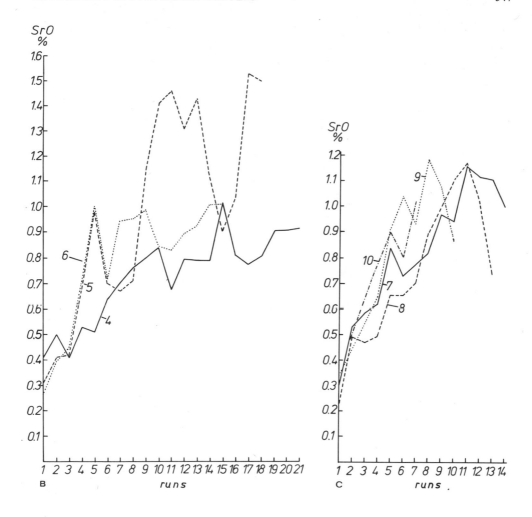

Fig. 7 (continued).

the deposit, and becomes steadily less in content higher in the section. Barite in the lower beds is low but steadily increases higher in the section. This relationship may depend on whether the available sulfur remains as sulfide or becomes oxidized to sulfate.

Calcite, as shown in Fig. 5K, shows greater abundances where the content of barite is lower.

The Sullivan deposit in Canada, like those of Rammelsberg and Meggen, is also a massive sulfide type deposit, with a thick section of banded iron sulfides, sphalerite, galena and silicate gangue. Observations of 403 layers (sulfide and gangue) in contact, and

58 observations of sulfide in gangue or in another sulfide (all observations on the lowest of the "B" Band Triplets) resulted in the ratios given in Fig. 6a and 6b. Barite has still not been identified in the Sullivan deposit.

RHYTHMICITY OF Sr IN BARITE NODULES AND IN BEDDED BARITE SEQUENCES

Barite nodules from a sample near the upper part of the barite section, about 18 m up in the section of the Chamberlain Creek syncline deposit were grouped into 10 sizes, ranging from 10 mm (sample No. 1) to 2 mm (sample No. 10); the powders derived from differential abrasion were analyzed for Sr by X-ray fluorescence. The abrasion was effected in a shaker and the nodules were "milled" together with glass balls of similar sizes in water-filled flasks. The largest nodules were run from 15 to 20 times and powders ranging up to 200 mg were collected from each run; the smallest nodules were run from 10 to 15 times (see Fig. 7). The analyses were compared with a standard, also from the Chamberlain Creek syncline deposit.

The nodules possess a mainly radiating structure, with one concentric break about halfway between the center and the outer wall visible in cross-section with a binocular microscope. A minute pyrite grain lies in the center of a few of the nodules; however, pyrite grains are more commonly situated near the outer walls of the nodules.

Curves (see Fig. 7) which are quite similar include the following: 1 and 2; 3 and 4; 6, 7 and 8; 9 and 10. In most cases an increase is seen (from the outside, or wall, inwards) in Sr, followed by a small decrease (example: runs 1–10 of the curve for sample No. 1) which is then followed by a greater increase in Sr than in the outer shells. These correspond to a type of rhythmic deposition and the curves also are concave upward (as with Sr analyses of barite strata from the Gap Mountain deposit near Glenwood, Arkansas; see below) when turned upside down so that analyses of the innermost parts of the nodules lie to the lower left. Small nodules probably formed later than the large nodules, inasmuch as the peaks of the Sr contents of the interiors of the small nodules correspond with the Sr contents of shells not entirely in the interiors of the larger nodules (for example, between shells milled during runs between 2 and 8 of sample No. 1 in Fig. 7). The diagenetic pore solution was already partly depleted in Sr and no longer as rich in this element as when the interiors of the largest nodule sizes were formed.

Deposition of barite in veins (Tischendorf, 1962) likewise shows generations of sedimentation. In a vein of the Schneckenstein barite deposit in Vogtland (near Auerbach in Saxony, East Germany – $SrSO_4$ content in the 1–2% range), the $SrSO_4$ content decreases generally from the wall of the veins inwards (deposition is believed to have proceeded from the wall inwards), but within each generation of deposition, the $SrSO_4$ content increases as deposition proceeds.

A concentrically zoned, 3-cm barite nodule with radial structure and about six concentric bands, from the Eagles Nest deposit near Fancy Hill Mountain in the Fancy Hill

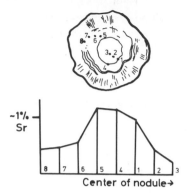

Fig. 8. Relative semi-quantitative Sr-content based on X-ray fluorescence analysis of the concentric bands of a 3 cm barite nodule from the Eagles Nest deposit near Fancy Hill Mountain, Arkansas.

Barite District near Glenwood, Arkansas (see Zimmermann, 1965), was analysed[1] by X-ray fluorescence and the rhythmic relationships (see Fig. 8) were found to be similar to those for the nodules from the Chamberlain Creek syncline. (Relative amounts were analysed here and not absolute amounts.)

RELATIONS OF THE Sr CONTENTS IN BARITE STRATA

Samples were taken from the different deposits in the Stanley Shale in Arkansas and from different places in the stratigraphic section of each deposit; those in Fig. 9 are from the Gap Mountain deposit. The analyses[2] of each sample represent 2—5 cm of the stratigraphic section which may be for an entire bed or only part of a bed. Here, relations are evident as seen in the graph for the Gap Mountain deposit (Fig. 9). The BaO/SrO ratio indicates a tendency to increase from lower to higher barite strata in the Gap Mountain deposit: the Sr content decreases from the lower to the upper part of the deposit.

INTERPRETATION OF GENESIS OF RHYTHMICITY

One process involved in accumulating barite layers and fissile shale seams quite likely involves a component of clastic deposition to account for the fissile layers interlayered with the barite, and the presence of the same black shale in sections of clastic sedimentary rocks above and below it. Deposition of barite, in general, can be accounted for only

---

[1] Analyses by Dr. Erik H. Schot.
[2] Flame photometer analyses available from the author on request (see also dissertation by Zimmermann, 1965).

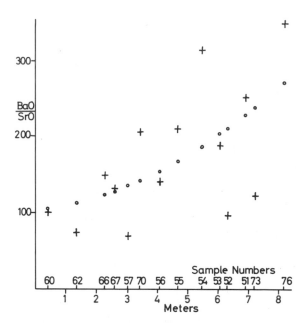

Fig. 9. BaO/SrO ratio of samples analyzed for Sr (flame photometer analyses) and Ba from the Gap Mountain deposit near Glenwood, Arkansas. The distance in meters represents the distance up from the base of the deposit. The crosses represent the actual results of the analyses; the circles represent the corresponding analyses calculated by the method of least squares.

by its very low solubility as a sulfate. Individual layers and even sub-layers of fissile shale in barite would be primary features of the barite and shale accumulated due to changes in deposition of clastics and chemical deposits of barite.

When considering the origin of rhythmic barite and shale (and one should point out that in the largest deposits the barite is in close association with both shale and chert, and in some cases like at Meggen, with limestone), one has to pay attention also to the genesis of sulfides and rhythmic precipitates of sulfides. There are at least two reasons for this: (1) most, if not all, barite deposits contain pyrite, commonly in the fissile-shale layers as in the Arkansas deposits; and (2) some deposits (especially of layered barite with shale or sandy shale, as in Meggen and Rammelsberg) contain enormous tonnages of sulfides in beds composed of thinly layered, interbanded pyrite and fissile shale or organic matter. Curiously, as in Meggen and Rammelsberg, the bulk of the barite is separated from the bulk of the pyrite (an exception is the association of lead and zinc sulfides in the Grauerz ore body at Rammelsberg), with lateral gradations at Meggen and Rammelsberg, but with a vertical separation, with barite overlying the sulfides, at Rammelsberg. At Meggen, the entire ore body of barite or pyrite is underlain by pyrite nodules with some barite (Zimmermann, 1970). Kuroko-type deposits in Japan also contain abundant barite in association with pyrite.

Formation of the sulfide rhythms would be as follows (see Lambert and Bubela,

1970): (1) direct precipitation of metallic sulfides with changes in pH, presence of mono-metallic or multi-metallic solutions, changes in ion concentrations and precipitation according to solubility products; and (2) sulfide ions diffuse into contact with metals which are in solution, or adsorbed to minerals such as clays; metals, initially precipitated as carbonates, were subsequently converted to sulfides. According to Ferguson et al. (1974), the Atlantis II Deep in the Red Sea (Degens and Ross, 1969) — where barite has also been found — provides evidence that metalliferous thermal pools, capable of precipitating stratiform bands of sulfides, can form from thermal conditions. The mechanism of base-metal sulfide precipitation in this case probably requires the availability of metals as complexes in the brine with precipitation on cooling (Bischoff, 1969; Ridge, 1973). A type of rhythmicity not resulting from fluctuations in composition (Fe) includes the fine banding of sphalerite with alternating yellow and brown bands found in the Upper Mississippi Valley Tri-State zinc—lead district. These deposits are the result of fluctuations in sulfur fugacity during formation (Browne and Lovering, 1973).

Directly or indirectly, oxidation—reduction processes leading to deposition of bands are due to bacterial activity. In oxidation—reduction reactions of manganese, for example, certain bacteria produce an enzyme system which catalyzes the reactions (Hammond, 1974). At Meggen, many of the sulfides are composed of replaced microfossils indicating, perhaps, the action of sulfate-reducing bacteria on organic matter. Bipartite bands in sedimentary rocks may likewise originate from oxidation—reduction reactions in non-baritic, non-sulfidic rocks: a decomposing, ammonia-producing, fetid black mud with organic matter grades up to calcium carbonate (precipitation of bicarbonate by ammonia) which, in turn, is abruptly overlain by a new fetid mud layer (owing to the sealing-off effect of the preceding calcium carbonate; Shrock, 1948). A similar type of rhythm is described by Seibold (1955, 1958, in Reineck and Singh, 1973, pp. 109—110) from the Adriatic Sea where precipitation of biogenic carbonate forms in the summer months: in the summer months biogenic carbonate is favored as a result of strong assimilation by photoplankton and increased evaporation. The terrigenous material, quartz, iron sulfide and organic matter are deposited during periods of higher rains in autumn, winter and spring.

In deep strata, where sulfate reduction may go on, the Ba ion concentration increases when $BaSO_4$ concentration drops below the solubility product. Concentration of Ba due to bacteriological activity in pore solutions at depth, with later precipitation in sulfate-rich sediments nearer the surface, appears to be well supported by observations in the coal mines in the Ruhr, Germany (Von Engelhardt, 1967; Puchelt, 1967). With increasing sulfate concentration, the Ba concentration drops (Puchelt, 1967, pp. 19—20).

Chukhrov and Ermilova (1970), Hanor (1969) and others investigated strontium-bearing nodules. Starke (1964) and Hanor (1966 in Hanor, 1968) pointed out that significant compositional variations over small distances are common in barite crystals and aggregates.

The rhythmicity in the strontium in the nodules indicates periodic deposition (even the presence of concentric shells suggests this) where the concentration of strontium in neighboring layers is similar but with a general decrease in concentration when viewed from the interior to exterior of the nodule.

Strontium is diadochic in the structure of barite; a difference in the free energy of formation for $BaSO_4$ and $SrSO_4$ is such that Ba goes into the structure first, or at a faster rate than Sr (Christ and Garrels, 1965). Sr goes to the carbonate phase. Investigations by Hanor (1968) indicate that barite apparently is unreactive in terms of Ba—Sr exchange in most sedimentary and hydrothermal environments.

According to Puchelt (1967, p. 177) barite crystals with $10-12\%$ $SrSO_4$ should form in dilute sodium-chloride solutions (for the solubility relations in the $20-63°$ temperature range see Scherp and Strübel, 1974). Experiments by Starke (1964, in Puchelt, 1967, p. 179) show that the content of Sr in barite precipitated from solution with constant Ba/Sr ratio decrease with a relatively higher salt content, and with lowering of temperature. In his laboratory experiments with dilute sulfate solutions, Starke (1964, in Puchelt, 1967, p. 177) found that more Sr precipitated with barium when $BaSO_4$ was precipitated over a short time than over long periods. As the precipitate aged, the strontium content decreased; the Sr was present as solution inclusions which were released on stirring the precipitate over a longer period.

It would appear from experimental work done that variations in Sr may arise from the original Ba—Sr content in seawater or pore solutions during diagenesis or a combination of these, and not from secondary ion exchange. The concentration of the pore solutions in the upper part of the barite section would also be affected by the rising pore solutions from lower barite strata during diagenesis.

The rather differing contents of barium which have been determined in many seas (Church and Wolgemuth, 1972) would indicate a variable Ba/Sr ratio. As shown by experiments by Puchelt (1967, p. 17), extremely variable concentrations of barium are possible in solutions with low salt concentrations in the $0-15\%$ range (and, therefore, variable Ba/Sr); in nature, such concentrations correspond to those found in and near esturaries.

REFERENCES

Bischoff, J.L., 1969. Red. Sea geothermal brine deposits: their mineralogy, chemistry and genesis. In: E. Degens and D.A. Ross (Editors), *Hot Brines and Recent Heavy Metal Deposits in the Red Sea.* Springer, New York, N.Y., pp. 368–401.

Browne, P.R.L. and Lovering, J.F., 1973. Composition of sphalerites from the Broadlands geothermal field and their significance to sphalerite geothermometry and geobarometry. *Econ. Geol.*, 68: 381–387.

Christ, C.L. and Garrels, R.M., 1965. *Solutions, Minerals and Equilibria.* Harper, New York, N.Y., 450 pp.

Chukhrov, F.V. and Ermilova, L.P, 1970. Zur Schwefel-Isotopenzusammensetzung in Konkretionen. *Ber. Dtsch. Ges. Geol. Reihe B.*, 15 (3/4): 255–267.

Church, T.M. and Wolgemuth, K., 1972. Marine barite saturation. *Earth Planet. Sci. Lett.*, 15: 35–44.

Degens, E.T. and Ross, D.A. (Editors), 1969. *Hot Brines and Recent Heavy Metal Deposits in the Red Sea*. Springer, New York, N.Y., 600 pp.

Ehrenberg, H., Pilger, A. and Schröder, F., 1954. Das Schwefelkies–"Zinkblende"–Schwerspatlager von Meggen (Westfalen). *Beih. Geol. Jahrb. Hannover*, 12: 352 pp.

Ferguson, J., Lambert, I.B. and Jones, H.E., 1974. Iron sulphide formation in an exhalative–sedimentary environment, Talasea, New Britain, P.N.G. *Miner. Deposita*, 9: 33–47.

Hammond, A.L., 1974. Manganese nodules (I): mineral resources on the deep seabed. *Science*, 183 (4123): 502–503.

Hanor, J.S., 1968. Frequency distribution of compositions in the barite–celestite series. *Am. Miner.*, 53: 1215–1222.

Hanor, J.S., 1969. Compositional zoning in concretions. *Geol. Soc. Am. Abstr.*, Part 7: p. 89.

Kraume, E., 1955. Die Erzlager des Rammelsberges bei Goslar. *Beih. Geol. Jahrb. Hannover*, 18: 194 pp.

Krumbein, W.C. and Sloss, L.L., 1963. *Stratigraphy and Sedimentation*. Freeman, San Francisco, Calif., 660 pp.

Lambert, I.B. and Bubela, B., 1970. The experimental production of monomineralic sulphide bands in sediments. *Miner. Deposita*, 5: 97–102.

Pettijohn, F.J., 1957. *Sedimentary Rocks*. Harper, New York, N.Y., 718 pp.

Puchelt, H., 1967. Zur Geochemie des Bariums im exogenen Zyklus. *Sitzungsber. Heidelb. Akad. Wiss. Math.-Naturwiss. Kl.*, 4: 205 pp.

Reineck, H.E. and Singh, I.B., 1973. *Depositional Sedimentary Environments with Reference to Terrigenous Clastics*. Springer, New York, N.Y., 439 pp.

Ridge, J.D., 1973. Volcanic exhalations and ore deposition in the vicinity of the sea floor. *Miner. Deposita*, 8: 332–348.

Shrock, R.R., 1948. *Sequence in Layered Rocks*. McGraw, New York, N.Y., 507 pp.

Scherp, A. and Strübel, G., 1974. Zur Barium–Strontium Mineralisation. *Miner. Deposita*, 9: 155–168.

Starke, R., 1964. Die Strontiumgehalte der Baryte. *Freiberg. Forschungsh.*, C (150): 86 pp.

Tischendorf, G., 1962. Fortschritte bei Untersuchungen über die Spurenelementgehalte im Baryt. *Geologie*, 11 (9): 1052–1058.

Von Engelhardt, W., 1967. Interstitial solutions and diagenesis in sediments. In: G. Larsen, and G.V. Chilingar, (Editors), *Diagenesis in Sediments*. Elsevier, Amsterdam, pp. 503–521.

Zimmermann, R.A., 1965. *The Origin of the Bedded Arkansas Barite Deposits (with Special Reference to the Genetic Value of Sedimentary Features in the Ore)*. Thesis, Univ. Missouri, Rolla, M.O., 367 pp., unpublished.

Zimmermann, R.A., 1967. Sedimentary features in the layered barite deposits of Nevada, Wisconsin and Meggen (Germany). In: *Int. Sediment. Congr., 7th, Great Britain, 1967*. Reprint.

Zimmermann, R.A., 1969. Stratabound barite deposits in Nevada: rhythmic layering, diagenetic features, and a comparison with similar deposits of Arkansas. *Miner. Deposita*, 4: 401–409.

Zimmermann, R.A., 1970. Sedimentary features in the Meggen barite–pyrite–sphalerite deposit and a comparison with the Arkansas barite deposits. *Neues Jahrb. Miner. Abh.*, 113: 179–214.

Zimmermann, R.A. and Amstutz, G.C., 1961. Sedimentary features in the Arkansas barite belt. *Geol. Soc. Am. Bull.*, 68: 306–307.

Zimmermann, R.A. and Amstutz, G.C., 1964a. Small scale sedimentary features in the Arkansas barite district. In: G.C. Amstutz (Editor), *Sedimentology and Ore Genesis (Developments in Sedimentology, 2)*. Elsevier, Amsterdam, pp. 157–163.

Zimmermann, R.A. and Amstutz, G.C., 1964b. Die Arkansas-Schwerspatzone: neue sedimentpetrographische Beobachtungen und genetische Umdeutung. *Erzmetall*, pp. 365–371.

Zimmermann, R.A. and Amstutz, G.C., 1964c. Genesis of layered barite deposits in Arkansas and Germany. *Geol. Soc. Am. Spec. Pap.*, 82: p. 234.